Signals and
Linear Systems

Signals and Linear Systems

ROBERT A. GABEL
RICHARD A. ROBERTS

Department of Electrical Engineering
University of Colorado

JOHN WILEY & SONS
New York   London   Sydney   Toronto

Library of Congress Cataloging in Publication Data

Gabel, Robert A.
  Signals and linear systems.

  Includes bibliographical references.
  1. Electric networks. 2. Electric engineering—
Mathematics. 3. System analysis. I. Roberts,
Richard A., 1935–   joint author. II. Title.

TK454.2.G22   621.3'01'51   621.381'042   72-8770
ISBN 0-471-28900-0

Printed in the United States of America

10 9

Preface

When viewing the number of linear system texts available to students and teachers, any author adding to this collection must feel a need to justify his work. This text grew out of several editions of classroom notes which we developed for a senior-level course at the University of Colorado. While organizing the course, we found no single text which covered the material we thought important at a level suitable for undergraduate students. In particular, material on discrete-time systems was scattered, and most texts were either directed toward a particular type of system—such as electrical circuits, control systems, or communication systems—or were arranged in a topical order which the students found difficult to follow.

Our intent in writing this book is to present, in an organizational format designed for the student, the basic general techniques for analyzing linear systems. We treat both the usual continuous-time systems, and also discrete-time systems, which are finding widespread use in modern communication and control systems. The material is so ordered that the topics covered reinforce one another. The student is led naturally from basic techniques of time-domain analysis to the more abstract, although computationally simpler, transform-domain techniques. Extensive use is made of examples to illustrate the use of these techniques in solving problems in many diverse areas. Several examples are treated using two or more techniques in order to aid the student in comparing and relating the different solution methods. The material is general, and is chosen to lead, if desired, into a second-semester course in communication systems, control systems, or other areas which use these basic techniques in specific advanced applications.

Prerequisites assumed of the student are a sophomore-level mathematics course covering differential equations, and a course in the student's major area (such as electrical circuits or mechanics) which treats the derivation of mathematical models of physical systems. A course in probability theory is also helpful; examples and problems in the text which illustrate applications to probability problems are identified with the symbol * and may be omitted by those without this background.

This text is used for a one-semester elective course in linear system analysis at the University of Colorado. The course meets for a total of forty lecture hours; in this time we cover most of Chapters 1–6 and some of Chapter 7 if time permits. There is enough material in the text to give the instructor a choice of emphasis. The following notes indicate, as a guide, how we treat the various chapters.

Chapter 1 is covered rapidly, since for most students this is a review. The classical solution of linear constant coefficient differential equations, involving sums of functions of the type ce^{rt}, is readily extended to difference equation solutions, with sums of sequences of the type cr^k

Chapter 2 represents a departure from the usual order found in most linear systems texts. We cover the concept of convolution for both discrete-time and continuous-time systems in some depth. We feel that convolutional methods are so important that they should not be avoided by the use of transform methods. The convolutional sum for discrete-time systems and the corresponding convolutional integral for continuous-time systems furnish the student with a simple and effective way of picturing the operation of a linear system. We introduce convolutional methods in terms of discrete-time systems because the impulse sequence $\{\delta_k\}$ is easily defined and understood. After covering discrete-time systems we progress to continuous-time systems. In using convolutional methods for analysis of linear systems, the impulse response sequence or function of the system must be known. We present a method of calculating the impulse response sequence or function based directly on the difference or differential equation modeling the system, which is more general and usually simpler than the corresponding transform methods.

The state variable material of Chapter 3 is included for two reasons: (1) the larger, more complex systems now treated by control, power, and communication engineers require this compact general description; (2) computer solution methods are well-suited to the matrix manipulation involved. Again, we find that the solution of discrete-time problems is simpler and more direct, and hence it is treated first. The solution of continuous-time problems then follows quite naturally.

Chapter 4 is placed deliberately to introduce students to transform-domain solutions. Since students are probably least familiar with Z-transforms, they are not distracted by mechanical techniques (such as those they have used previously with Laplace transforms) and can concentrate on the central ideas involved in transforming from one domain to another. The analogous use of logarithms discussed in the chapter introduction is useful in this respect.

Chapter 5 introduces transform methods for continuous-time systems by considering the Fourier series and Fourier integral as methods of representing continuous-time functions. The generalized Fourier series is covered in depth in order to suggest why one might wish to have alternate representations of functions available. Walsh functions, for example, are shown to represent rectangular waveforms more "naturally" than the usual sinusoids. The Fourier transform is introduced as a generalization of the exponential Fourier series. The properties of the Fourier integral are introduced, pointing out the similarity with the Z-transform properties, and these properties are then used to obtain transforms of "energy signals" and then so-called "power signals." Power signals are time functions which are not absolutely integrable but do have finite power. By using the definition of the impulse function and the properties of the transform, one can easily generate very useful transform pairs for power signals, with a minimum of "hand-waving" arguments. Chapter 5 concludes with a discussion of the discrete Fourier transform as a method of numerically calculating the Fourier transform by a digital computer. Throughout the chapter we emphasize the physical nature of the transform by the use of examples taken primarily from communication theory applications.

Chapter 6 continues the discussion of transform methods for continuous-time signals by considering the Laplace transform. The Laplace transform is presented as a generalization of the Fourier transform. Both the bilateral and one-sided transforms are covered. In order to emphasize the unity inherent in transform methods, several examples are worked using both the Fourier and the Laplace transforms. These examples serve to point out that in most cases of a practical nature one does not need the Laplace transform, although numerical calculations with the Laplace transform are in some cases less complex. The Z-transform and the Fourier and Laplace transforms are integrated through a discussion of sampled time functions.

Our main purpose in including Chapter 7 is to further illustrate the relationships between continuous-time and discrete-time systems. In particular, we treat the problem of obtaining a desired continuous-time transfer function through the use of a discrete-time system (a digital filter). This chapter probably cannot be included in a one-semester course unless some earlier material is omitted. However, after the material treating the equivalent transfer function has been covered, students should be able to read through the remaining sections without undue difficulty. Alternatively, this chapter could be covered in a following semester or in a laboratory section.

Most new books have a way of building on the works of previous authors. In our case we are indebted to those authors mentioned in the references.

During the preparation of this book we have had many discussions with colleagues and graduate students. In particular, Min-Yen Wu and Jack Koplowitz offered many comments on the manuscript. Comments by Henry Hermes clarified many of our discussions in this book. The publisher's reviewers, John Thomas and Mac VanValkenburg, were most helpful. Teaching this material over the past several years has been an exciting and rewarding experience, due in part to the interest of our students. Their suggestions and comments have improved the presentation markedly.

The typing was accomplished through the efforts of Mrs. Marie Krenz and Miss Judy Price. Their efforts are greatly appreciated. Finally, we would like to thank any kind readers who forward to us corrections and suggestions for improvements in this book.

Denver, Colorado   ROBERT A. GABEL
Boulder, Colorado   RICHARD A. ROBERTS

Contents

ix

3 State Variables   110

Part Two   TRANSFORM-DOMAIN TECHNIQUES

4 The Z-Transform   169

5 The Fourier Transform   214

6 The Laplace Transform   306

Signals and
Linear Systems

PART ONE

Time-Domain Techniques

1

Linear Systems

1.1 INTRODUCTION

The study of linear systems has been an essential part of formal undergraduate training for many years. Linear system analysis is important primarily because of its utility, for even though physical systems are never completely linear, a linear model is often appropriate over certain ranges of application. Also, a large body of mathematical theory is available for engineers and scientists to use in the analysis of such systems. In contrast, the analysis of nonlinear systems is essentially *ad hoc:* that is, each nonlinear system must be studied as one of a kind. There are no general methods of analysis and no general solutions.

The analysis of a given linear system is often facilitated by use of a particular class of input signals or a particular signal representation. Thus it is natural to include a study of signals and their properties in a study of linear systems. In later chapters we shall find this study especially fruitful.

As engineers, we are interested not only in the analysis but also in the synthesis of systems. In fact, it is the synthesis or design of systems that is the really creative portion of engineering. Yet, as in so many creative efforts, one must learn first how to analyze a system before one can proceed with system design. This work is directed primarily toward the analysis of certain classes of linear systems, although, because design and analysis are so intimately connected, this material will provide a basis for simple design.

We can divide the analysis of systems into three aspects:

(1) The development of a suitable mathematical model for the physical problem of interest. This portion of the analysis is concerned with obtaining the "equations of motion," boundary or initial conditions, parameter values, etc. This is the process wherein judgment, experience,

3

and experiments are combined to develop a suitable model. In some sense this first step is the hardest to develop formally.

(2) After a suitable model is obtained, one then solves the resultant equations to obtain solutions in various forms.

(3) One then relates or interprets the solution to the mathematical model in terms of the physical problem. It is to be hoped that the development in (1) has been accurate enough so that meaningful interpretations and predictions concerning the physical system can be made.

The primary emphasis of this work is on the second and third aspects mentioned above. The first step is essential but is probably better and more completely accomplished within a particular discipline. Thus, chemical engineers will learn to write equations of motion for chemical processes, electrical engineers for electrical circuits, and so on. After a model is obtained, one can consider various techniques for its solution and provide a basis for its mathematical interpretation.

Because linear models are so often used in all disciplines of engineering and science, this material is very useful. Perhaps the best way to point this fact out is to present examples from various physical problems. The only drawback to this method is that the reader may not always possess the necessary background to perform the first step in the analysis, to write the equations of motion. This problem is to be expected. As one gains familiarity with a given discipline, this first step becomes natural. Thus, without further apologies, we shall attempt to give physical examples from many fields without always attempting to develop a complete basis for the derivation of the equations of motion for a given system.

This material is presented as a summarizing work which brings together techniques and concepts that can be used to analyze a great variety of physical phenomena. This unity that one obtains is most useful and satisfying.

1.2 DEFINITIONS

We are primarily concerned with linear systems apart from their inherent physical structure. Thus, we often shall represent a linear system schematically as a box with inputs $x_1(t), x_2(t), \ldots, x_n(t)$ and outputs $y_1(t), y_2(t), \ldots, y_m(t)$ as in Figure 1.1. The inputs $x_i(t)$, $i = 1, 2, \ldots, n$ and

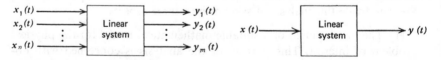

FIGURE 1.1
Schematic representations of a linear system.

outputs $y_j(t)$, $j = 1, 2, \ldots, m$ will, in general, be time signals: i.e., any physical variables of interest which vary with time. For the moment, let us focus on a single input, single output linear system. We shall consider multiple input-output systems in detail in Chapter 3.

Definition of a Linear System

The word linear suggests something pertaining to a straight line relationship. Thus, we might suspect that a linear system is one in which the output is in some sense proportional to the input. That is, if $x(t)$ gives rise to $y(t)$, then $\alpha x(t)$ gives rise to $\alpha y(t)$ for any constant α. In symbols, if

$$x(t) \to y(t)$$

then

$$\alpha x(t) \to \alpha y(t) \tag{1.1}$$

This property, called *homogeneity*, is a property of all linear systems. However, a linear system involves much more than (1.1). It must also possess the property of *superposition*. That is, if

$$x_1(t) \to y_1(t)$$

and

$$x_2(t) \to y_2(t)$$

then

$$x_1(t) + x_2(t) \to y_1(t) + y_2(t) \tag{1.2}$$

for some class of inputs $\{x(t)\}$. A system is linear if and only if superposition and homogeneity hold. We can combine (1.1) and (1.2) into a single equation. We define a system to be *linear* if and only if

$$\alpha x_1(t) + \beta x_2(t) \to \alpha y_1(t) + \beta y_2(t) \tag{1.3}$$

where α and β are constants. A convenient notation for the arrows of (1.1)–(1.3) is to use functional notation and represent the transformation of inputs into outputs by

$$y(t) = H[x(t)] \tag{1.4}$$

The system represented by (1.4) is linear if and only if H is a linear transformation: i.e., $H[\alpha x_1(t) + \beta x_2(t)] = \alpha H[x_1(t)] + \beta H[x_2(t)]$ Although the property of homogeneity can be inferred from the superposition property if α is a rational number, there are mathematical transformations which satisfy superposition and not homogeneity. However, these are really pathological examples, and they would not arise physically. Thus, we shall be content to verify the linearity of an input-output relation by checking superposition only.

EXAMPLE 1.1.   Suppose a system has an input-output relation given by the linear equation

$$y(t) = ax(t) + b \qquad (1.5)$$

Graphically, $x(t)$ and $y(t)$ are related as shown in Figure 1.2. Does Figure 1.2 represent a linear system? Consider the superposition property. If we apply an input $x_1(t)$, the corresponding output is $y_1(t) = ax_1(t) + b$. Similarly the input $x_2(t)$ gives as an output $y_2(t) = ax_2(t) + b$. If we now apply the input $x_1(t) + x_2(t)$, we obtain as an output $a(x_1(t) + x_2(t)) + b$ which is not equal to $(y_1(t) + y_2(t)) = a(x_1(t) + x_2(t)) + 2b$ unless $b = 0$. Thus, the system of Figure 1.2 is not linear even though the equation that relates $x(t)$ and $y(t)$ is a linear equation. It seems rather discouraging to have a system described by a linear equation and yet not be able to use linear analysis to analyze the system. We shall show in Section 1.5 how one can deal with this problem so that linear analysis can be applied to this system.

EXAMPLE 1.2.   Consider the circuit shown in Figure 1.3. In this system, suppose that $x(t)$ is an input voltage and that $y(t)$ is an output voltage. As long as point A is less than 3 V, then

$$y(t) = x(t)/2 \qquad (1.6)$$

Thus, $x_1(t)$ gives rise to $y_1(t) = x_1(t)/2$ and $x_2(t)$ gives rise to $y_2(t) = x_2(t)/2$. This means that $(x_1(t) + x_2(t))$ gives rise to $y_3(t) = (x_1(t) + x_2(t))/2$, which is $y_1(t) + y_2(t)$ (provided that $(x_1(t) + x_2(t)) < 3$). Thus, the system is linear as long as the diode is not conducting. This fact implies that point A must be less than 3 V: i.e., the system is linear for the class of signals $\{x(t)\}$, where $x(t)$ or any combination of $x(t)$'s is

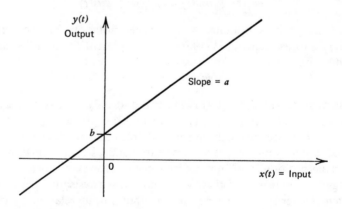

FIGURE 1.2

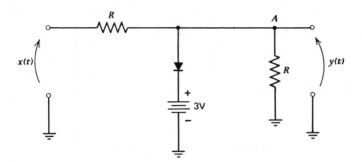

FIGURE 1.3

always less than 3 V. If any input or combination of inputs in the class of signals $\{x(t)\}$ exceeds 3 V, then the system is nonlinear. Notice that to be precise, we must specify not only the system, but also the class of input signals. Usually the latter is implicitly specified.

Continuous-Time and Discrete-Time Systems

As we have implied, we are interested in inputs and responses that may vary with time. If the input and output of a system are capable of changing at any instant of time, we call the system a continuous-time system. The input $x(t)$ and output $y(t)$ are functions of the continuous-time variable t. Notice that the input and output do not themselves have to be continuous functions.

Other systems are such that the signals associated with the system change only at discrete instants—for example, the contents of an arithmetic unit of a digital computer, or the nerve impulses from a sensory organ in an animal. The discrete time intervals may change or be constant. The values of the signals may change from one time interval to the next. The signals may not be defined between intervals, or they may be constant. These systems are called discrete-time systems. To indicate the value of the signals at the time intervals t_k, we use the notations x_{t_k}, x_k, or $x(k)$ interchangeably.

Continuous-time systems usually are modeled by the use of differential equations. The analogous models for discrete-time systems are so-called difference equations.

EXAMPLE 1.3. The system shown in Figure 1.4 takes an input sequence of values denoted by $x_1, x_2, x_3, \ldots$ and transforms them into an output sequence of values $y_1, y_2, y_3, \ldots$. The output at time $t = k$ is given by $y_k = Gx_k + x_{k-1}$, where G is a constant. Suppose that we

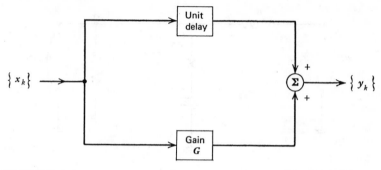

FIGURE 1.4

have two input sequences $\{x_k\}$ and $\{x_k'\}$. The two corresponding output sequences are $\{y_k\}$ and $\{y_k'\}$, where

$$y_k = Gx_k + x_{k-1}, \qquad y_k' = Gx_k' + x_{k-1}'$$

If we have an input $\{x_k\} + \{x_k'\}$, we obtain the output

$$G(x_k + x_k') + (x_{k-1} + x_{k-1}') = Gx_k + x_{k-1} + Gx_k' + x_{k-1}'$$

Thus, the system satisfies superposition and so is linear.

Time-Invariant and Time-Varying Systems

The class of all linear systems can be subdivided into subclasses. For example, there are linear time-invariant and time-varying systems. Systems with parameters that vary with time are called time-varying systems. These systems are most often modeled by means of linear differential or difference equations with time-varying coefficients. We shall deal only briefly with such systems. Time-invariant or constant-parameter systems are most often modeled by linear differential or difference equations with constant coefficients.

A simple characterization of time-invariant systems can be given by shifting the input signal in time. Suppose that $x(t)$ gives rise to an output $y(t)$. Now consider the output due to $x(t - T)$: i.e., the original input signal delayed by T. If a system is time invariant, then $x(t - T)$ gives rise to an output $y(t - T)$. In symbols, for a time-invariant system, if

$$x(t) \rightarrow y(t)$$

then

$$x(t - T) \rightarrow y(t - T) \tag{1.7}$$

for any values of t and T.

The discrete analog of (1.7) is that if

$$x(n) \to y(n)$$

then

$$x(n - k) \to y(n - k) \tag{1.8}$$

for any values of n and k.

Equations (1.7) and (1.8) are another way of saying that the system response is independent of the time origin and depends only on the shape of the input.

Lumped and Distributed Parameter Systems

A physical system is a collection of interconnected elements. When we apply an input $x(t)$ to a system, then either the input stimulus reaches each element essentially simultaneously or there is some propagation time for the input to affect various elements within the system. If the input stimulus is propagated instantaneously throughout the system, the system is called a lumped parameter system. These systems are modeled by ordinary differential equations. In electrical systems, this fact means that the wavelength of the stimulus is large compared to the dimension of the elements.

Distributed parameter systems can be modeled by use of partial differential equations. In these systems, the input stimuli affect elements of the system in some nonzero time depending on the propagation speed of the stimulus through the system. A chemical system involving the diffusion of one material throughout another material is an example of a distributed parameter system. Propagation of electrical energy through transmission lines is another example. Distributed parameter models are also useful in modeling biological systems—eg., understanding the electrical activity of a heart involves the analysis of a distributed parameter system.

Deterministic and Nondeterministic Systems

In many physical systems, it often occurs that one cannot control certain parameters within the system; or it may be that in the analysis of a system, certain parameters within the system are generated by a process too complicated to be completely understood. In these cases, there are system parameters whose values are unknown, or at least uncertain. Usually we attempt to model such situations probabilistically. For example, when we model communication through the atmosphere, disturbances or noises introduced are specified statistically. A system with one or more unknown or uncertain parameter elements we call nondeterministic. If the system parameters are known exactly, the system is called deterministic.

Memory and Memoryless Systems

Another distinction which is often useful concerns the system's memory of an input stimulus. If the output at any time t or t_k depends only on the input at the same time, the system is called memoryless. If the output at time t or t_k depends on input values over some interval, say $(t - T, t)$, then the system has memory of length T. For example, a system consisting of a capacitor with the input defined as current and output defined as voltage has a memory of infinite duration, because

$$e(t) = \frac{1}{C} \int_{-\infty}^{t} i(t') \, dt' \tag{1.9}$$

Equation (1.9) indicates that the output depends on the input over the entire interval $(-\infty, t)$.

In contrast, a system consisting of a water pipe with an input equal to flow rate $f(t)$ and an output equal to pressure $P(t)$ is a memoryless system. The pressure $P(t)$ is affected only by the immediate flow rate and is given by

$$P(t) = R f(t) \tag{1.10}$$

where R is the resistance of the pipe to water flow.

We have classified systems in a variety of ways:

(1) linear versus nonlinear
(2) continuous time versus discrete time
(3) time-invariant versus time-varying
(4) lumped parameter versus distributed parameter
(5) deterministic versus nondeterministic
(6) memoryless versus memory

We introduce these classifications so that we can conveniently lump together models and techniques that apply to particular classes of systems. It is probably as important to realize what techniques are not available for a given class of systems as it is to know those techniques that can be used. In other words, it is important to know the boundaries of our knowledge. These classifications permit us to conveniently specify such boundaries. In the material which follows we shall be primarily concerned with models for *deterministic, time-invariant, lumped parameter, linear* systems.

1.3 MODELS FOR PHYSICAL SYSTEMS

Linear models for systems arise naturally as a convenient formulation for a wide variety of physical phenomena. We present some examples taken from various disciplines to motivate the use of linear models as a description of physical systems.

EXAMPLE 1.4. Robert Hooke studied mechanical springs quantitatively and found by experiment that over a limited range, the deflection of a spring is proportional to the applied force F. If we let y be the spring deflection, then a model describing this aspect of the spring is

$$\text{Force} = k \cdot \text{Deflection} \tag{1.11}$$

where k is a constant characteristic of the spring. Equation (1.11), known as Hooke's law, is a good model for the spring so long as the deflection y is small. If we identify the force F as the input of the system and y as the output, then the input-output equation for this system is

$$y(F) = F/k \tag{1.12}$$

Application of the definition of linearity to (1.12) quickly confirms that this system is linear. If we apply F_1 and F_2 separately, the corresponding outputs are F_1/k and F_2/k, respectively. If we now add the inputs and apply this to our system, the output is $(F_1 + F_2)/k$, which equals the sum of the two outputs obtained previously. Thus, the system satisfies superposition, and so is linear.

EXAMPLE 1.5. Population studies attempt to predict future growth based on previous growth data. Populations are, of course, discrete, but if we look at large populations, we can consider the population at time t, $P(t)$, a continuous variable. The simplest growth model is obtained by noting that experimental data indicate that the rate of growth of population is proportional to the total population for limited population ranges. Thus, if k is the net per capita growth rate, then we can write that

$$\frac{dP(t)}{dt} = kP(t) \quad \text{or} \quad \frac{dP(t)}{dt} - kP(t) = 0, \quad t > 0 \tag{1.13}$$

This simple model, in the form a linear differential equation, results in a model that predicts the future population as

$$P(t) = P_0 e^{kt}, \quad t \geq 0 \tag{1.14}$$

where P_0 is the initial population. In this system, we identify the output of the system as the population at time t. What is the input to this system? Usually in such models the input is found on the right-hand side of (1.13), which in this case is zero. However, if we consider the initial population P_0 as the input to the system, we again see that our model is linear. If we have inputs P_0 and P_0', the outputs are $P_0 e^{kt}$ and

$P_0'e^{kt}$, respectively. If we have the input $(P_0 + P_0')$, then the output is $(P_0 + P_0')e^{kt}$. This quantity equals the sum of the outputs P_0e^{kt} and $P_0'e^{kt}$ resulting from the separate inputs, and thus verifies superposition. This simplified model is, of course, accurate only over a limited range of t, because ultimately such factors as food supplies, disease, and so on will limit the validity of (1.13) as a model.

EXAMPLE 1.6. A simple RC network shown in Figure 1.5 has an input $i(t)$ and an output $e(t)$ as indicated. Assuming that the initial energy storage in the system is zero (this assumption means there is initially no net charge in the capacitor), we can use Kirchhoff's voltage equation to write

$$e(t) = Ri(t) + \frac{1}{C} \int_{-\infty}^{t} i(t') \, dt' \tag{1.15}$$

One can again easily verify that (1.15) represents the input-output relation for a linear system. Suppose that we have two separate inputs $i_1(t)$ and $i_2(t)$. The corresponding outputs are

$$e_1(t) = Ri_1(t) + \frac{1}{C} \int_{-\infty}^{t} i_1(t') \, dt'$$

$$e_2(t) = Ri_2(t) + \frac{1}{C} \int_{-\infty}^{t} i_2(t') \, dt'$$

If we now apply as an input $(i_1(t) + i_2(t))$, the corresponding output is $e_3(t)$, given by

$$e_3(t) = R(i_1(t) + i_2(t)) + \frac{1}{C} \int_{-\infty}^{t} (i_1(t') + i_2(t')) \, dt'$$

$$= Ri_1(t) + \frac{1}{C} \int_{-\infty}^{t} i_1(t') \, dt' + Ri_2(t) + \frac{1}{C} \int_{-\infty}^{t} i_2(t') \, dt'$$

$$= e_1(t) + e_2(t)$$

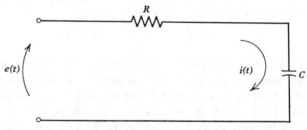

FIGURE 1.5

Thus, the system of Figure 1.5 satisfies the property of superposition and so is linear. Notice that in this example we have assumed the initial voltage on the capacitor to be zero. Suppose instead that there is an initial charge on the capacitor. This initial charge divided by the capacitance C is then some initial voltage across the capacitor. Assume the sign of this initial voltage to be a voltage drop in the clockwise direction. The resultant Kirchhoff voltage equation is

$$e(t) = Ri(t) + \frac{1}{C} \int_0^t i(t') \, dt' + v_0, \qquad t \geq 0 \qquad (1.16)$$

where

$$v_0 = \frac{q}{C} = \frac{\int_{-\infty}^0 i(t') \, dt'}{C} \qquad (1.17)$$

is the initial voltage on the capacitor. Because of this constant v_0, which represents the initial energy storage, (1.16) does not represent a linear input-output relation between $i(t)$ and $e(t)$. It is for a similar reason that Example 1.1 does not represent a linear system. Section 1.5 points out how one might handle this problem of initial energy storage. Unless we state otherwise, we shall assume initial energy storage within a system to be zero.

EXAMPLE 1.7.   In mammal bones, calcium is continually being exchanged between the bone and the interstitial fluids that surround the bones. The bones act as a reservoir for calcium for use in the rest of the body. Experimental evidence[1] indicates that this bone-calcium system can be modeled by the use of first-order rate processes: i.e., the rate of diffusion of material is proportional to the total amount of material present. Suppose in Figure 1.6 that $x(t)$ is the amount of

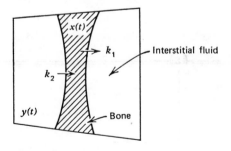

FIGURE 1.6

[1] Solomons, C., Ernisse, D., and R. A. Roberts, "The Effects of Mechanical Stress on Ca^{45} Transport in Male Rat Bone *in Vitro*," *Mathematical Biosciences* **9**, 17–27 (July, 1970).

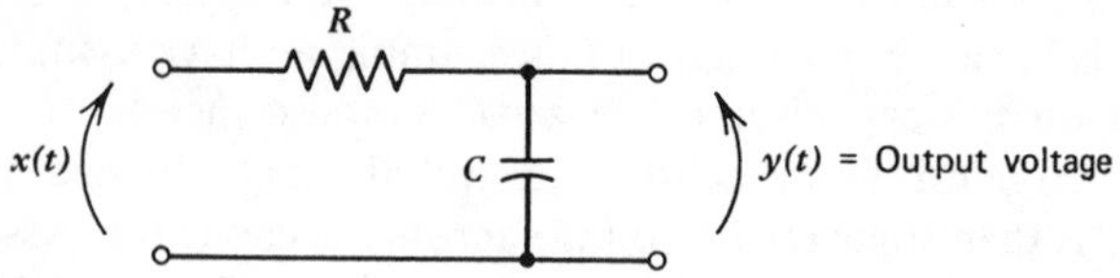

R

$x(t)$ ⎛ C ⎞ $y(t)$ = Output voltage

FIGURE 1.7

calcium in the bone at time t and $y(t)$ is the amount of calcium in the interstitial fluids at time t. Identify $x(t)$ as the input to this system and $y(t)$ as the output. Let k_1 and k_2 be the rate constants governing the transport of calcium out of and into the bone, respectively. Assuming that no net calcium is being either destroyed or created in the body, we find that the rate of change of the amount of calcium in the bone is

$$\frac{dy(t)}{dt} = k_1 x(t) - k_2 y(t) \qquad (1.18)$$

subject to the condition that $y(t) + x(t) =$ constant. Equation (1.18) is a linear model describing the variation of calcium within the bone and the interstitial fluids. With this model, for example, we can measure $y(t)$ and thereby predict the value of $x(t)$, which is essentially unmeasurable directly. This very simple model has been used with some success to characterize calcium transport in rat bones.

EXAMPLE 1.8.   A simple analog filter which has the effect of smoothing input voltage signals $x(t)$ is a single stage RC filter shown in Figure 1.7. One can obtain a similar system for discrete-time signals by using an adder, an amplifier, and a delay as shown in Figure 1.8. In this system, the output at time $t = k$ is

$$y_k = G y_{k-1} + x_k \qquad (1.19)$$

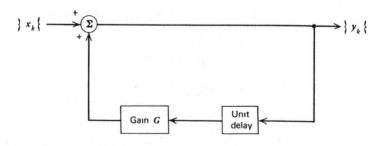

FIGURE 1.8

Equation (1.19) is a linear first-order difference equation which models the system of Figure 1.8. The system is linear, as we can verify by checking superposition. This discrete-time system performs a smoothing operation on the sequence $\{x_k\}$ in a manner analogous to the smoothing operation performed by the RC filter on $x(t)$.

EXAMPLE 1.9. Consider the mechanical system shown in Figure 1.9. Take as the coordinate y the distance of the mass from the rest position. The equation of motion for this system can be obtained from Newton's second law, $F = ma$. In this system, the restoring forces are the spring force $ky(t)$ and the viscous friction force $\beta\, dy(t)/dt$. Thus, the equation that describes the variation of $y(t)$ is

$$M\frac{d^2y(t)}{dt^2} = -ky(t) - \beta\frac{dy(t)}{dt} \tag{1.20}$$

If we interpret $y(t)$ as the output of a system, we see that (1.20) is a linear differential equation which relates $y(t)$ to an input consisting of the initial energy storage. If we forced the system with an external force $x(t)$, this term would appear on the right-hand side of (1.20). One can again check that superposition holds and so confirm that (1.20) is a linear system model.

The preceding examples illustrate the wide application of linear models. Notice that all the models evolved as some form of linear equation—either algebraic, differential, or difference equations. In what follows, we shall be primarily interested in systems we can model with linear differential or difference equations with constant coefficients. This class of models is important in applications, and so we turn now to a brief review of classical time-domain methods for their solution. We first consider the solution of linear differential equations with constant coefficients.

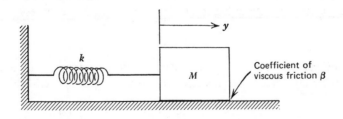

FIGURE 1.9

1.4   TIME-DOMAIN SOLUTION OF LINEAR DIFFERENTIAL EQUATIONS

An ordinary linear differential equation with constant coefficients is characteristic of linear lumped time-invariant systems. In this section, we review briefly the theory of solving this class of differential equations. We restrict our discussion here to single input-output systems. The multidimensional case is the subject of Chapter 3. Suppose the differential equation is of order n

$$b_n \frac{d^n y(t)}{dt^n} + b_{n-1} \frac{d^{n-1} y(t)}{dt^{n-1}} + \cdots + b_0 y(t) = x(t) \qquad (1.21)$$

It is convenient to write (1.21) in operational form as

$$\left(b_n \frac{d^n}{dt^n} + b_{n-1} \frac{d^{n-1}}{dt^{n-1}} + \cdots + b_0 \right)[y] = x(t) \qquad (1.22)$$

or more briefly as

$$L[y] = x(t) \qquad (1.23)$$

where L is the operator

$$L = b_n \frac{d^n}{dt^n} + b_{n-1} \frac{d^{n-1}}{dt^{n-1}} + \cdots + b_0 \qquad (1.24)$$

It is easy to verify that L is a linear operator, that is

$$L[c_1 y_1 + c_2 y_2] = c_1 L[y_1] + c_2 L[y_2]$$

Thus, if $y_1(t)$ and $y_2(t)$ are solutions to $L[y] = 0$, then so also is $c_1 y_1(t) + c_2 y_2(t)$ a solution to $L[y] = 0$.

The solution to (1.21) is made up of two components: (1) the source-free (transient, natural, complementary, homogeneous) solution, and (2) the component resulting from the source (forced, steady-state, nonhomogeneous, particular solution). The transient solution $y_c(t)$ is found from the homogeneous equation corresponding to (1.21). That is, $y_c(t)$ must satisfy

$$L[y_c(t)] = 0 \qquad (1.25)$$

For the special case of constant coefficients, the solution to (1.25) has been completely solved.[2] In this case, the solution has a general form

$$y_c(t) = c_1 y_1(t) + c_2 y_2(t) + \cdots + c_n y_n(t) \qquad (1.26)$$

[2] Wylie, C. R., *Advanced Engineering Mathematics*, McGraw-Hill, New York, 1966.

The functions $y_1(t), y_2(t), \ldots, y_n(t)$ depend on the roots of the associated characteristic equation,

$$f(r) = b_n r^n + b_{n-1} r^{n-1} + \cdots + b_0 = 0 \qquad (1.27)$$

where $b_n, \ldots, b_0$ are the same coefficients as in (1.24). If (1.27) has n distinct roots $r_1, r_2, \ldots, r_n$, the functions $y_i(t)$, $i = 1, 2, \ldots, n$ are $y_i(t) = e^{r_i t}$ and the transient or complementary solution is

$$y_c(t) = c_1 e^{r_1 t} + c_2 e^{r_2 t} + \cdots + c_n e^{r_n t} \qquad (1.28)$$

Depending on the multiplicity of the roots in (1.27), the functions $y_i(t)$ take on various forms. We summarize the results below. After finding the n roots of the characteristic equation (1.27), assign the $y_i(t)$, $i = 1, 2, \ldots, n$ as follows:

(1) for each real root r, the function e^{rt}
(2) for each real root r of multiplicity k, the functions $e^{rt}, te^{rt}, \ldots, t^{k-1} e^{rt}$
(3) for each simple complex pair of roots $a \pm jb$, the functions $e^{at} \cos bt$ and $e^{at} \sin bt$
(4) for each complex pair of roots $a \pm jb$ of multiplicity k, the functions $e^{at} \cos bt$, $e^{at} \sin bt$, $te^{at} \cos bt$, $te^{at} \sin bt$, $\ldots$, $t^{k-1} e^{at} \cos bt$, $t^{k-1} e^{at} \sin bt$.

The constants c_i, $i = 1, 2, \ldots, n$ in (1.26) are determined from the boundary or initial conditions of the problem.

EXAMPLE 1.10. It is a simple matter to verify that $y_c(t)$ given in (1.28) is a solution to the homogeneous equation of (1.25). Suppose that the roots of the characteristic equation are distinct. The solution is then given by (1.28). Substituting (1.28) into (1.25) yields

$$b_n \frac{d^n}{dt^n} y_c(t) + b_{n-1} \frac{d^{n-1}}{dt^{n-1}} y_c(t) + \cdots + b_0 y_c(t) \overset{?}{=} 0$$

Performing the indicated differentiations gives

$$b_n(c_1 r_1^n e^{r_1 t} + c_2 r_2^n e^{r_2 t} + \cdots + c_n r_n^n e^{r_n t})$$
$$+ b_{n-1}(c_1 r_1^{n-1} e^{r_1 t} + \cdots + c_n r_n^{n-1} e^{r_n t})$$
$$+ \cdots + b_0(c_1 e^{r_1 t} + \cdots + c_n e^{r_n t}) \overset{?}{=} 0$$

Now rearrange terms to obtain

$$c_1 e^{r_1 t}[b_n r_1^n + b_{n-1} r_1^{n-1} + \cdots + b_0]$$
$$+ c_2 e^{r_2 t}[b_n r_2^n + b_{n-1} r_2^{n-1} + \cdots + b_0] + \cdots \overset{?}{=} 0 \qquad (1.29)$$

Now consider the bracketed expressions. They are all zero because each of the roots $r_1, r_2, \ldots, r_n$ satisfies the characteristic equation

$$b_n r_i^n + b_{n-1} r_i^{n-1} + \cdots + b_0 = 0, \qquad i = 1, 2, \ldots, n$$

Therefore the left-hand side of (1.29) is zero and $y_c(t)$ of (1.28) is a solution to the homogeneous equation (1.25).

EXAMPLE 1.11.   Consider the differential equation

$$\frac{d^3y(t)}{dt^3} - \frac{d^2y(t)}{dt^2} + \frac{dy(t)}{dt} - y(t) = 0$$

The characteristic equation is

$$f(r) = r^3 - r^2 + r - 1 = 0$$

which has roots j, $-j$, and 1. The homogeneous solution function is therefore

$$
\begin{aligned}
y_c(t) &= c_1 e^t + c_2 e^{-jt} + c_3 e^{jt} \\
&= c_1 e^t + c_2 \cos t - c_2 j \sin t + c_3 \cos t + c_3 j \sin t \\
&= c_1 e^t + c_2' \cos t + c_3' \sin t
\end{aligned}
$$

We call $y_c(t)$ a *solution function* if the constants c_i, $i = 1, 2, \ldots, n$ are not specified.

The solution resulting from the forced response is somewhat more involved. There are several methods used in obtaining $y_p(t)$ including educated guessing. The method of undetermined coefficients can be used if the derivatives of $x(t)$ result in a finite number of independent functions. We outline the method of undetermined coefficients briefly.

Let $D \triangleq d/dt$ so that the operator L can be written

$$L = b_n D^n + b_{n-1} D^{n-1} + \cdots + b_0 \tag{1.30}$$

i.e., as a polynomial operator in the operator D. We wish to find an operator which "annihilates" $x(t)$. That is, given $x(t)$, we wish to find an operator L_A such that

$$L_A[x(t)] = 0 \tag{1.31}$$

Such an operator can always be found if $x(t)$ is a solution of a homogeneous equation with constant coefficients. For example, if $x(t) = e^{at}$, then the annihilator operator is $L_A = D - a$. If $x(t) = A \cos bt + B \sin bt$, then the annihilator operator is $L_A = D^2 + b^2$. For a sum of such terms, the annihilator operator is the product of the operators for each term in the sum.

Once the annihilator operator has been found, applying it to both sides of the original nonhomogeneous equation results in a homogeneous equation. That is, suppose L_A is the annihilator for $x(t)$ and we wish to solve

$$L[y(t)] = x(t) \tag{1.32}$$

Then

$$L_A\{L[y(t)]\} = L_A[x(t)] = 0 \tag{1.33}$$

We can solve (1.33) by the method outlined previously. The solution function is

$$y(t) = c_1 y_1(t) + \cdots + c_n y_n(t) + c_{p_1} y_{p_1}(t) + \cdots + c_{p_r} y_{p_r}(t).$$

The first n solution functions satisfy $L[y(t)] = 0$. Therefore, if we substitute $y(t)$ into (1.32), these terms will sum to zero on the left-hand side. The only terms left will be $c_{p_1} y_{p_1}(t) + \cdots + c_{p_r} y_{p_r}(t)$ which arise from the forcing function. If we now equate coefficients of like functions on both sides of the equation, we can evaluate the constants $c_{p_1}, \ldots, c_{p_r}$, and thus obtain the forced solution. The following examples demonstrate these ideas.

EXAMPLE 1.12.   Consider the differential equation

$$L[y(t)] = (D^2 + 1)[y(t)] = e^t$$

In this case, the annihilator for e^t is $(D - 1)$ because $(D - 1)[e^t] = 0$. Thus we operate on both sides by $(D - 1)$ and obtain the homogeneous equation

$$(D - 1)(D^2 + 1)[y(t)] = 0$$

We can solve this equation by means of the characteristic equation

$$(r - 1)(r^2 + 1) = 0$$

$$\alpha \pm \beta j$$

This equation has roots $1, j, -j$. Thus, the solution function is

$$0 \pm j$$

$$y(t) = c_1 \cos t + c_2 \sin t + c_3 e^t$$

If we now substitute this solution function into the original equation, the first two terms are zero because they are solutions to $L[y] = 0$. We obtain one equation for c_3, the undetermined coefficient,

$$(D^2 + 1)[c_1 \cos t + c_2 \sin t + c_3 e^t] = e^t$$

Thus

$$0 + (D^2 + 1)[c_3 e^t] = e^t$$

$$c_3 e^t + c_3 e^t = e^t$$

or

$$2c_3 e^t = e^t$$

Hence, $c_3 = 1/2$, and so $c_p y_p(t) = e^t/2$.

EXAMPLE 1.13.   Consider the differential equation

$$L[y(t)] = (D^2 + 1)[y(t)] = \sin t$$

In this problem, the annihilator is $(D^2 + 1)$. Thus, the homogeneous equation we wish to solve is

$$(D^2 + 1)(D^2 + 1)[y(t)] = 0$$

The corresponding characteristic equation is

$$(r^2 + 1)(r^2 + 1) = 0$$

which has roots j, $-j$ each of multiplicity two. Thus, the solution function is

$$y(t) = c_1 \cos t + c_2 \sin t + c_3 t \cos t + c_4 t \sin t$$

Substituting this solution function into the original equation, we have

$$(D^2 + 1)[c_1 \cos t + c_2 \sin t + c_3 t \cos t + c_4 t \sin t] = \sin t$$

Performing the indicated differentiation and equating coefficients of like functions on both sides of the equation, we find that $c_4 = 0$ and $c_3 = -\frac{1}{2}$. Hence, $c_p y_p(t) = -\frac{1}{2} t \cos t$. If the initial conditions are $y(0) = 1$ and $y'(0) = 0$, we can solve for c_1 and c_2 using our knowledge of the forced solution. We have

$$y(t) = c_1 \cos t + c_2 \sin t - \tfrac{1}{2} t \cos t$$

And so

$$y(0) = 1 = c_1$$
$$y'(0) = 0 = c_2 - \tfrac{1}{2}$$

Thus, we know that

$$c_1 = 1$$
$$c_2 = \tfrac{1}{2}$$

and the complete solution is

$$y(t) = \cos t + \tfrac{1}{2} \sin t - \tfrac{1}{2} t \cos t$$

Transient and Steady-State Components

At this point, it is useful to view the mathematical solution for (1.21) in terms of the system and the input stimulus $x(t)$. We have decomposed the total response of the system $y(t)$ into two components. The source-free or transient solution is obtained for a zero input stimulus. Hence, it depends only on the character of the system and not on any external signals. Because this response occurs for $x(t) \equiv 0$, it is also known as the natural response of the system. The terms source-free, transient, natural, homogeneous, and complementary are all used to describe this response that results only from the character of the system. The term $\cos t + \frac{1}{2} \sin t$ in Example 1.13 is an example of a transient solution.

In contrast, the forced solution is characteristic of both the system and the input stimulus. If we change either the system or the input, we change the forced response. The forced response is also known as the steady-state response because this response is a driven response and can exist after the transient response has died away. The terms steady-state, forced, driven, component resulting from the source, nonhomogeneous, and particular are all used to describe this response. The term $-\frac{1}{2}t\cos t$ in Example 1.13 is an example of a forced solution.

To summarize: The transient response is characteristic of the system and is obtained by solving the homogeneous differential equation that models the system; the steady-state response depends on both the system and the nature of the input stimulus. It is the output response that is forced upon the system by the input stimulus and must be found by solving the nonhomogeneous differential equation.

1.5 INITIAL ENERGY STORAGE IN LINEAR SYSTEMS

There is another decomposition of the total response that is sometimes useful in linear systems analysis. This decomposition involves separating the response resulting from the initial energy storage and the response resulting from the system input. By separating the initial energy storage from the remaining system response, we often obtain a simpler way of handling and interpreting initial energy storage in a system.

The decomposition shown in Figure 1.10 allows us to resolve the problem we encountered in Examples 1.1 and 1.6, where a seemingly linear system was described by an input-output relation which did not satisfy the superposition property and hence was not linear. In the decomposition shown in

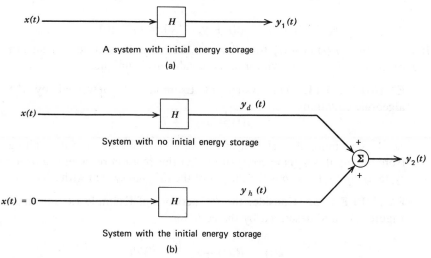

$x(t) \longrightarrow \boxed{H} \longrightarrow y_1(t)$

A system with initial energy storage

(a)

$x(t) \longrightarrow \boxed{H} \xrightarrow{y_d\ (t)}$

System with no initial energy storage

$x(t) = 0 \longrightarrow \boxed{H} \xrightarrow{y_h\ (t)}$

System with the initial energy storage

$\xrightarrow{+}\ \Sigma\ \xrightarrow{} y_2(t)$

$+$

(b)

FIGURE 1.10

Figure 1.10, each of the systems labeled H is identical to the given system in the top of the figure. We write the system output $y_2(t)$ as the sum of $y_d(t)$, the output of an initially relaxed system driven by the input $x(t)$, plus $y_h(t)$, the output of an unforced system with initial conditions identical to those of our given system. Then we have

$$y_2(t) = y_d(t) + y_h(t)$$

where

$$L[y_d(t)] = x(t)$$

$$y_d(0) = y'_d(0) = \cdots = y_d^{(n-1)}(0) = 0$$

and

$$L[y_h(t)] = 0$$

$$y_h(0) = y_1(0), \qquad y'_h(0) = y'_1(0), \ldots, \qquad y_h^{(n-1)}(0) = y_1^{(n-1)}(0).$$

Here $y_1(0), y'_1(0), \ldots, y_1^{n-1}(0)$ are the given initial conditions for our system. To demonstrate that $y_2(t)$ of the decomposed system equals $y_1(t)$ of the original system, we note that the two functions satisfy the same differential equation

$$L[y_2(t)] = L[y_h(t) + y_d(t)] = x(t)$$

$$L[y_1(t)] = x(t)$$

and that they have the same initial conditions

$$y_2(0) = y_h(0) + y_d(0) = y_1(0)$$
$$y'_2(0) = y'_h(0) + y'_d(0) = y'_1(0)$$
$$\cdots$$
$$y_2^{(n-1)}(0) = y_h^{(n-1)}(0) + y_d^{(n-1)}(0) = y_1^{(n-1)}(0)$$

It follows that $y_1(t) = y_2(t)$ for all $t > 0$, because the solution to an nth order differential equation with n initial conditions is unique.

EXAMPLE 1.14. The system of Example 1.1, described by the algebraic equation

$$y(t) = ax(t) + b$$

can be decomposed as shown in Figure 1.11. Here the output resulting from the input is $y_d(t) = ax(t)$ (note that this relation represents a *linear* system), and $y_h(t) = b$ is the output of the original system with $x(t) = 0$.

EXAMPLE 1.15. Consider the RC network of Example 1.6 shown in Figure 1.12 and described by the equation

$$e(t) = Ri(t) + \frac{1}{C} \int_{-\infty}^{t} i(t') \, dt'$$

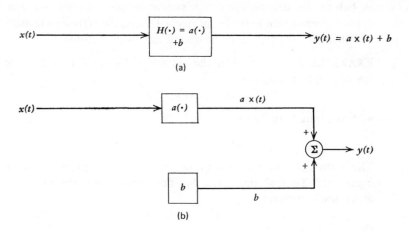

(a)

(b)

FIGURE 1.11

where $i(t)$ is the system input and $e(t)$ the system output. In our system decomposition, we set $y_a(t)$ equal to the output of the same system, but with no energy storage at $t = 0$. Thus

$$y_a(t) = Ri(t) + \frac{1}{C} \int_0^t i(t') \, dt'$$

The term $y_h(t)$ equals the output of the original system with zero input, but with the given initial condition $e(0) = v_0$.

$$y_h(t) = v_0$$

Thus we have

$$e(t) = y_a(t) + y_h(t)$$

$$= Ri(t) + \frac{1}{C} \int_0^t i(t') \, dt' + v_0$$

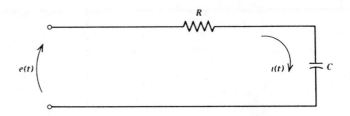

FIGURE 1.12

as before. By decomposing our system in this manner, we can now apply superposition to the linear part—that is, the $x(t) \to y_d(t)$ relation—and then add the term $y_h(t)$ to obtain the entire output.

EXAMPLE 1.16. Consider the system of Example 1.13 described by the differential equation

$$(D^2 + 1)[y(t)] = x(t)$$

with the initial conditions

$$y(0) = 1 \qquad y'(0) = 0$$

The term $y_h(t)$ in the system decomposition is independent of the input $x(t)$. To find this term, we need only solve the homogeneous differential equation

$$(D^2 + 1)[y_h(t)] = 0$$

with initial conditions $\pm j$

$$y_h(0) = 1, \qquad y_h'(0) = 0$$

$2c$

Solving this equation, we have

$$y_h(t) = c_1 \cos t + c_2 \sin t$$

where

$$y_h'(t) = c_2 \cos t - c_1 \sin t$$

$$y_h(0) = 1 \quad \text{gives} \quad c_1 = 1$$

$$y_h'(0) = 0 \quad \text{gives} \quad c_2 = 0$$

Thus $y_h(t) = \cos t$, and we can sketch the equivalent system as in Figure 1.13. To find $y_d(t)$, we must solve the nonhomogeneous equation

$2s$

$$(D^2 + 1)[y_d(t)] = x(t)$$

with $x(t) = \sin t$ and zero initial conditions. As before, we obtain

$$y_d(t) = c_1 \cos (t) + c_2 \sin t + c_3 t \cos t + c_4 t \sin t$$

$2s$

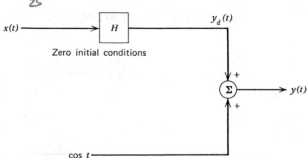

FIGURE 1.13

and substituting in the original differential equation, we find

$$c_3 = -\tfrac{1}{2}$$
$$c_4 = 0$$

as before. Applying the initial conditions

$$y_d(0) = 0$$
$$y_d'(0) = 0$$

we find that

$$c_1 = 0$$
$$c_2 = 1/2$$

Thus

$$y_d = \tfrac{1}{2}(\sin t - t \cos t)$$

Summing the terms $y_h(t)$ and $y_d(t)$, we find the solution

$$y(t) = y_h(t) + y_d(t)$$
$$= \cos t + \tfrac{1}{2}(\sin t - t \cos t)$$

as before.

1.6 LINEAR DIFFERENCE EQUATIONS

The theory of linear difference equations closely parallels the theory of linear differential equations. In fact, the major distinction between the two is that a differential equation expresses a relationship between *continuous-time* functions, whereas a difference equation expresses a relationship between *discrete-time* functions or *sequences*. Before considering the solution of difference equations, we shall diverge briefly to define what we mean by a sequence and to look at three examples which are naturally modeled by the use of difference equations.

Sequences

A sequence of numbers $\{x_k\}$ is merely a collection of numbers which are indexed by a set of integers $\{k\}$. We denote the sequence itself by $\{x_k\}$ and the values it assumes for the index k by x_k. Thus, if the index k takes on the values $0, 1, \ldots$, then the values of the sequence are $x_0, x_1, x_2, \ldots$, respectively. We can define $\{x_k\}$ in two ways:

(1) We can specify a rule or formula for calculating x_k. For example, if $x_k = a^k$, we generate the sequence $\{\cdots a^{-1}, 1, a^1, a^2, \ldots\}$.

(2) We can also list the values of the sequence explicitly. For example, $\{x_k\} = \{\ldots, 0, 0, 1, 5, -3, 4, 9, 0, 0, \ldots\}$. We use the arrow to denote the x_0 term.

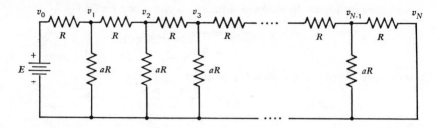

FIGURE 1.14

We define the sum of two sequences, $\{x_k\} + \{z_k\}$, as the sequence $\{y_k\}$, where

$$y_k = x_k + z_k$$

Similarly, the product of two sequences $\{x_k\}\{z_k\}$ is the sequence $\{y_k\}$, with

$$y_k = x_k z_k$$

The product of a constant α and a sequence $\alpha\{x_k\}$ is another sequence $\{y_k\}$, with

$$y_k = \alpha x_k$$

Difference Equations

The following three examples are intended to illustrate the occurrence of difference equations in various applications and to motivate the solution of these equations.

EXAMPLE 1.17. Suppose that we wished to find the voltages $v_1, v_2, v_3, \ldots, v_N$ at the nodes of the resistive network in Figure 1.14. These voltages may be related using Kirchhoff's current law, as shown in Figure 1.15. Summing currents at the node for v_k, we have $i_1 = i_2 + i_3$: i.e.,

$$\frac{v_{k-1} - v_k}{R} = \frac{v_k - v_{k+1}}{R} + \frac{v_k}{aR}$$

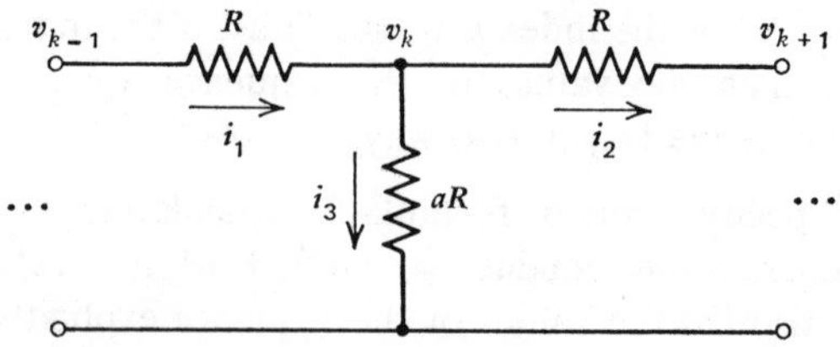

FIGURE 1.15

to yield the difference equation

$$av_{k+1} - (2a + 1)v_k + av_{k-1} = 0, \qquad k = 1, 2, \ldots, N - 1 \quad (1.34)$$

with the boundary conditions

$$v_0 = E$$
$$v_N = 0$$

In the case of an infinite ladder, the appropriate second boundary condition is

$$\lim_{k \to \infty} v_k = 0$$

In the next section we solve (1.34) to obtain the voltage v_k at any node k. Notice in this example that the index k is not a time index, but rather an index on the location of a node in a network.

*EXAMPLE 1.18. Two players, A and B, are flipping coins which land head up with probability p and tail up with probability $1 - p$. If a coin comes up heads, A gives B one chip; if tails, B gives A one chip. The game ends when either player wins all the coins. Suppose that A and B initially have a total of N coins, and that A now has k coins, $0 \le k \le N$. What is the probability, denoted by p_k, that A will eventually win the game? To set up the difference equation for this problem, assume A has k coins. He can win in two ways. He can lose one coin (probability p) and then eventually win (probability p_{k-1}). The probability of this joint event is thus pp_{k-1}. Similarly, he can win one coin (probability $1 - p$) and then eventually win (probability p_{k+1}). The probability of this joint event is $(1 - p)p_{k+1}$. Thus, we have

$$p_k = pp_{k-1} + (1 - p)p_{k+1}$$

or

$$(1 - p)p_{k+1} - p_k + pp_{k-1} = 0 \quad (1.35)$$

The boundary conditions are

$$p_0 = 0 \quad \text{and} \quad p_N = 1$$

If the coin is fair, $p = 1/2$. Then (1.35) can be written as

$$p_{k+1} - 2p_k + p_{k-1} = 0 \quad (1.36)$$

EXAMPLE 1.19. Once every cycle in a chemical process, x_k liters of chemical A and $100 - x_k$ liters of chemical B are added to 900 liters of mixture in a large vat, where $0 \le x_k \le 100$; $k = 1, 2, 3, \ldots$. The vat contents are thoroughly mixed and 100 liters of mixture are drawn off. Let y_k be the fractional concentration of chemical A in the mixture drawn off: i.e., $1000y_k$ is the amount of chemical A in the mixture. The total amount of chemical A at cycle k is also equal to the amount of chemical A added at cycle k plus the amount of chemical A at cycle $k - 1$, i.e.,

$$1000y_k = 900y_{k-1} + x_k$$

or

$$y_k - .9y_{k-1} = .001x_k, \qquad k = 1, 2, 3, \ldots \qquad (1.37)$$

with y_0 equal to the initial concentration of chemical A in the vat. In this system, the volume x_k could be considered the input, and the concentration y_k the output. The values of x_k and y_k for $k = 1, 2, 3, \ldots$ constitute the input and output sequences.

We see in the examples above that the sequences related or defined by a difference equation may be either finite or infinite, that the index need not necessarily be related to time, and that the input and output of a discrete-time system, as with a continuous-time system, may be measured in different units. We turn now to a concise method for solving difference equations like the ones we have derived in the examples above. Because the theory of differential equations is probably more familiar to most readers, we shall often use an analogy between the two to aid the discussion.

1.7   SOLUTION OF LINEAR DIFFERENCE EQUATIONS

To illustrate the solution of a difference equation, consider the equation

$$y_{k+1} - \alpha y_k = 1, \qquad k = 0, 1, 2, \ldots$$

where α is a constant and the initial condition is $y_0 = 1$. We can, by setting $k = 0, 1, 2, \ldots$, obtain that $y_1 = 1 + \alpha, y_2 = 1 + \alpha + \alpha^2, \ldots$. Reasoning inductively, we can write that

$$y_k = 1 + \alpha + \alpha^2 + \cdots + \alpha^k \qquad (1.38)$$

The above represents an explicit solution: however, it is not a closed form, because the dots must be included unless k is specified. We recognize that the right-hand side of (1.38) forms a geometric progression. Recalling a result of elementary algebra for the partial sum of a geometric series,[3] we have that

$$y_k = \frac{1 - \alpha^{k+1}}{1 - \alpha}, \qquad \alpha \neq 1, \qquad k = 0, 1, 2, \ldots$$

$$= k + 1, \qquad \alpha = 1, \qquad k = 0, 1, 2, \ldots$$

Thus, we have an explicit closed form expression for y_k.

---

[3] See Appendix B.

We are primarily interested in solution methods which result in closed-form solutions. We consider first the general homogeneous linear difference equation of nth order (with constant coefficients).

$$b_n y_{k+n} + b_{n-1} y_{k+n-1} + \cdots + b_0 y_k = 0, \qquad k = 0, 1, 2, \ldots \quad (1.39)$$

Define the shift operator S as

$$S[y_k] = y_{k+1}$$

Then we can write (1.39) as

$$L[y_k] = 0$$

where L is the linear difference operator

$$L = b_n S^n + b_{n-1} S^{n-1} + \cdots + b_0$$

As in the case of the analogous differential equations, we obtain a general solution to (1.39) as

$$y_k = c_1 y_k^1 + c_2 y_k^2 + \cdots + c_n y_k^n, \qquad k = 0, 1, 2, \ldots \quad (1.40)$$

where $\{y_k^i\}$, $i = 1, 2, \ldots, n$ are n linearly independent sequences which are obtained by use of the characteristic equation

$$f(r) = b_n r^n + b_{n-1} r^{n-1} + \cdots + b_0 = 0 \quad (1.41)$$

If (1.41) has n distinct roots $r_1, r_2, \ldots, r_n$, the sequences $\{y_k^i\}$, $i = 1, 2, \ldots, n$ are given by

$$y_k^i = r_i^k, \qquad i = 1, 2, \ldots, n$$

We can again verify that (1.40) is a solution of the homogeneous equation (1.39) by substitution. Assume that the characteristic equation (1.41) has n distinct roots. The solution function is then

$$y_k = c_1 r_1^k + c_2 r_2^k + \cdots + c_n r_n^k \quad (1.42)$$

Substituting (1.42) into (1.39) gives

$$b_n(c_1 r_1^{k+n} + c_2 r_2^{k+n} + \cdots + c_n r_n^{k+n})$$
$$+ b_{n-1}(c_1 r_1^{k+n-1} + \cdots + c_n r_n^{k+n-1}) + \cdots + b_0(c_1 r_1^k + \cdots + c_n r_n^k) \overset{?}{=} 0$$

Rearranging terms yields

$$c_1 r_1^k [b_n r_1^n + b_{n-1} r_1^{n-1} + \cdots + b_0]$$
$$+ c_2 r_2^k [b_n r_2^n + b_{n-1} r_2^{n-1} + \cdots + b_0] + \cdots \overset{?}{=} 0 \quad (1.43)$$

Again notice the bracketed terms. By virtue of the fact that the $r_1, r_2, \ldots, r_n$ are roots of the characteristic equation, these bracketed terms are all zero.

Thus, the left-hand side of (1.43) is zero. This fact verifies that (1.42) is a solution to (1.39).

EXAMPLE 1.20.   Consider the linear second-order difference equation

$$y_{k+2} - 5y_{k+1} + 6y_k = 0, \qquad k = 0, 1, 2, \ldots$$

The characteristic equation is

$$f(r) = r^2 - 5r + 6 = 0$$

This equation has roots of 2 and 3. Hence, the solution function y_k is

$$y_k = c_1 2^k + c_2 3^k, \qquad k = 0, 1, 2, \ldots$$

Again depending on the multiplicity of the roots in the characteristic equation, the functions y_k^i take on various forms. We summarize the results below. Assign the functions y_k^i according to the following rules:

(1) For each simple real root r, assign the function r^k.

(2) For each multiple real root r of multiplicity m, assign m functions $r^k, kr^k, \ldots, k^{m-1}r^k$.

(3) For each pair of complex roots $a \pm jb$, assign the functions $\zeta^k \cos \phi k$, $\zeta^k \sin \phi k$ where $\zeta = (a^2 + b^2)^{1/2}$ and $\phi = \tan^{-1}(b/a)$.

(4) For each multiple pair of complex roots $a \pm jb$ of multiplicity m, assign the functions $\zeta^k \cos \phi k$, $\zeta^k \sin \phi k$; $k\zeta^k \cos \phi k$, $k\zeta^k \sin \phi k$; $\ldots$ $k^{m-1}\zeta^k \cos \phi k$, $k^{m-1}\zeta^k \sin \phi k$.

As in the case of differential equations, the constants are evaluated by means of the initial or boundary conditions.

EXAMPLE 1.21.   The second-order difference equation

$$y_{k+1} - 2ay_k + y_{k-1} = 0 \qquad (1.44)$$

occurs rather frequently in applications, for example, in the resistor ladder network of Example 1.17 and the coin flipping game of Example 1.18. To solve this equation in general, we form the characteristic equation

$$f(r) = r - 2a + r^{-1} = 0$$

which may be rewritten as

$$r^2 - 2ar + 1 = 0$$

The roots of this equation are

$$r_1 = a + \sqrt{a^2 - 1}$$

and

$$r_2 = a - \sqrt{a^2 - 1}$$

Depending on the value of a, these roots may be simple and real-valued, simple and complex-valued, or repeated. We consider these cases separately.

(1) If $|a| > 1$, the roots r_1 and r_2 are real-valued and distinct. Thus the solution to (1.44) may be written as

$$y_k = c_1(r_1)^k + c_2(r_2)^k$$
$$= c_1(a + \sqrt{a^2 - 1})^k + c_2(a - \sqrt{a^2 - 1})^k$$

(2) If $|a| < 1$, the roots r_1 and r_2 are complex-valued and distinct. Thus the solution sequence is

$$y_k = c_1(r_1)^k + c_2(r_2)^k$$
$$= c_1(a + j\sqrt{1 - a^2})^k + c_2(a - j\sqrt{1 - a^2})^k$$
$$= \hat{c}_1 \cos \phi k + \hat{c}_2 \sin \phi k$$

where $\phi = \tan^{-1}\left(\dfrac{\sqrt{1 - a^2}}{a}\right) = \cos^{-1}(a)$

$$\zeta = \sqrt{a^2 + (1 - a^2)} = 1$$

(3) If $a = 1$, the characteristic equation

$$r^2 - 2ar + 1 = 0$$

has a repeated root at $r = 1$. Thus the solution sequence is

$$y_k = c_1 + c_2 k$$

(4) If $a = -1$, the characteristic equation has a repeated root at $r = 1$. Thus

$$y_k = c_1(-1)^k + c_2 k(-1)^k$$
$$= (c_1 + c_2 k)(-1)^k$$

EXAMPLE 1.22. In Example 1.17, assume $a = 1$ in (1.34). To find the voltage on the kth node, we must solve the difference equation

$$v_{k+1} - 3v_k + v_{k-1} = 0, \qquad k = 0, 1, 2, \ldots, N \qquad (1.45)$$

Using the previous example, $a = 1.5$ and so

$$r_1 = 2.62$$
$$r_2 = 0.38$$

Thus, the solution to (1.45) is

$$v_k = c_1(2.62)^k + c_2(.38)^k, \qquad k = 0, 1, 2, \ldots, N$$

For an infinite ladder, $\lim v_k = 0$. This implies that $c_1 = 0$, and using $v_0 = E$, we have as the solution

$$v_k = (.38)^k E$$

*EXAMPLE 1.23.   In Example 1.18, to find the probability that A eventually wins the game with fair coins ($p = 1/2$), we can solve (1.36). Again referring to Example 1.21, we see that $a = 1$, and so

$$p_k = c_1 + c_2 k \qquad (1.46)$$

The initial conditions are $p_0 = 0$ and $p_N = 1$. Using these conditions we obtain

$$p_0 = 0 = c_1 + c_2(0) \Rightarrow c_1 = 0$$

$$p_N = 1 = c_1 + c_2(N) \Rightarrow c_2 = \frac{1}{N}$$

Hence,

$$p_k = \frac{k}{N}, \qquad 0 \le k \le N$$

EXAMPLE 1.24.   In Example 1.19, with $x_k = 0$ in (1.37), we must solve the difference equation

$$y_k - .9y_{k-1} = 0, \qquad k = 0, 1, 2, \ldots \qquad (1.47)$$

The corresponding characteristic equation is

$$1 - .9r^{-1} = 0$$

which has a single root

$$r = .9$$

Thus, the solution of (1.47) is

$$y_k = c_1(.9)^k, \qquad k = 0, 1, 2, \ldots$$

If the initial concentration is y_0, then the constant c_1 is

$$y_0 = c_1(.9)^0 = c_1$$

and so

$$y_k = y_0(.9)^k, \qquad k = 0, 1, 2, \ldots$$

EXAMPLE 1.25.   Consider the evaluation of the integral

$$I_k(\phi) = \int_0^\pi \frac{\cos k\theta - \cos k\phi}{\cos \theta - \cos \phi} \, d\theta, \qquad k = 0, 1, 2, \ldots$$

whose value is a function of the discrete parameter k and the continuous parameter ϕ. One can show that for fixed ϕ, this integral satisfies the difference equation

$$I_{k+1} - 2I_k \cos \phi + I_{k-1} = 0, \qquad k = 1, 2, \ldots$$

with initial conditions $I_0 = 0$, $I_1 = \pi$. From the results of Example 1.21, it follows that I_k has the general form

$$I_k = c_1 \cos k\phi + c_2 \sin k\phi$$

Using the initial conditions, we see that $c_1 = 0$, $c_2 \sin \phi = \pi$, so that

$$I_k = c_2 \sin k\phi = \pi \cdot \frac{\sin k\phi}{\sin \phi}$$

1.8 NONHOMOGENEOUS SOLUTION OF LINEAR DIFFERENCE EQUATIONS

We now turn our attention to finding the steady-state solution to the difference equation

$$L[y_k] = x_k \tag{1.48}$$

As in the case of differential equations, let us explore the method of undetermined coefficients for obtaining the forced solution to a difference equation. You will recall that the method depends on finding an annihilator operator for the forcing function. This operator is then applied to both sides of the equation, and the resulting homogeneous equation is solved by the use of techniques previously explained. The solution function is then substituted into the original difference equation, and the undetermined coefficients are evaluated.

EXAMPLE 1.26. Let us take $x_k = 50$ in (1.37). That is, we add a 50% solution of chemical A to the vat at each step. Now we have

$$y_k - .9y_{k-1} = .05, \qquad k = 0, 1, 2, \ldots \tag{1.49}$$

The annihilator operator L_A is $S - 1$. Hence, we must solve the homogeneous equation

$$(S - 1)(1 - .9S^{-1})[y_k] = (S - 1)[.05] = 0$$

The roots of the corresponding characteristic equation

$$(r - 1)(1 - .9r^{-1}) = 0$$

are $r_1 = 1$ and $r_2 = .9$. The complete solution has the form

$$y_k = c_1 + c_2(.9)^k$$

Substituting into the original difference equation (1.49), we have

$$c_1(1 - .9) + 0 = .05$$

from which $c_1 = .5$.

Now, applying the initial condition, we have

$$.5 + c_2 \cdot 1 = y_0$$

giving the solution

$$y_k = .5 + (y_0 - .5)(.9)^k, \qquad k = 0, 1, 2, \ldots$$

The forced and transient parts of the solution should be evident in the expression above.

EXAMPLE 1.27. Suppose we have the difference equation

$$L[y_k] = y_{k+1} - 3y_k + 2y_{k-1} = e^{\alpha k}, \qquad k = 0, 1, 2, \ldots$$

The appropriate operator in this case is $(S - e^{\alpha})$, because $(S - e^{\alpha})$ $[e^{\alpha k}] = e^{\alpha(k+1)} - e^{\alpha}e^{\alpha k} = 0$. Thus, applying $(S - e^{\alpha})$ to both sides of the original equation, we have

$$(S - e^{\alpha})(S - 3 + 2S^{-1})[y_k] = (S - e^{\alpha})[e^{\alpha k}] = 0$$

The corresponding characteristic equation is

$$(r - e^{\alpha})(r - 3 + 2r^{-1}) = 0$$

This equation has roots e^{α}, 2, 1, from which we obtain the solution function as

$$y_k = c_1(1)^k + c_2(2)^k + c_3(e^{\alpha})^k, \qquad k = 0, 1, 2, \ldots$$

Substituting this solution function into the original difference equation allows us to find c_3. The resulting equation is

$$(S - 3 + 2S^{-1})[c_1 + c_2 2^k + c_3 e^{\alpha k}] = e^{\alpha k}$$

We know that $c_1 + c_2 2^k$ satisfies the homogeneous equation, and so $(S - 3 + 2S^{-1})[c_1 + c_2 2^k] = 0$. We can use the above equation to find c_3. Thus,

$$c_3 e^{\alpha(k+1)} - 3c_3 e^{\alpha k} + 2c_3 e^{\alpha(k-1)} = e^{\alpha k}$$

or

$$c_3 = \frac{1}{e^{\alpha} - 3 + 2e^{-\alpha}}$$

The forced solution to the original difference equation is therefore

$$y_k\Big|_{\text{forced}} = \frac{e^{\alpha k}}{e^{\alpha} - 3 + 2e^{-\alpha}}, \qquad k = 0, 1, 2, \ldots$$

EXAMPLE 1.28. Suppose we take the same difference equation as in the previous example, but with a forcing function $1 + a^k$: i.e., consider

$$y_{k+1} - 3y_k + 2y_{k-1} = 1 + a^k, \qquad k = 0, 1, 2, \ldots$$

In this case, we can annihilate $(1 + a^k)$ with the operator $(S - 1)(S - a)$. This operator is merely the product of the annihilators for each term in the sum. Applying this operator to both sides of the difference equation, we obtain

$$(S - 1)(S - a)(S - 3 + 2S^{-1})[y_k] = 0$$

The corresponding characteristic equation has roots 1, a, 2, 1. Note the multiplicity of the root 1. Thus, the solution function is

$$y_k = c_1 + c_2 2^k + c_3 k + c_4 a^k, \qquad k = 0, 1, 2, \ldots$$

This function is now substituted into the original difference equation, which leads to

$$-c_3 + \left(a - 3 + \frac{2}{a}\right)c_4 a^k = 1 + a^k$$

Thus, it follows that

$$c_3 = -1, \qquad c_4 = \frac{a}{(a - 1)(a - 2)}, \qquad a \neq 1, 2$$

With this restriction on a, the forced solution is then

$$y_k\bigg|_{\text{forced}} = -k + \frac{a^{k+1}}{(a - 1)(a - 2)}, \qquad k = 0, 1, 2, \ldots$$

We can again repeat our discussion of the transient and steady-state components as we did for linear differential equations. The transient or homogeneous solution is characteristic of the system, while the steady-state or forced response depends on both system and the driving function.

This brief explanation of linear difference and differential equations completes our discussion of the classical time-domain methods of solution. We shall return to the solution process for these equations in the material which follows.

1.9 APPLICATION TO LINEAR SYSTEMS

Now that we have developed some mathematics that can be used to solve linear difference and differential equations, let us apply these techniques to some specific examples.

EXAMPLE 1.29. In biological problems, one often is interested in modeling the transfer of material across membranes. Cells must metabolize for their own existence and to perform specialized functions that occur in various cells such as neuron cells, liver cells, and so on. The metabolic process in the cells includes an inflow of food for the cell and an output of waste and other products. Membrane transfer involves both passive diffusion processes and active processes involving electrical and chemical potentials. For our discussion, we shall begin by assuming that only passive diffusion is involved in membrane transport.

Suppose we have two lumped compartments separated by a membrane and between which material flows at a rate proportional to its concentration difference. For example, the two compartments might model water and electrolyte flow in extracellular and intracellular fluids, or perhaps flows and storage of carbon dioxide and oxygen in the lungs and arterial-venous systems. We represent these two compartments schematically as shown in Figure 1.16. We assume that there is a reservoir of material being used by a consuming compartment. The reservoir has a constant concentration c_1 moles/cm³ of material. This constant concentration is regulated by some external system. The user compartment has a concentration c_2 moles/cm³ of material and consumes material at a rate of r moles/sec. We denote the net flow rate into compartment 2 as f_{12} in moles/sec. We can write the flow f_{12} as

$$v_2 \frac{dc_2(t)}{dt} + r = f_{12} = k[c_1(t) - c_2(t)] \qquad (1.50)$$

where k is the diffusion constant in cm³/sec and v_2 is the volume of the user compartment. We assume that the concentrations are small enough

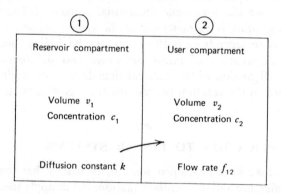

①	②
Reservoir compartment	User compartment
Volume v_1 Concentration c_1	Volume v_2 Concentration c_2
Diffusion constant k	Flow rate f_{12}

FIGURE 1.16

so that the flow does not change the volumes of the compartments. We can rewrite (1.50) as

$$\tau \frac{dc_2(t)}{dt} + c_2(t) = c_1(t) - \frac{r}{k} \tag{1.51}$$

where $\tau = v_2/k$. Because $c_1(t)$ is assumed to be constant, we can easily solve for the concentration $c_2(t)$. The characteristic equation corresponding to (1.51) is

$$\tau r + 1 = 0$$

Thus, the single root is $r = -1/\tau$, from which the transient function is

$$c_2(t)|_{\text{transient}} = \alpha_1 e^{-t/\tau}, \qquad t \geq 0$$

where α_1 is some constant dependent on initial conditions. The steady-state solution to (1.51) is found easily, because the annihilator operator in this case is merely (D). Thus, we examine the equation

$$D(D\tau + 1)[c_2(t)] = 0$$

This equation has a characteristic equation with roots 0, $1/\tau$. Thus the solution function for determining the forced or steady-state solution is

$$c_2(t) = \alpha_1 e^{-t/\tau} + \alpha_2 \tag{1.52}$$

We substitute (1.52) into (1.51) to obtain

$$(D\tau + 1)[\alpha_2] = c_1 - \frac{r}{k}$$

Hence,

$$\alpha_2 = c_1 - \frac{r}{k}$$

The steady-state solution is therefore

$$c_2(t)\Big|_{\text{steady-state}} = c_1 - \frac{r}{k}$$

This result agrees with our intuition. It merely states that after a long time, the concentration in the user compartment is the reservoir concentration minus a value proportional to the rate of consumption in the user compartment.

We can extend our model to a more meaningful level by assuming that there are active flow processes involved and that the separating membrane has different permeabilities in each direction. Suppose k_{12} is the diffusion constant from 1 to 2 and k_{21} is the diffusion constant from 2 to 1. Assume that there is an active source or pump which pumps material out of the user compartment at a rate $k_{21}R$. Then the net flow f_{12} is

$$f_{12} = k_{12}(c_1 - c_2(t)) - k_{21}(c_2(t) - c_1) - k_{21}R$$

The equation for the concentration $c_2(t)$ is therefore

$$v_2 \frac{dc_2(t)}{dt} + r = (k_{12} + k_{21})(c_1 - c_2(t)) - k_{21}R$$

which we can rewrite as

$$\tau \frac{dc_2(t)}{dt} + c_2(t) = c_1 - \frac{k_{21}R + r}{k_{12} + k_{21}} \qquad (1.53)$$

τ in this case is $v_2/(k_{12} + k_{21})$. Equation (1.53) can be solved in the same manner as before. The steady-state solution is $c_2(t) = c_1 - R$ for $k_{21} \gg k_{12}$.

EXAMPLE 1.30. Electrical networks are a classical example of systems which are modeled by linear differential equations. Consider the network of Figure 1.17. Kirchhoff's voltage law taken about the single loop leads to the equation

$$e(t) = Ri(t) + L\frac{di(t)}{dt} + \frac{1}{C}\int_{-\infty}^{t} i(t')\,dt', \qquad t > 0 \qquad (1.54)$$

We can convert (1.54) into an equation involving only derivatives by differentiating both sides with respect to t. Thus

$$\frac{de(t)}{dt} = 0 = R\frac{di(t)}{dt} + L\frac{d^2i(t)}{dt^2} + \frac{1}{C}i(t) \qquad (1.55)$$

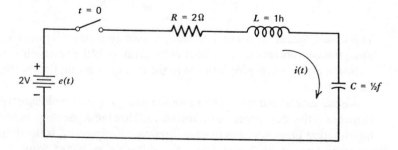

FIGURE 1.17

The initial conditions must be specified to obtain a solution. They can be obtained from the network. If we assume that the switch closes at time $t = 0$, then $i(0^+)$[5] must be zero because current cannot change instantaneously through an inductor. Thus, no current flows at $t = 0^+$, and the entire voltage appears across the inductor. Thus,

$$L \frac{di(0^+)}{dt} = e(0^+)$$

or

$$\frac{di(0^+)}{dt} = \frac{e(0^+)}{L} = 2 \text{ amp/sec} \qquad (1.56)$$

Equation (1.56) furnishes us with our second initial condition. Our problem is thus to solve

$$\frac{d^2 i(t)}{dt^2} + 2 \frac{di(t)}{dt} + 2i(t) = 0$$

with
$$i(0^+) = 0, \qquad \frac{di(t)}{dt}\bigg|_{t=0^+} = 2 \qquad (1.57)$$

The characteristic equation corresponding to (1.57) is

$$r^2 + 2r + 2 = 0$$

This equation has roots $-1 \pm j$. The transient solution function is therefore

$$i(t) = c_1 e^{(-1+j)t} + c_2 e^{(-1-j)t}$$
$$= e^{-t}(c_1' \cos t + c_2' \sin t)$$

where $c_1' = c_1 + c_2$ and $c_2' = i(c_1 - c_2)$. The initial conditions can now be used to find c_1' and c_2'.

$$i(0^+) = 0 = e^{-0}(c_1' \cos 0 + c_2' \sin 0) = c_1'$$

Because c_1' is zero, we can write

$$\frac{di(0^+)}{dt} = 2 = c_2'(\cos t e^{-t} - e^{-t} \sin t)\big|_{t=0^+} = c_2'$$

The transient solution is thus

$$i(t) = 2e^{-t} \sin t, \qquad t \geq 0 \qquad (1.58)$$

This solution is also the general solution to (1.57), because (1.57) is a homogeneous equation. The solution corresponds to our physical intuition. The switch causes a step of unchanging voltage to be placed

---

[5] The notation 0^+ means time $t = 0$ just after the switch is closed.

across the series network. Because the capacitor is in series and responds only to time-varying voltages, the current in the loop must eventually die away. If in (1.55) the term $de(t)/dt$ were not zero, then, of course, one would obtain a forced or steady-state solution in addition to the transient solution of (1.58). Notice that the process of determining initial conditions from the physical problem involves a basic understanding of the physical problem in question. It is not simply a matter of turning a convenient crank.

EXAMPLE 1.31. As a second example of solving network equations, consider the circuit in Figure 1.18. The switch is closed at $t = 0$. We wish to determine the voltage $y(t)$.

Taking $i_1(t)$, $i_2(t)$, and $i_3(t)$ as shown in Figure 1.19, and applying Kirchhoff's current law, we arrive at the appropriate differential equation

$$C\frac{d^2y(t)}{dt^2} + \left(\frac{1}{R_1} + \frac{R_2C}{L}\right)\frac{dy(t)}{dt} + \left(\frac{1}{L} + \frac{R_2}{R_1L}\right)y(t) = \frac{R_2}{R_1L}u(t) + \frac{1}{R_1}\frac{du(t)}{dt}$$

with

$$y(t)\big|_{t=0} = 0$$

$$\frac{dy(t)}{dt}\bigg|_{t=0} = \frac{u(t)}{R_1C}\bigg|_{t=0}$$

To simplify the following algebra, we take $R_1 = R_2 = L = C = 1$, yielding the differential equation

$$(D^2 + 2D + 2)[y(t)] = (D + 1)[u(t)] \tag{1.59}$$

with

$$y(0) = 0$$

$$y'(0) = u(0)$$

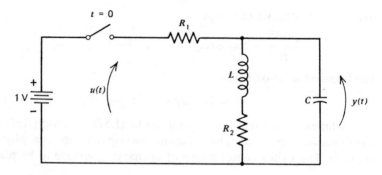

FIGURE 1.18

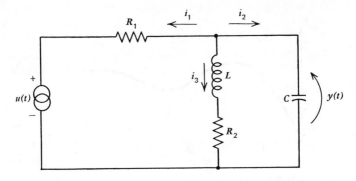

FIGURE 1.19

and

$$u(t) = \begin{cases} 0, & t < 0 \\ 1, & t \geq 0 \end{cases}$$

Assuming that the circuit is initially relaxed (i.e., no energy is stored in the inductor or capacitor) we have $y(t) = 0$ for $t < 0$. For $t \geq 0$, $u(t) = 1$ and the annihilator operator for this driving function is $L_A = D$. Thus, we must solve the homogeneous equation

$$D(D^2 + 2D + 2)[y(t)] = 0$$

whose characteristic equation

$$r(r^2 + 2r + 2) = 0$$

has the roots

$$r_1 = 0$$
$$r_2 = -1 + i$$
$$r_3 = -1 - i$$

The general solution has the form

$$y(t) = c_1 + c_2 e^{-t} \sin t + c_3 e^{-t} \cos t$$

Substituting in (1.59), we obtain for the forced solution (in this case, the constant c_1)

$$2c_1 = 0 + 1$$

from which $c_1 = \frac{1}{2}$. With $y(0) = 0$,

$$\tfrac{1}{2} + c_2 \cdot 0 + c_3 \cdot 1 = 0$$

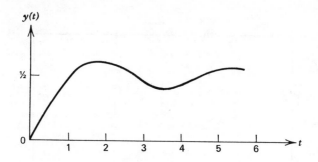

FIGURE 1.20

giving $c_3 = -\frac{1}{2}$. Finally, from $y'(0) = 1$,

$$[c_2(-e^{-t}\sin t + e^{-t}\cos t) - c_3(e^{-t}\cos t + e^{-t}\sin t)]\big|_{t=0}$$
$$= c_2 - c_3 = 1$$

from which $c_2 = 1 + c_3 = \frac{1}{2}$. The final solution is

$$y(t) = \begin{cases} 0, & t < 0 \\ \frac{1}{2}(1 + e^{-t}\sin t - e^{-t}\cos t), & t \geq 0 \end{cases}$$

as sketched in Figure 1.20. In the following chapters, we shall reexamine this circuit and determine its response using other approaches.

EXAMPLE 1.32. Suppose we wished to find the response of the discrete-time system in Figure 1.21 to a step input (this system might, for example, represent a digital filter or a sampled chemical process). The appropriate difference equation is

$$y_{k+2} = -6y_k + 5y_{k+1} + u_k$$

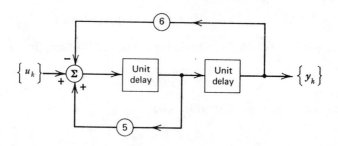

FIGURE 1.21

that is,

$$y_{k+2} - 5y_{k+1} + 6y_k = u_k \qquad (1.60)$$

with

$$u_k = \begin{cases} 0, & k < 0 \\ 1, & k \geq 0 \end{cases}$$

Assume the initial conditions are

$$y_0 = y_1 = 0$$

Since the system is initially relaxed, we have $y_k = 0$ for $k < 0$. For $k \geq 0$, the annihilator operator for $u_k = 1$ is $S - 1$. Hence, we must solve the homogeneous equation

$$(S - 1)(S^2 - 5S + 6)[y_k] = 0$$

The general solution is

$$y_k = c_1 + c_2 2^k + c_3 3^k$$

Substituting into (1.60), we find that $c_1 = \frac{1}{2}$. From the initial conditions, we determine that $c_2 = -1$ and that $c_3 = \frac{1}{2}$. Thus, the response of our system to a step input is given by

$$y_k = \begin{cases} 0, & k < 0 \\ \frac{1}{2} - 2^k + \dfrac{3^k}{2}, & k \geq 0 \end{cases}$$

As with the preceding example, we shall reexamine this system in later chapters and solve for its output using several other methods.

EXAMPLE 1.33. To illustrate another use of linear difference equations, consider the problem of determining small deflections of a tightly stretched string loaded at n equally spaced points as shown in Figure 1.22.

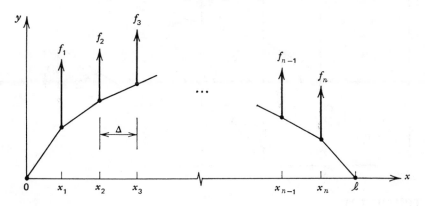

FIGURE 1.22

Let x_k, $k = 1, 2, \ldots, n$ be the n equally spaced points and f_k, $k = 1, 2, \ldots, n$ the forces on the string. The end points of the string are attached to the x axis at $x = 0$ and $x = \ell = (n + 1)\Delta$, where Δ is spacing between x points. In Figure 1.23, we illustrate three points x_{k-1}, x_k, and x_{k+1}. We assume that the string is under a uniform tension T and that the slope of each segment of the deflected string is small. Referring to Figure 1.23, we can write that the deflections at x_k and x_{k+1} differ by

$$y_{k+1} - y_k = \Delta \tan \phi_k \tag{1.61}$$

Because the forces are in equilibrium at (x_k, y_k), we see that

$$T(\sin \phi_k - \sin \phi_{k-1}) + f_k = 0 \tag{1.62}$$

For small angles, we use the approximations that the sine and tangent of an angle are equal to the angle. Thus eliminating the angles between (1.61) and (1.62), we obtain

$$y_{k+1} - 2y_k + y_{k-1} = -\frac{\Delta f_k}{T}, \qquad k = 0, 1, 2, \ldots, n \tag{1.63}$$

with boundary conditions $y_0 = y_{n+1} = 0$. Suppose the forces are beads of mass m attached to the string at the points x_k. The forces are thus the gravity forces acting on the n mass points: i.e., $f_k = -mg$, so that (1.63) is

$$y_{k+1} - 2y_k + y_{k-1} = \frac{mg\Delta}{T}, \qquad k = 0, 1, 2, \ldots, n \tag{1.64}$$

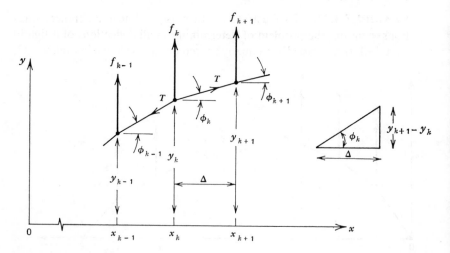

FIGURE 1.23

The homogeneous solution function for (1.64) is found from the roots of the characteristic function.

$$r^2 - 2r + 1 = 0$$

This equation has roots 1, 1. Thus, the homogeneous solution function is

$$y_k = c_1 + c_2 k, \qquad k = 0, 1, 2, \ldots, n$$

The forced solution is found by applying the annihilator operator $(S - 1)$ to both sides of (1.64). We obtain

$$(S - 1)(S^2 - 2S + 1)[y_k] = 0 \qquad (1.65)$$

The characteristic equation corresponding to (1.65) has three roots at $+1$. Thus, the forced solution is of the form

$$y_k = c_3 k^2 \qquad (1.66)$$

We find c_3 by substituting (1.66) into (1.64) and then equating coefficients. This procedure leads to

$$c_3 = \frac{mg\Delta}{2T}$$

The general solution is of the form

$$y_k = c_1 + c_2 k + \frac{mg\Delta}{2T} k^2, \qquad k = 0, 1, 2, \ldots, n$$

We find c_1 and c_2 by use of the boundary conditions. Thus

$$y_0 = 0 = c_1$$

and

$$y_{n+1} = 0 = c_2(n + 1) + \frac{mg\Delta}{2T}(n + 1)^2$$

or

$$c_2 = -\frac{mg\Delta}{2T}(n + 1)$$

The general solution is therefore

$$y_k = \frac{mg\Delta}{2T} k(k - (n + 1)), \qquad k = 0, 1, 2, \ldots, n \qquad (1.67)$$

Notice that $x_k = k\Delta$ and that $\ell = (n + 1)\Delta$. Hence, we can rewrite (1.67) as

$$y(x_k) = -\frac{mg}{2T\Delta} x_k(\ell - x_k) \qquad (1.68)$$

From (1.68), we see that the segments of the deflected string are chords of the parabola

$$y(x) = -\frac{mg}{2T\Delta} x(\ell - x)$$

If the mass is spread uniformly over the string in the x-coordinate by letting $\Delta \rightarrow 0$ with $mg/\Delta = \xi$, the string takes on the well known parabolic form

$$y(x) = -\frac{\xi}{2T} x(\ell - x) \tag{1.69}$$

PROBLEMS

1. Are the following systems linear? Show.

a. $H[x(t)] = \begin{cases} 0, & t < T \\ \alpha x(t), & t \geq T \end{cases}$

b.

$$x(t) = \begin{bmatrix} x_1(t) \\ x_2(t) \end{bmatrix} \longrightarrow \boxed{H} \longrightarrow \min\left\{ x_1(t), x_2(t) \right\}$$

Vector input

c.

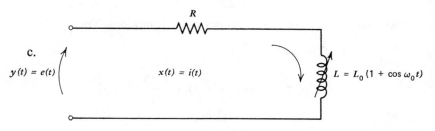

$y(t) = e(t)$ $x(t) = i(t)$ R $L = L_0(1 + \cos \omega_0 t)$

d. $H[x(t)] = 3x(t)(t + 4)$

e.

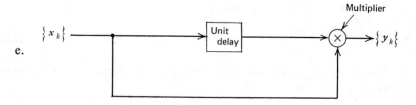

$\left\{ x_k \right\} \longrightarrow$ Unit delay $\longrightarrow \otimes \longrightarrow \left\{ y_k \right\}$ Multiplier

f. $y(t) = \alpha x(t) + x(t^\alpha), \qquad t > 0$

g. $y_k = \sum\limits_{i=k}^{k+2} i^2 x_i$

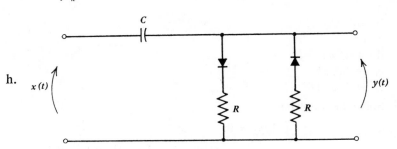

h. $x(t)$ $y(t)$

2. Let H be a linear memoryless time-invariant system. If for some particular input $\hat{x}(t) \neq 0$, $H[\hat{x}(t)] = 0$, show that $H[x(t)] = 0$ for all $x(t)$.

3. a. Time-invariant linear systems forced by sinusoidal inputs are an important class of problems. One model of such systems is made by means of a linear differential equation. Show that the forced solution of the following differential equation must be of the form αe^{kt}:

$$\frac{d^2 y(t)}{dt^2} + a\,\frac{dy(t)}{dt} + by(t) = ce^{kt}$$

where a, b, c, and k are known constants and k is not a root of the characteristic equation.

   b. In view of the result of part a, what must the form of the forced solution be for a forcing function $c \sin \omega t$ (rather than ce^{kt})?

   c. What is the form of the forced solution if the forcing function is ce^{kt} and k satisfies the equation $k^2 + ak + b = 0$? What physical interpretation can you give to this result?

4. Give some examples of a system that satisfies the property of homogeneity but not the principle of superposition.

5. Find the appropriate annihilator for each of the following forcing functions:

   a. Ae^{at}

   b. $Bt^2 e^{at}$

   c. $A \sin \omega t$

   d. $A \sin \omega t + B \cos \omega t$

   e. $A \sin (\omega t + \phi)$

   f. $Ae^{at} + Be^{bt} + Ce^{ct}$

   g. $t^3 + B \sin t$

6. Find the appropriate annihilator for each of the following discrete-time forcing functions:

  a. a^k
  b. ka^k
  c. e^{bk}
  d. $A \sin ak$
  e. $B \sinh ak$
  f. $k^2 a^k + A e^{bk}$

7. Solve the following:

  a. $y_{k+2} + 7y_{k+1} + 12y_k = 0, \qquad k \geq 0$

  $$Ans.\ y_k = c_1(-3)^k + c_2(-4)^k, \qquad k \geq 0$$

  b. $y_{k+2} + 2y_{k+1} + 2y_k = 0, \qquad k \geq 0$

  $$Ans.\ y_k = 2^{k/2}\left[c_1 \cos \frac{3\pi k}{4} + c_2 \sin \frac{3\pi k}{4}\right], \qquad k \geq 0$$

  c. $y_{k+2} + y_k = \sin k, \qquad k \geq 0$

  $$Ans.\ y_k = c_1 \cos \frac{\pi k}{2} + c_2 \sin \frac{\pi k}{2} + \frac{\sin k + \sin (k-2)}{2(1 + \cos 2)}, \qquad k \geq 0$$

  ✓ d. $(D^4 + 8D^2 + 16)[y(t)] = -\sin t$

  $$Ans.\ y(t) = c_1 \cos 2t + c_2 \sin 2t + c_3 t \cos 2t + c_4 t \sin 2t - \frac{\sin t}{9}$$

  e. $(D^3 - 2D^2 + D - 2)[y(t)] = 0 \qquad y(0) = y'(0) = y''(0) = 1$

  $$Ans.\ y(t) = \tfrac{1}{5}(2e^{2t} + 3 \cos t + \sin t)$$

  ✓ f. $(D^4 - D)[y(t)] = t^2$

  $$Ans.\ y(t) = c_1 + c_2 e^t + e^{-t/2}\left(c_3 \cos \frac{\sqrt{3}\,t}{2} + c_4 \frac{\sin\sqrt{3}\,t}{2}\right) - \frac{t^3}{3}$$

8. Consider the difference equation $y_{k+2} - 2\tau y_{k+1} + y_k = 0$ and find the solution for the following cases:

  a. $\tau < -1$
  b. $\tau = -1$
  c. $-1 < \tau < 1$
  d. $\tau = 1$
  e. $\tau > 1$

9. Derive a model for nanoplankton respiration which accounts for gross community photosynthesis, storage, and respiration.[6] The following analogies may be helpful.

---

[6] Odum, H. T., et al., "Consequences of Small Storage Capacity in Nanoplankton Pertinent to Primary Production in Tropical Waters," *Journal of Marine Research* (June 1963).

The input to the system is sunlight, which can be represented by a voltage, say e_b. The production rate p of material by photosynthesis is proportional to the difference in input sunlight and the material already in the system (a "back potential" e_m). The constant of proportionality between production rate p and this difference potential is a conductance G_b. The community respiration rate p_r is assumed to be proportional to e_m. The storage rate p_c of material in the system is assumed to be proportional to the rate of change of material already in the system, and the constant of proportionality is C^{-1}, representative of the storage capacity of the system. In symbols, $p_c = (1/C)\,de_m/dt$. Also, the total production rate p is equal to the sum of the respiration and storage rates. (Notice that production rates are analogous to currents.)

a. Obtain an electrical schematic model for this system described.

b. Using your model, find the dependence of the respiration rate p_r upon a step input of sunlight.

10. Consider a continuous unloaded uniform beam resting on N equally spaced supports with separation d. The deflection $y(x)$ of the beam is governed by the equation $EI\,d^2y(x)/dx^2 = M$, where EI is the constant flexural rigidity and M the bending moment. Let x be the distance to the right of the kth support as shown in Figure P1.1. Using the requirements

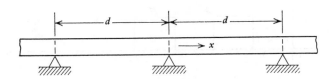

FIGURE P1.1

that $y(x)$ vanish at $x = 0$ and $\pm d$ and that $y'(x)$ be continuous at $x = 0$, show that

$$M(x) = M_k + (M_k - M_{k-1})\frac{x}{d}, \qquad -d \leq x \leq 0$$

$$= M_k + (M_{k+1} - M_k)\frac{x}{d}, \qquad 0 \leq x \leq d$$

and hence show that the bending moment M_k at the kth supports satisfies

$$M_{k+1} + 4M_k + M_{k-1} = 0, \qquad k = 2, 3, \ldots, N - 1$$

11. A system is described by the following matrix equation

$$\begin{bmatrix} y_1(t) \\ y_2(t) \end{bmatrix} = \begin{bmatrix} 3 & 4 & 2 \\ -1 & t^2 & 0 \end{bmatrix} \begin{bmatrix} x_1(t) \\ x_2(t) \\ x_3(t) \end{bmatrix}$$

   a. Is this system linear?
   b. Is this system time invariant?

12. In Figure P1.2 is shown a modulating system for amplitude modulation of the waveform $A(t)$. Is this system linear? Is this system time invariant?

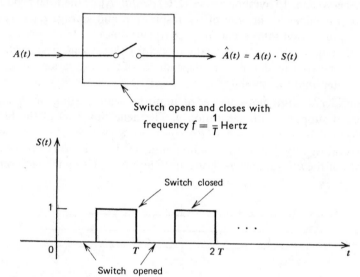

$A(t) \longrightarrow \qquad \longrightarrow \hat{A}(t) = A(t) \cdot S(t)$

Switch opens and closes with
frequency $f = \dfrac{1}{T}$ Hertz

$S(t)$

Switch closed

1

0   T $2T$ t

Switch opened

FIGURE P1.2

13. a. In the RC circuit shown in Figure P1.3, the initial voltage on the capacitor is 2 V: i.e., $y(0) = 2$. If the input voltage $x(t) = 0$ for $t \geq 0$ (obtained by placing a short across the input), what is the output voltage $y(t)$ for $t \geq 0$? Ans.: $y(t) = 2e^{-t/RC}$.

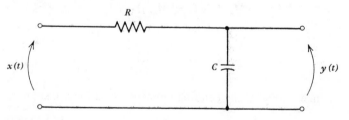

R

$x(t)$ C $y(t)$

FIGURE P1.3

b. Let $y_0 = 2$ be the initial voltage output of the unit delay in the discrete-time system shown in Figure P1.4. If the input $x_k = 0$ for $k \geq 0$, what is the output y_k for $k \geq 0$? *Ans.:* $y_k = 2a^k$.

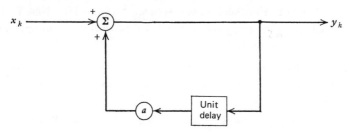

FIGURE P1.4

c. How should the constant a be chosen in part b so that the discrete-time output sequence $\{y_k\}$ for $k \geq 0$ is identical to the sequence $\{y(kT)\}$ formed by sampling the continuous-time output $y(t)$ in part a every T seconds? *Ans.:* $a = e^{-T/RC}$.

d. With the constant a as found in part c, compute the outputs $y(t)$ and $\{y_k\}$ for step inputs $x(t) = u(t)$ and $x_k = u_k$ with the initial conditions as before. Is the discrete-time system output $\{y_k\}$ still the same as the sampled continuous output $\{y(kT)\}$?

14. Take n nonparallel lines in a plane. Let y_n be the number of separate regions into which these n lines divide the plane. Derive a first-order difference equation that can be used to find the number of separate regions. No three lines may intersect at one point.

15. Consider the system in Figure P1.5, consisting of two balls of masses m and M, where $m < M$. The two masses slide on a horizontal plane between two perfectly elastic walls. The coefficient of restitution between the balls is e. If u_n and v_n are the velocities of m and M, respectively, before the nth impact, then derive a difference equation model which one can solve to find u_n and v_n.

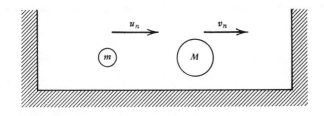

FIGURE P1.5

SUGGESTED READINGS

Schwarz, R. J., and B. Friedland, *Linear Systems*, McGraw-Hill, New York, 1965.
This book is a complete treatment of linear systems at somewhat higher than the junior-senior level.

Cheng, D. K., *Analysis of Linear Systems*, Addison-Wesley, Reading, Mass., 1959.
This book contains several chapters on mathematical models for linear systems.

Gardner, M. F., and J. L. Barnes, *Transients in Linear Systems*, Wiley, New York, 1961.
A classic in the area of linear analysis. Chapter 2 has a good treatment of mathematical models for mechanical and electromechanical systems.

Hildebrand, F. B., *Finite Difference Equations and Simulations*, Prentice-Hall, Englewood Cliffs, N.J., 1968.
Chapter 1 of this book contains a complete discussion of linear difference equations and applications of linear difference equations.

Wylie, C. R., *Advanced Engineering Mathematics*, McGraw-Hill, New York, 1966.
A good general reference on the solution of differential equations with some discussion of difference equations.

2

Convolution

2.1 INTRODUCTION

In the analysis of linear systems, we are interested primarily in the system response or output resulting from various input signals. In the preceding chapter we characterized or modeled the system by means of linear differential or difference equations. The output of the system to any input signal was then obtained by solving the appropriate equation. Another method of analysis, based on the superposition property of linear systems, uses the convolution operation. This method is the subject of this chapter.

The idea of convolution occurs again and again in many different disciplines. This fact accounts for the multiplicity of terms used to describe convolution—superposition integral, Duhamel integral, smoothing or scanning integral, weighted mean, and Faltung integral (folding in German). In physics, for example, one usually encounters convolution as describing the effect of an instrument in a measurement process. An observing instrument must always perform some weighted averaging of the observed physical quantity. The resultant instrument value is a convolution of the instrument weighting function and the distribution of the physical quantity rather than the exact quantity itself. In engineering, convolution arises as another method of characterizing linear systems. Convolution gives us another viewpoint from which to view these systems; and with this alternate viewpoint come new insight and intuition.

2.2 SUPERPOSITION AND CONVOLUTION— DISCRETE-TIME SYSTEMS

Linear systems, as we have seen in Chapter 1, possess the property of superposition. That is, if we know the response for the input sequence

53

$\{x_1(k)\}$ and $\{x_2(k)\}$ separately, then we also know the response for the input $\{x_1(k) + x_2(k)\}$. It is merely the sum of the two individual responses. If the linear system is also time invariant, then we can shift these inputs to any place along the time axis and obtain the outputs by making a corresponding time shift. In symbols, if $x_1(k)$ gives rise to $y_1(k)$, then $x_1(k \pm n)$ gives an output of $y_1(k \pm n)$. In this chapter we use these two properties of linear time-invariant systems to develop an alternate formulation for describing the input-output relationship. This formulation makes use of a function characteristic of the system. This characteristic function is the response of the system to an *impulse function*, or in the case of discrete-time systems an *impulse sequence*. The response of a discrete-time system to an impulse sequence is commonly called the *impulse response sequence*.

The impulse sequence $\{\delta_k\}$ is the sequence defined by (2.1) and shown in Figure 2.1.

$$\delta_k = \begin{cases} 1, & k = 0 \\ 0, & k \neq 0 \end{cases} \tag{2.1}$$

The response of a discrete-time system to an input impulse sequence $\{\delta_k\}$ is by definition the *impulse response sequence* $\{h_k\}$: i.e.,

$$\{\delta_k\} \rightarrow \{h_k\}$$

If we multiply the impulse sequence by a constant c, then because of linearity the output is also multiplied by c: i.e.,

$$c\{\delta_k\} \rightarrow c\{h_k\}$$

If we shift the position of the impulse sequence in time, then because of the time invariance of the system, we also shift the output by the same amount: i.e.,

$$c\{\delta_{k \pm n}\} \rightarrow c\{h_{k \pm n}\}$$

These relationships are shown schematically in Figure 2.2.

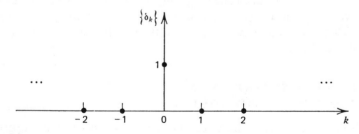

FIGURE 2.1

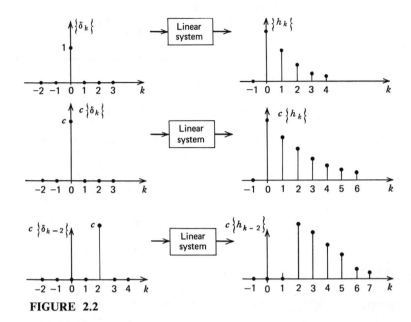

FIGURE 2.2

We are, of course, interested in the response resulting from arbitrary input signal sequences, say $\{x_k\}$. How can we use the impulse response sequence to help us? Suppose we represent the input sequence as follows:

$$\{x_k\} = \cdots + x_{-2}\{\delta_{k+2}\} + x_{-1}\{\delta_{k+1}\} + x_0\{\delta_k\} + x_1\{\delta_{k-1}\} + \cdots$$

In other words, we take each value of the sequence x_i and multiply it by a shifted version of the impulse sequence $\{\delta_{k-i}\}$. Because $\{\delta_{k-i}\}$ has value 1 only for $k = i$, this procedure allows us to represent an arbitrary input sequence as a weighted sum of shifted unit impulse sequences. Thus,

$$\{x_k\} = \sum_{i=-\infty}^{\infty} x_i\{\delta_{k-i}\} \tag{2.2}$$

Figure 2.3 depicts an example of the representation of (2.2).

Now we use the linearity of the system to find the output response. We first find the output that results from each term in the input, $x_i\{\delta_{k-i}\}$, and then sum these outputs to obtain the total response. The response to $x_0\{\delta_k\}$ is, for example,

$$x_0\{\delta_k\} \rightarrow x_0\{h_k\}$$

Similarly, the response to $x_1\{\delta_{k-1}\}$ is

$$x_1\{\delta_{k-1}\} \rightarrow x_1\{h_{k-1}\}$$

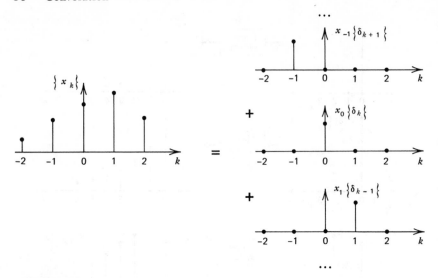

FIGURE 2.3

The general term $x_j\{\delta_{k-j}\}$ gives rise to a response $x_j\{h_{k-j}\}$. The total response is, because of the superposition property, merely the sum of responses of the form $x_j\{h_{k-j}\}$. Thus, the output sequence $\{y_k\}$ is

$$\{y_k\} = \sum_{j=-\infty}^{\infty} x_j\{h_{k-j}\}$$

The kth term of the output sequence is thus

$$y_k = \sum_{j=-\infty}^{\infty} x_j h_{k-j} \qquad (2.3)$$

The sum of (2.3) is known as a *convolution sum*. The shorthand notation used to indicate the calculation of (2.3) is

$$\{y_k\} = \{x_k\} * \{h_k\} \qquad (2.4)$$

If we let $m = k - j$ in (2.3), then we can rewrite (2.3) as

$$y_k = \sum_{m=-\infty}^{\infty} x_{k-m} h_m \qquad (2.5)$$

which means convolution is commutative: i.e.,

$$\{y_k\} = \{h_k\} * \{x_k\} = \{x_k\} * \{h_k\}$$

The characterization of a linear system in terms of a convolution operation is an extremely important concept. Even though the sum in (2.3) or (2.5) may be difficult to calculate, this formulation is a strong conceptual aid in understanding linear systems. The use of superposition and the impulse response of a system is also a more general characterization of linear systems than the corresponding and often used transform characterization, which we shall treat in Chapters 4–6. For example, if a linear system contains time-varying coefficients, the usual transform approach is no longer valid. However, a characterization in terms of a time-varying impulse response $\{h(k, n)\}$ and a superposition sum or integral is valid. Another example of the general nature of this approach occurs in the study of linear systems excited by random input signals. One can use the impulse response and superposition integral to study a larger class of random input signals than is possible with transform methods.

2.3   THE CONVOLUTION OPERATION—DISCRETE-TIME SYSTEMS

The convolution sum of (2.3) can be interpreted graphically. Consider two sequences $\{x(k)\}$ and $\{h(k)\}$. The convolution of these two sequences is another sequence $\{y(k)\}$, given by

$$\{y(k)\} = \{x(k)\} * \{h(k)\}$$

where

$$y(k) = \sum_{n=-\infty}^{\infty} x(n)h(k - n) \qquad (2.6)$$

For example, suppose $\{h(k)\}$ is the sequence $\{1, 2, 1\}$,[1] $k = 0, 1, 2$ and $\{x(k)\}$ is the sequence $\{\ldots, 1, 3, 1, 3, 1, \ldots\}$. Consider the calculation of $y(1)$.
$$\uparrow$$

$$y(1) = \sum_{n=-\infty}^{\infty} x(n)h(1 - n)$$

To calculate $y(1)$, we need $h(1 - n)$. We first take the mirror image of $\{h(n)\}$ about the vertical axis through the origin to obtain $\{h(-n)\}$ as shown in Figure 2.4a and 2.4b. Now shift $\{h(-n)\}$ one unit to the right to obtain $\{h(1 - n)\}$, shown in Figure 2.4c. This shifted sequence is multiplied with $\{x(n)\}$, depicted in 2.4d, and the resultant sequence values, shown in Figure 2.4e, are then added together to obtain one term, $y(1)$, of the sequence $\{y(k)\}$.

$$y(1) = \sum_{n=-\infty}^{\infty} x(n)h(1 - n) = 3 + 2 + 3 = 8$$

---

[1] If no arrow is included in the sequence listing, assume the first term within the brackets is the $k = 0$ term.

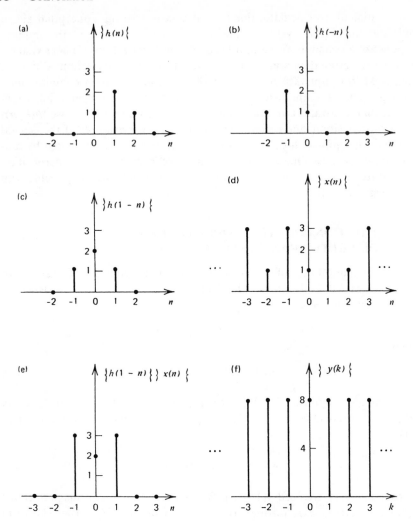

FIGURE 2.4

To calculate $y(2)$, one shifts $\{h(-n)\}$ two steps to the right to obtain $\{h(2 - n)\}$. This sequence and $\{x(n)\}$ are then multiplied together. The resultant sequence values are summed to find $y(2)$. The other values of the sequence $\{y(k)\}$ are obtained in a similar manner. The output sequence for this example is shown in Figure 2.4f. Notice that the convolution operation has produced a smoothing effect on the input sequence $\{x(n)\}$.

To summarize, we can view the convolution sum as composed of four basic operations:

(1) Take a mirror image of $\{h(n)\}$ about vertical axis through the origin to obtain $\{h(-n)\}$.

(2) Shift $\{h(-n)\}$ by an amount equal to the value of k where the output sequence is to be evaluated to find $\{h(k-n)\}$.

(3) Multiply this shifted sequence $\{h(k-n)\}$ and the input sequence $\{x(n)\}$.

(4) Add the resultant sequence values to obtain the value of the convolution at k.

There is another algorithm which one can use to evaluate discrete convolutions. Suppose that we wish to convolve $\{h(k)\}$ and $\{x(k)\}$, where

$$h(k) = \begin{cases} (\tfrac{1}{2})^k, & k \geq 0 \\ 0, & k < 0 \end{cases}$$

and

$$x(k) = \{3 \quad 2 \quad 1\}$$

Construct a matrix with $\{h(k)\}$ bordering the top of the matrix and $\{x(k)\}$ the left side of the matrix, as shown in Figure 2.5. In this case, the matrix is infinite because $\{h(k)\}$ is infinite. The entries in the matrix are the product of the corresponding row and column headers. To find the convolution of the

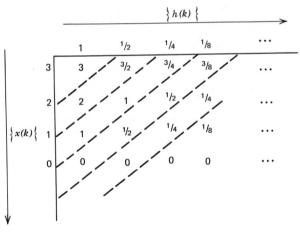

FIGURE 2.5

two sequences, we need only "fold and add" according to the dotted diagonal lines. The first term, $y(0)$, is thus 3. The second term, $y(1)$, is equal to $2 + 3/2 = 7/2$, which is the sum of the terms contained between the first and second diagonal lines. We can continue in this way and obtain the output sequence as

$$\{y_k\} = \left\{3 \quad \tfrac{7}{2} \quad \tfrac{11}{4} \quad \tfrac{11}{8} \quad \tfrac{11}{16} \quad \cdots \quad \frac{11}{2^k} \cdots \right\}$$

The reader can verify this result in terms of the formal calculations indicated in (2.3). In the case of two-sided sequences, the zeroth term in the output is contained between the diagonals containing the intersection term of the zeroth indices for the row and column sequences.

EXAMPLE 2.1. Convolve the impulse sequence $\{\delta(k)\} = \{\ldots, 0, 0, 1, 0, 0, \ldots\}$ with any arbitrary sequence $\{x(k)\} = \{\ldots, x(-1), x(0), x(1), \ldots\}$. From (2.3), we know that the kth term of the resultant sequence $\{y(k)\}$ is

$$y(k) = \sum_{n=-\infty}^{\infty} x(n)\, \delta(k - n) \tag{2.7}$$

Now, each term in the $\{\delta(k - n)\}$ is zero except for $k = n$. In this case, we have $\delta(0) = 1$. Thus, the only nonzero term in the sum of (2.7) occurs for $n = k$, and the evaluation of (2.7) yields

$$y(k) = x(k)$$

In other words, the convolution of $\{x(k)\}$ and $\{\delta(k)\}$ reproduces the sequence $\{x(k)\}$.

EXAMPLE 2.2. Convolve the sequences $\{x_k\}$ and $\{h_k\}$, where

$$x_k = \begin{cases} a^k, & k \geq 0 \\ 0, & k < 0 \end{cases}$$

and

$$h_k = \begin{cases} b^k, & k \geq 0 \\ 0, & k < 0 \end{cases}$$

The resultant sequence is $\{y_k\}$, given by

$$y_k = \sum_{n=-\infty}^{\infty} x_n h_{k-n} = \sum_{n=0}^{k} x_n h_{k-n} \tag{2.8}$$

The limits on the sum in (2.8) result because $x_n = 0$ for $n < 0$ and $h_{k-n} = 0$ for $n > k$. Thus,

$$y_k = \begin{cases} 0, & k < 0 \\ \displaystyle\sum_{n=0}^{k} a^n b^{k-n}, & k \geq 0 \end{cases}$$

For $k \geq 0$, we can evaluate the sum using the formula for the partial sum of a geometric sequence. That is,

$$y_k = \sum_{n=0}^{k} a^n b^{k-n}$$

$$= b^k \sum_{n=0}^{k} \left(\frac{a}{b}\right)^n$$

$$= \begin{cases} b^k \cdot \dfrac{1 - \left(\dfrac{a}{b}\right)^{k+1}}{1 - \left(\dfrac{a}{b}\right)}, & a \neq b \\ b^k \cdot (k+1), & a = b \end{cases} \qquad k \geq 0$$

EXAMPLE 2.3. Using the convolution sum, determine the output of the digital circuit shown in Figure 2.6 to an input sequence $\{x_k\} = \{3 \quad -1 \quad 3\}$. Assume that the gain G is $\frac{1}{2}$.

The difference equation which describes this system can be obtained by equating the output of the summer y_k to the two inputs into the summer: i.e.,

$$y_k = \tfrac{1}{2} y_{k-1} + x_k \tag{2.9}$$

Assume that the system is initially at rest, so that $y_{-1} = 0$. To use the convolution sum, we must first calculate the impulse response function $\{h_k\}$. One method of finding this response is to use the difference equation

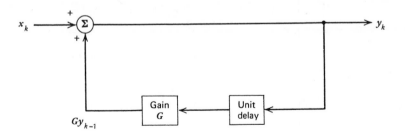

FIGURE 2.6

and calculate the output iteratively. This is what we might term a brute-force approach. From (2.9), we have

$$h_0 = \delta_0 + \tfrac{1}{2}h_{-1} = 1 + 0 = 1$$

$$h_1 = \delta_1 + \tfrac{1}{2}h_0 \quad = 0 + \tfrac{1}{2} \cdot 1 = \tfrac{1}{2}$$

$$h_2 = \delta_2 + \tfrac{1}{2}h_1 \quad = 0 + \tfrac{1}{2} \cdot \tfrac{1}{2} = \tfrac{1}{4}$$

$$\cdot$$
$$\cdot$$
$$\cdot$$

$$h_k = \delta_k + \tfrac{1}{2}h_{k-1} = (\tfrac{1}{2})^k$$

The impulse response function is

$$h_k = \begin{cases} (\tfrac{1}{2})^k, & k \geq 0 \\ 0, & k < 0 \end{cases} \tag{2.10}$$

The output is therefore given by

$$\{y_k\} = \{3 \quad -1 \quad 3\} * \{(\tfrac{1}{2})^k\}, \qquad k \geq 0 \tag{2.11}$$

One simple way to calculate this convolution is to use the matrix with the "fold and add" method as shown in Figure 2.7. Thus, the output sequence is

$$\{y_k\} = \left\{ 3 \quad \tfrac{1}{2} \quad \tfrac{13}{4} \quad \tfrac{13}{8} \quad \tfrac{13}{16} \quad \cdots \quad \frac{13}{2^k} \quad \cdots \right\}$$

In this example, we were able to obtain a closed-form solution for the impulse-response sequence. This iterative approach to finding the impulse-response sequence has the disadvantage that it may not always be possible to

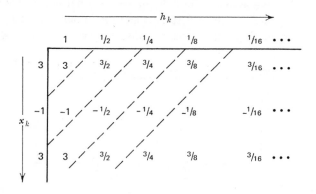

FIGURE 2.7

recognize the form of the general term. In such cases, the solution for $\{h_k\}$ is not in closed form and may not be an acceptable answer.

2.4   FINDING THE IMPULSE-RESPONSE SEQUENCE

In order to use the convolution sum as a method of calculating the response of a linear system, we must be able to find the impulse response of the system. The previous example demonstrates one method: namely, use the difference equation for the system with a forcing function equal to the impulse sequence and merely solve for the output iteratively. As mentioned above, this method has one major disadvantage: the answer may not be in a closed form. This problem will limit the usefulness of this method in some applications.

There is a second method which does not suffer from this drawback. This method is based on knowledge of the homogeneous solutions to the difference equation that models the system. Suppose that we have a system modeled by the following difference equation.

$$L[y_k] = y_{k+n} + b_{n-1}y_{k+n-1} + b_{n-2}y_{k+n-2} + \cdots + b_0 y_k = x_k \quad (2.12)$$

where L is the difference operator

$$L = S^n + b_{n-1}S^{n-1} + b_{n-2}S^{n-2} + \cdots + b_0$$

By definition, the impulse-response sequence results when the input is $\{\delta_k\}$. If we apply an impulse sequence on the right-hand side of (2.12), we see that

$$L[h_k] = 0, \quad k > 0$$

because for $k > 0$, $\{\delta_k\} = 0$. Thus, the impulse response $\{h_k\}$ must satisfy the homogeneous difference equation corresponding to (2.12) for $k > 0$. This means that $\{h_k\}$ can be expressed as a sum of n linearly independent solutions, $\{\phi_{1_k}\}, \{\phi_{2_k}\}, \ldots, \{\phi_{n_k}\}$ of the equation $L[\phi_k] = 0$. That is,

$$h_k = c_1\phi_{1_k} + c_2\phi_{2_k} + \cdots + c_n\phi_{n_k}, \quad k > 0 \quad (2.13)$$

The constants c_i, $i = 1, 2, \ldots, n$ are evaluated based on initial conditions for h_k. These initial conditions can be evaluated with reference to (2.12). Let $k = -n$ in (2.12). We obtain for an impulse input

$$h_0 + b_{n-1}h_{-1} + b_{n-2}h_{-2} + \cdots + b_0 h_{-n} = \delta_{-n} = 0 \quad (2.14)$$

Assuming that the system is initially at rest means $h_k = 0$, $k < 0$. Thus (2.14) implies that $h_0 = 0$. Now let $k = -n + 1$ in (2.12). We have

$$h_1 + b_{n-1}h_0 + b_{n-2}h_{-1} + \cdots + b_0 h_{-n+1} = \delta_{-n+1} = 0 \quad (2.15)$$

Again, we see that $h_1 = 0$. Continuing in like manner, we obtain the initial conditions

$$h_1 = h_2 = h_3 = \cdots = h_{n-1} = 0, \qquad h_n = 1 \tag{2.16}$$

The initial conditions of (2.16) can also be obtained by drawing a block diagram of a system represented by the difference equation of (2.12). Such a diagram is shown in Figure 2.8. Referring to Figure 2.8, we see that for an impulse input $\{\delta_k\}$, the unit value at $k = 0$ is delayed n times before appearing at the output. Thus, if the system is initially at rest, the first nonzero output occurs for $k = n$ and has value 1. In other words, $h_1 = h_2 = \cdots = h_{n-1} = 0$, and $h_n = 1$, which are precisely the initial conditions of (2.16). The impulse-response sequence is

$$h_k = \begin{cases} c_1\phi_{1_k} + c_2\phi_{2_k} + \cdots + c_n\phi_{n_k}, & k > 0 \\ 0, & k \leq 0 \end{cases} \tag{2.17}$$

The constants c_i, $i = 1, 2, \ldots, n$ are evaluated from the n equations obtained from (2.16), namely

$$c_1\phi_{1_1} + c_2\phi_{2_1} + \cdots + c_n\phi_{n_1} = 0$$
$$\vdots \tag{2.18}$$
$$c_1\phi_{1_{n-1}} + c_2\phi_{2_{n-1}} + \cdots + c_n\phi_{n_{n-1}} = 0$$
$$c_1\phi_{1_n} + c_2\phi_{2_n} + \cdots + c_n\phi_n = 1$$

This method of finding the impulse-response sequence for a discrete-time system is probably the most general and powerful method available. It can be used whenever the homogeneous solutions can be found. It is not restricted to time-invariant systems.

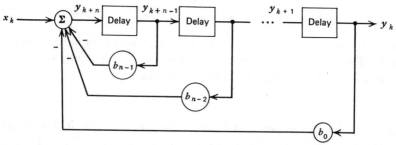

FIGURE 2.8

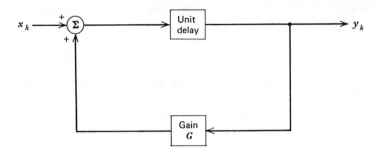

FIGURE 2.9

EXAMPLE 2.4.   Find the impulse-response sequence of the first-order system in Figure 2.9. The difference equation for this system is

$$y_{k+1} - Gy_k = x_k$$

To find the impulse response, we solve $(S - G)[h_k] = 0$ with initial condition $h_1 = 1$. Thus

$$h_k = \begin{cases} c_1 G^k, & k > 0 \\ 0, & k \le 0 \end{cases}$$

The constant c_1 is obtained by using the condition $h_1 = 1$.

$$h_1 = 1 = c_1 G^1; \quad c_1 = \frac{1}{G}$$

The impulse-response sequence is thus

$$h_k = \frac{1}{G} G^k = \begin{cases} G^{k-1}, & k > 0 \\ 0, & k \le 0 \end{cases}$$

A plot of this impulse response is shown in Figure 2.10 for $G < 1$.

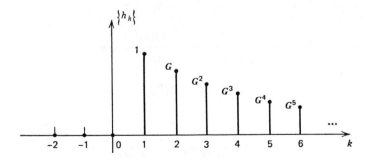

FIGURE 2.10

EXAMPLE 2.5. Find the impulse-response sequence for a system modeled by the difference equation

$$(S^3 - 3S^2 + 3S - 1)[y_k] = x_k, \qquad k \geq 0$$

The homogeneous difference equation has a characteristic equation with a three-fold repeated root: i.e., $r_1 = r_2 = r_3 = 1$. Thus, the homogeneous solutions are 1, k and k^2. Hence,

$$h_k = \begin{cases} c_1 + c_2 k + c_3 k^2, & k > 0 \\ 0, & k \leq 0 \end{cases} \qquad (2.19)$$

The constants c_1, c_2, and c_3 are obtained from

$$h_1 = 0 = c_1 + \ c_2 + \ c_3$$
$$h_2 = 0 = c_1 + 2c_2 + 4c_3$$
$$h_3 = 1 = c_1 + 3c_2 + 9c_3$$

Solving for the constants in the above equations yields $c_1 = 1$, $c_2 = -\frac{3}{2}$, and $c_3 = \frac{1}{2}$. Thus, the impulse response is

$$h_k = \begin{cases} \dfrac{2 - 3k + k^2}{2}, & k > 0 \\ 0, & k \leq 0 \end{cases} \qquad (2.20)$$

The output sequence $\{y_k\}$ can now be found using the convolution sum as

$$y_k = \begin{cases} \dfrac{1}{2} \displaystyle\sum_{m=0}^{k-1} [2 - 3(k - m) + (k - m)^2] x_m, & k > 0 \\ 0, & k \leq 0 \end{cases} \qquad (2.21)$$

EXAMPLE 2.6. Let us return to Example 1.32 and solve this problem using convolution techniques. The difference equation is

$$(S^2 - 5S + 6)[y_k] = x_k$$

Hence, the impulse response is given by

$$h_k = \begin{cases} c_1 2^k + c_2 3^k, & k > 0 \\ 0, & k \leq 0 \end{cases}$$

with

$$h_1 = 0 = 2c_1 + 3c_2$$
$$h_2 = 1 = 4c_1 + 9c_2$$

from which $c_1 = -\frac{1}{2}$ and $c_2 = \frac{1}{3}$. Thus, the output is given by

$$y_k = \sum_{m=0}^{k-1} [-\tfrac{1}{2}2^{k-m} + \tfrac{1}{3}3^{k-m}]x_m$$

With $\{x_k\}$ a unit step, we have for the output

$$y_k = \sum_{m=0}^{k-1} [-2^{k-m-1} + 3^{k-m-1}] \cdot 1$$

$$= \sum_{p=0}^{k-1} [-2^p + 3^p] \cdot 1$$

$$= \begin{cases} \frac{1}{2} - 2^k + \dfrac{3^k}{2}, & k > 0 \\ 0, & k \leq 0 \end{cases} \qquad (2.22)$$

as we found previously.

A Generalization of the Difference-Equation Model

There are many applications in which the driving function of a discrete-time system is not merely x_k, but is rather some functional of x_k which we can denote as $L_D[x_k]$. For example, $x_{k+1} - 3x_k$ might be the driving sequence rather than x_k. The impulse response of the system represented by $L[y_k] = x_{k+1} - 3x_k$ is not the same as that of the system modeled by $L[y_k] = x_k$.

In order to explain how we obtain the impulse response for the system $L[y_k] = L_D[x_k]$, we need some additional notation. In block diagram form, suppose we represent the system $L[y_k] = x_k$ as shown in Figure 2.11.

We use the inverse operator L^{-1} to represent the system in the block diagram because we obtain the output y_k as

$$y_k = L^{-1}[x_k] \qquad (2.23)$$

The inverse operator L^{-1} is defined implicitly as

$$L^{-1}[L[y_k]] = y_k \qquad (2.24)$$

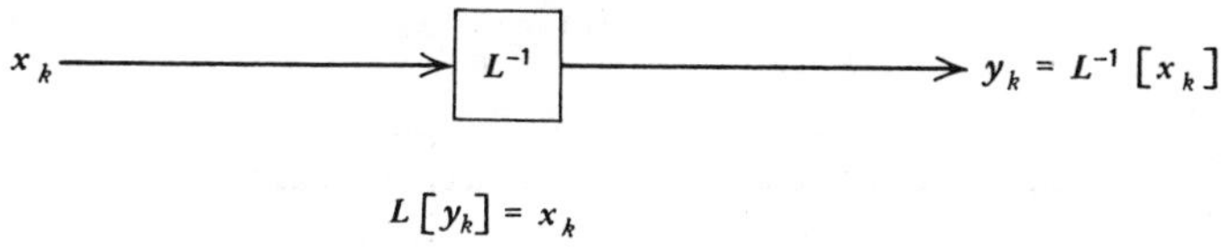

$$L[y_k] = x_k$$

FIGURE 2.11

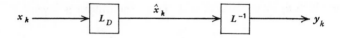

FIGURE 2.12

In Figure 2.11, we obtain the output by operating on the input with the operator within the block diagram. We shall adhere to this convention in what follows. Consider now the system

$$L[y_k] = L_D[x_k] \tag{2.25}$$

We obtain y_k by applying the inverse operator L^{-1} to both sides of (2.25). Thus, we obtain

$$y_k = L^{-1}[L_D[x_k]] \tag{2.26}$$

If we represent (2.26) in block diagram form, we obtain Figure 2.12, where $L_D[x_k]$ is defined as $\hat{x}_k$. For linear time-invariant systems, L_D and L^{-1} commute. Thus, we can rewrite (2.26) as

$$y_k = L_D[L^{-1}[x_k]] \tag{2.27}$$

This system is shown in Figure 2.13. This system suggests the following method for obtaining the impulse response for the system $L[y_k] = L_D[x_k]$. First find the impulse response of the first part of the system, represented by $L[y_k] = x_k$. Call this response $\hat{h}_k$. This problem is, of course, the one we have previously discussed. To obtain the overall impulse response, now apply L_D to $\hat{h}_k$. Notice that we do not have to solve a system represented by an inverse operator for the system represented by L_D. Thus, the second part of the solution is direct, and we obtain as the overall impulse response

$$h_k = L_D[\hat{h}_k] \tag{2.28}$$

One could also calculate the impulse response of a system represented by (2.25) by convolving two impulse responses, one obtained from the system $L[y_k] = x_k$ and the other obtained from the system $L_D^{-1}[y_k] = x_k$.

EXAMPLE 2.7.   A system is modeled by the difference equation

$$y_{k+2} - 5y_{k+1} + 6y_k = x_{k+2} - 3x_k \tag{2.29}$$

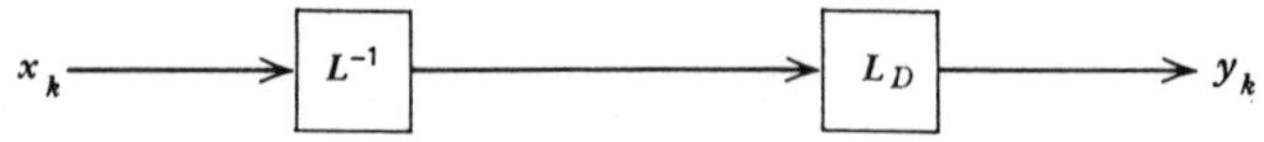

$$x_k \longrightarrow \boxed{L^{-1}} \longrightarrow \boxed{L_D} \longrightarrow y_k$$

FIGURE 2.13

What is the impulse response of this system? In this case, $L_D = S^2 - 3$, because $(S^2 - 3)[x_k] = x_{k+2} - 3x_k$. We first find the impulse-response sequence for the system modeled by

$$y_{k+2} - 5y_{k+1} + 6y_k = x_k \tag{2.30}$$

The characteristic equation has roots $r_1 = 3$ and $r_2 = 2$, which means that $\hat{h}_k$ is given by

$$\hat{h}_k = \begin{cases} c_1 3^k + c_2 2^k, & k > 0 \\ 0, & k \leq 0 \end{cases} \tag{2.31}$$

The initial conditions are $\hat{h}_1 = 0$ and $\hat{h}_2 = 1$. Thus, we evaluate c_1 and c_2 using

$$\hat{h}_1 = 0 = 3c_1 + 2c_2$$
$$\hat{h}_2 = 1 = 9c_1 + 4c_2$$

Solving for c_1 and c_2 gives $c_1 = \frac{1}{3}$ and $c_2 = -\frac{1}{2}$. The impulse response $\hat{h}_k$ is therefore

$$\hat{h}_k = \begin{cases} 3^{k-1} - 2^{k-1}, & k > 0 \\ 0, & k \leq 0 \end{cases} \tag{2.32}$$

We can write (2.32) in a compact form as

$$\hat{h}_k = (3^{k-1} - 2^{k-1})\{u_{k-2}\}$$

where $\{u_k\}$ is the unit step sequence defined as

$$u_k = \begin{cases} 1, & k \geq 0 \\ 0, & k < 0 \end{cases} \tag{2.33}$$

The overall impulse-response sequence is

$$h_k = (S^2 - 3)[\hat{h}_k] = (S^2 - 3)[(3^{k-1} - 2^{k-1})\{u_{k-2}\}]$$
$$= (3^{k+1} - 2^{k+1})\{u_k\} - 3(3^{k-1} - 2^{k-1})\{u_{k-2}\} \tag{2.34}$$

We can represent $\{u_k\}$ as

$$\{u_k\} = \{\delta_k\} + \{\delta_{k-1}\} + \{u_{k-2}\} \tag{2.35}$$

Thus, (2.34) can be written as

$$h_k = (3^{k+1} - 2^{k+1})(\{\delta_k\} + \{\delta_{k-1}\} + \{u_{k-2}\}) + (-3 \cdot 3^{k-1} + 3 \cdot 2^{k-1})\{u_{k-2}\}$$
$$= \{\delta_k\} + 5\{\delta_{k-1}\} + (9 \cdot 3^{k-1} - 4 \cdot 2^{k-1} - 3 \cdot 3^{k-1} + 3 \cdot 2^{k-1})\{u_{k-2}\}$$

And so

$$h_k = \{\delta_k\} + 5\{\delta_{k-1}\} + (2 \cdot 3^k - 2^{k-1})\{u_{k-2}\} \tag{2.36}$$

A plot of this impulse response is shown in Figure 2.14.

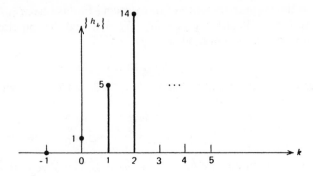

FIGURE 2.14

EXAMPLE 2.8. The difference equation model of (2.12) is often modified so that the output, instead of being delayed by n time units (where n is the order of the difference equation), is not delayed. In other words, the difference equation is written in the form

$$y_k + b_{n-1}y_{k-1} + b_{n-2}y_{k-2} + \cdots + b_0y_{k-n} = x_k \qquad (2.37)$$

Schematically, we can represent (2.37) in block diagram form as shown in Figure 2.15. Figures 2.8 and 2.15 differ only in the place where we obtain the output y_k. What is the impulse-response sequence of the system shown in Figure 2.15? We can easily find $\{h_k\}$ for this system by observing that (2.37) is (2.12) with the right-hand side equal to $L_D[x_k]$ where $L_D = S^n$. That is, (2.37) is of the form

$$y_{k+n} + b_{n-1}y_{k+n-1} + \cdots + b_0y_k = L_D[x_k], \qquad L_D = S^n$$

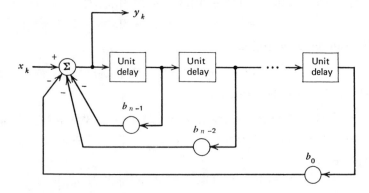

FIGURE 2.15

Thus, the impulse-response sequence is given by applying S^n to (2.17). This procedure yields as the impulse-response sequence for (2.37) the following with $c_1 \ldots c_n$ as in (2.17):

$$
h_k = \begin{cases} c_1 \phi_{1_{k+n}} + c_2 \phi_{2_{k+n}} + \cdots + c_n \phi_{n_{k+n}}, & k \geq 0 \\ 0, & k < 0 \end{cases} \tag{2.38}
$$

In other words, the form of the impulse response for (2.37) is precisely the same as for the difference equation (2.12). The only change is that the impulse response has been shifted forward in time n time units.

EXAMPLE 2.9.   Consider the system of Example 2.4, shown in Figure 2.9. Suppose we take the output in front of the delay unit as shown in Figure 2.16. What is the impulse-response sequence for this system? The difference equation is

$$
y_k - G y_{k-1} = x_k
$$

To find the impulse response, we use the impulse response of Example 2.4 and apply the operator $L_D = S$ to h_k obtained in Example 2.4. Thus

$$
h_k = S\,[G^{k-1}\{u_{k-1}\}] = G^k\{u_k\} \tag{2.39}
$$

or

$$
h_k = \begin{cases} G^k, & k \geq 0 \\ 0, & k < 0 \end{cases} \tag{2.40}
$$

This is the same response shown in Figure 2.10, except that it is shifted forward one unit in time.

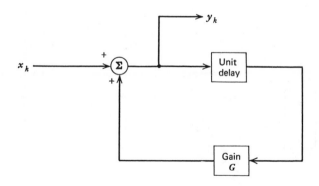

FIGURE 2.16

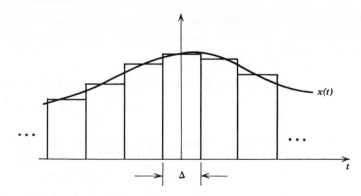

FIGURE 2.17

2.5 SINGULARITY FUNCTIONS AND THE
REPRESENTATION OF CONTINUOUS-TIME SIGNALS

We turn now to the development of convolution techiques for the analysis of continuous-time systems. As with discrete-time systems, we shall decompose the input $x(t)$ into a sum of impulse functions and then express the output $y(t)$ as a sum of the responses resulting from the individual impulses. First, however, we must define an appropriate continuous-time impulse function.

We begin by approximating an arbitrary function $x(t)$ by a series of pulses, as shown in Figure 2.17. The approximation for $x(t)$ is

$$x(t) \cong \sum_{n=-\infty}^{\infty} x(n\Delta)p_\Delta(t - n\Delta) \tag{2.41}$$

where $p_\Delta(t)$ is a pulse of unit height and width Δ as shown in Figure 2.18. This approximation of Figure 2.17 becomes better as Δ is decreased and more

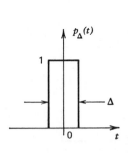
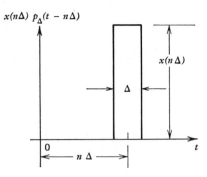

FIGURE 2.18

pulses are used in the representation of $x(t)$. In the limit as Δ goes to zero, we write the sum as an integral. In symbols we have

$$x(t) = \lim_{\Delta \to 0} \sum_{n=-\infty}^{\infty} x(n\Delta) p_\Delta(t - n\Delta)$$

$$= \lim_{\Delta \to 0} \sum_{n=-\infty}^{\infty} x(n\Delta) \left[\frac{1}{\Delta} p_\Delta(t - n\Delta) \right] \cdot \Delta$$

$$= \int_{-\infty}^{\infty} x(\tau)\, \delta(t - \tau)\, d\tau \tag{2.42}$$

Here $n\Delta$ is replaced by the continuous variable τ, Δ becomes $d\tau$, and we define

$$\delta(t - \tau) = \lim_{\Delta \to 0} \frac{1}{\Delta} \cdot p_\Delta(t - n\Delta)$$

or equivalently,

$$\delta(t) = \lim_{\Delta \to 0} \frac{1}{\Delta} \cdot p_\Delta(t) \tag{2.43}$$

The limiting process of (2.43) is shown schematically in Figure 2.19.

The *delta function* or *impulse function*, denoted by $\delta(t)$, is, roughly speaking, a pulse of unbounded amplitude and zero duration. It is the same kind of abstraction as a point charge or a point mass. This impulse function must be treated as a so-called *generalized function* because we cannot define its value point by point as with ordinary functions. Referring to Figure 2.19, we can

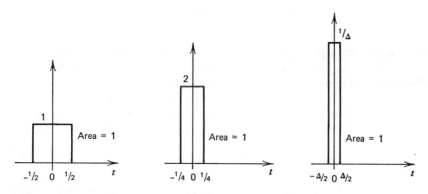

FIGURE 2.19

infer that $\delta(t)$ possesses the following properties:

(1) $\delta(t) = 0$ for $t \neq 0$

(2) $\delta(t)$ is undefined for $t = 0$

(3) $\displaystyle\int_{-\infty}^{\infty} \delta(t)\, dt = 1$

One can obtain the impulse function by using a limiting process on other functions. In fact, any sequence of non-negative functions $\{s_k(t)\}$ which has $\int_{-\infty}^{\infty} s(t)\, dt = 1$ can be used in the limiting form as an impulse function. We merely define the sequence as[2]

$$s_k(t) = ks(kt)$$

In other words,

$$\delta(t) = \lim_{k \to \infty} ks(kt) \tag{2.44}$$

For example, consider the function

$$s(t) = \frac{1}{\pi(1 + t^2)}$$

This function is non-negative for all t and has unit area. Thus the limit of the sequence of functions $s_k(t)$ given by

$$s_k(t) = ks(kt) = \frac{k}{\pi(1 + k^2 t^2)}$$

is an impulse function. That is,

$$\delta(t) = \lim_{k \to \infty} s_k(t) = \lim_{k \to \infty} \left(\frac{k}{\pi(1 + k^2 t^2)} \right)$$

The particular sequence of functions we use to obtain an impulse function is not really critical. Our main interest concerns the properties of the limiting form: i.e., $\delta(t)$.

The three properties we listed previously for the impulse function can be conveniently summarized into one defining equation for $\delta(t)$. This equation is

$$\int_{-\infty}^{\infty} f(t)\, \delta(t - t_0)\, dt = f(t_0) \tag{2.45}$$

---

[2] Greenberg, M. D., *Application of Green's Function in Science and Engineering*, Prentice-Hall, Englewood Cliffs, N.J., 1971.

provided $f(t)$ is continuous at $t = t_0$. One might reason (2.45) as follows. Let $f(t)$ be any function continuous at $t = t_0$. Consider a narrow pulse centered at $t = t_0$ as shown in Figure 2.20. For small Δ, we can write

$$\int_{-\infty}^{\infty} f(t)\left[\frac{1}{\Delta}p_\Delta(t - t_0)\right] dt = \frac{1}{\Delta}\int_{t_0-\Delta/2}^{t_0+\Delta/2} f(t)\, dt \cong f(t_0) \qquad (2.46)$$

The approximation occurs because $f(t)$ may not have constant slope in the neighborhood of t_0. As Δ is decreased, however, the approximation becomes better, and in the limit (2.46) becomes

$$\lim_{\Delta \to 0} \int_{-\infty}^{\infty} f(t)\left[\frac{1}{\Delta}\, p_\Delta(t - t_0)\right] dt = f(t_0) \qquad (2.47)$$

Hence, we arrive at (2.45) for our implicit definition of $\delta(t)$.

Generalized Functions

We digress briefly here to develop certain properties of the impulse function and related generalized functions. We shall use these properties in our discussion of convolution for continuous-time systems.

If we are restricted to using $\delta(t)$ only within an integral as part of integrand, then how can we determine the properties of $\delta(t)$?

One method is to use the concept of testing functions. A testing function $\theta(t)$ is a function which is continuous, has continuous derivatives of all orders, and is zero outside a finite interval. One class of testing functions is

$$\theta(t) = \begin{cases} e^{-\alpha^2/(\alpha^2-t^2)}, & |t| < \alpha \\ 0, & \text{otherwise} \end{cases} \qquad (2.48)$$

One uses testing functions as a method of "examining" the impulse function by integrating $\theta(t)$ and $\delta(t)$ on the interval $(-\infty, \infty)$. This procedure is,

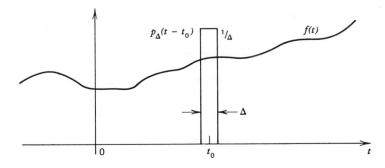

FIGURE 2.20

roughly speaking, analogous to using the output of a measuring instrument to deduce properties about what is being measured. Consider the result of integrating $\theta(t)$ and $\delta(t)$. Call this result $F(\theta)$.

$$F(\theta) = \int_{-\infty}^{\infty} \theta(t)\, \delta(t)\, dt \qquad (2.49)$$

$F(\theta)$ is a generalization of the concept of a function and is called a linear functional on the space of testing functions $\theta(t)$. Equation (2.49) assigns the number $F(\theta)$ to each function $\theta(t)$. In this case, we assign to each function $\theta(t)$ the value $F(\theta) = \theta(0)$. Let us see how this concept can be used to aid us in determining the properties of $\delta(t)$.

The Equivalence Property

One of the basic definitions we use again and again in studying impulse functions by means of testing functions is the so called equivalence property. Suppose that $d_1(t)$ and $d_2(t)$ are expressions involving impulse functions and other functions. We define $d_1(t) = d_2(t)$ if and only if

$$\int_{-\infty}^{\infty} \theta(t)\, d_1(t)\, dt = \int_{-\infty}^{\infty} \theta(t)\, d_2(t)\, dt \qquad (2.50)$$

for all testing functions $\theta(t)$ for which the integral exists. Roughly speaking, $d_1(t) = d_2(t)$ if the "examining instrument" can detect no differences between them.

EXAMPLE 2.10.   To illustrate the use of (2.50), let us demonstrate that $\delta(t)$ can be written as the derivative of the unit step function $u(t)$. The unit step function is defined as

$$u(t) = \begin{cases} 1, & t \geq 0 \\ 0, & t < 0 \end{cases} \qquad (2.51)$$

By the equivalence property, we must show that

$$\int_{-\infty}^{\infty} \delta(t)\theta(t)\, dt = \int_{-\infty}^{\infty} u'(t)\theta(t)\, dt \qquad (2.52)$$

Integrating (2.52) on the right by parts, we have

$$\int_{-\infty}^{\infty} u'(t)\theta(t)\, dt = u(t)\theta(t)\Big|_{-\infty}^{\infty} - \int_{-\infty}^{\infty} u(t)\theta'(t)\, dt$$

$$= 0 - \int_{0}^{\infty} \theta'(t)\, dt$$

$$= -\theta(t)\Big|_{0}^{\infty} = \theta(0)$$

Because the left-hand side is by definition also equal to $\theta(0)$, the equivalence of $\delta(t)$ and $u'(t)$ is proved.

EXAMPLE 2.11.   Another useful equivalence is

$$f(t)\,\delta(t) = f(0)\,\delta(t) \tag{2.53}$$

provided $f(t)$ is continuous at $t = 0$. Equation (2.53) can be seen from

$$\int_{-\infty}^{\infty} f(t)\,\delta(t)\theta(t)\,dt = \int_{-\infty}^{\infty} \delta(t)f(t)\theta(t)\,dt$$

$$= f(0)\theta(0)$$

$$= \int_{-\infty}^{\infty} f(0)\,\delta(t)\theta(t)\,dt$$

Using (2.53), we find, for example, that

$$Ae^t\,\delta(t) = A\,\delta(t)$$

$$e^t \cos t\,\delta(t) = \delta(t)$$

$$A \sin t\,\delta(t) = A \sin 0\,\delta(t) = 0$$

Higher-Order Derivatives of $\delta(t)$

Denote the nth derivative of $\delta(t)$ as $\delta^{(n)}(t)$. We claim that

$$\int_{-\infty}^{\infty} \delta^{(n)}(t)\theta(t)\,dt = (-1)^n\theta^{(n)}(t)\big|_{t=0} = (-1)^n\theta^{(n)}(0) \tag{2.54}$$

We obtain (2.54) by integrating by parts n times. Thus

$$\int_{-\infty}^{\infty} \delta^{(n)}(t)\theta(t)\,dt = \delta^{(n-1)}(t)\theta(t)\Big|_{-\infty}^{\infty} - \int_{-\infty}^{\infty} \delta^{(n-1)}(t)\theta^{(1)}(t)\,dt$$

$$= -\delta^{(n-2)}(t)\theta^{(1)}(t)\Big|_{-\infty}^{\infty} - \int_{-\infty}^{\infty} \delta^{(n-2)}(t)\theta^{(2)}(t)\,dt$$

$$\vdots$$

$$= (-1)^n\theta^{(n)}(0)$$

2.6   SUPERPOSITION AND CONVOLUTION FOR CONTINUOUS-TIME SYSTEMS

We turn our attention now to a discussion of convolution in continuous-time systems. We shall again show that we can characterize the input-output relationship for linear time-invariant systems by a convolution operation involving the input and the impulse-response function of the system.

The impulse response $h(t)$ is defined as the output resulting from an input of $\delta(t)$: i.e.,

$$\delta(t) \rightarrow h(t)$$

If we apply an input impulse of area k, then by the linearity of the system, the output is $kh(t)$.

$$k\,\delta(t) \rightarrow kh(t)$$

Because the system is time-invariant,

$$k\,\delta(t - t_0) \rightarrow kh(t - t_0)$$

These relationships are shown schematically in Figure 2.21.

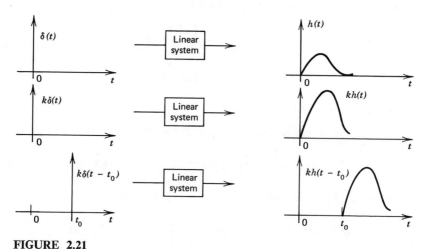

FIGURE 2.21

To find the response of a continuous-time system to an arbitrary input $x(t)$, we represent $x(t)$ by a train of impulse functions. Recall our representation of $x(t)$ in (2.42):

$$x(t) = \lim_{\Delta \to 0} \sum_{n=-\infty}^{\infty} x(n\Delta)\left[\frac{1}{\Delta} P_\Delta(t - n\Delta)\right] \cdot \Delta \qquad (2.55)$$

Let us replace $(1/\Delta)\,P_\Delta(t - n\Delta)$ by its limit, $\delta(t)$. The representation for $x(t)$ is then

$$x(t) = \lim_{\Delta \to 0} \sum_{n=-\infty}^{\infty} x(n\Delta)\,\delta(t - n\Delta) \cdot \Delta \qquad (2.56)$$

as before. Now we have the same sort of decomposition for the continuous-time signal $x(t)$ as we had for the sequence $\{x_k\}$ in (2.2).

If we apply the impulse train representation for $x(t)$ to the system, we can calculate the output response by calculating the response that results from each impulse separately and then adding these individual responses to obtain the complete output. This method of finding the output is possible because of the superposition property of linear systems.

The response to each impulse is quickly found as follows. The impulse at $t = 0$ gives rise to an output

$$\Delta x(0) \; \delta(t) \to \Delta x(0)h(t)$$

Similarly, the impulse at $t = \Delta$ gives rise to an output

$$\Delta x(\Delta) \; \delta(t - \Delta) \to \Delta x(\Delta)h(t - \Delta)$$

In general, the impulse at $t = n \, \Delta$ produces an output

$$\Delta x(n \, \Delta) \; \delta(t - n \, \Delta) \to \Delta x(n \, \Delta)h(t - n \, \Delta)$$

The complete response $y(t)$ is the sum of these individual responses: i.e.,

$$y(t) = \lim_{\Delta \to 0} \sum_{n=-\infty}^{\infty} \Delta x(n\Delta)h(t - n\Delta) \tag{2.57}$$

As we allow $\Delta \to 0$ and the number of impulses to grow, $n \to \infty$, so that $(n\Delta)$ becomes a continuous variable τ, the sum of (2.57) approaches an integral. The output response $y(t)$ resulting from an input $x(t)$ is thus

$$y(t) = \int_{-\infty}^{\infty} x(\tau)h(t - \tau) \, d\tau \tag{2.58}$$

Equation (2.58) is called the convolution of $x(t)$ and $h(t)$. We use the notation $y(t) = x(t)*h(t)$ to denote convolution.

We can also arrive at (2.58) by a somewhat different argument. The input waveform $x(t)$ can be approximated by

$$x(t) \cong \sum_{n=-\infty}^{\infty} \Delta x(n\Delta) \; \delta(t - n\Delta)$$

In the limit as $\Delta \to 0$, the sum becomes an integral, so that

$$x(t) = \int_{-\infty}^{\infty} x(\tau) \; \delta(t - \tau) \, d\tau \tag{2.59}$$

We interpret (2.59) as expressing $x(t)$ by a continuum of impulse functions of the appropriate area. The linearity of the system allows us to calculate the system reponse by adding the responses that result from the continuum of input impulse functions. If $h(t)$ is the response to $\delta(t)$, then it follows that

$$y(t) = \int_{-\infty}^{\infty} x(\tau) h(t - \tau) \, d\tau \qquad (2.60)$$

Equation (2.60) is not always easy to calculate. It is, in fact, often so involved that we use alternate solution methods based on transform techniques. Nevertheless, (2.60) is a strong conceptual aid in understanding linear systems.

2.7   THE CONVOLUTION OPERATION— CONTINUOUS-TIME SYSTEMS

As in the discrete-time case, a graphical interpretation of the convolution operation is often an aid to understanding how the system modifies input signals to obtain the output signal $y(t)$. Consider the convolution of $x(t)$ and $h(t)$

$$y(t) = x(t) * h(t)$$

$$= \int_{-\infty}^{\infty} x(\tau) h(t - \tau) \, d\tau \qquad (2.61)$$

Suppose for the moment that $h(t)$ is known. Equation (2.61) then gives a function $y(t)$ for each function $x(t)$ that we substitute into this equation. To calculate a single point on $y(t)$, say $y(t_1)$, we must know $x(t)$ over its complete range of t, because

$$y(t_1) = \int_{-\infty}^{\infty} x(\tau) h(t_1 - \tau) \, d\tau$$

We can demonstrate these ideas graphically as shown in Figure 2.22. Figure 2.22 depicts a given $h(\tau)$ and $x(\tau)$. The convolution as written in (2.61) uses $h(t - \tau)$ in the integrand. The function $h(-\tau)$, shown in Figure 2.22b, is merely the mirror image of $h(\tau)$ about the line $\tau = 0$; and $h(t - \tau)$ is, for $t > 0$, the function $h(-\tau)$ shifted to the right by t. This shift is shown in Figure 2.22c. To calculate $y(t)$, we multiply $h(t - \tau)$ and $x(\tau)$ and integrate the product function, shown shaded in Figure 2.22d. The area under this shaded function is then $y(t)$. Notice that we obtain only a single value for $y(t)$ by this process. To obtain the graph of $y(t)$ for all t, we must allow the

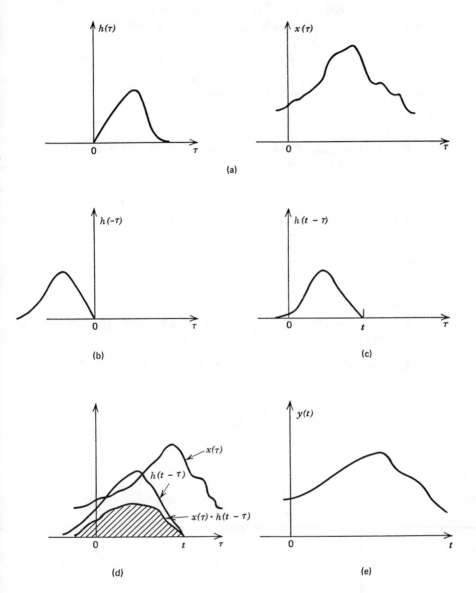

FIGURE 2.22

variable t in $h(t - \tau)$ to take on all values in the interval $(-\infty, \infty)$. The result is a smoothed version of $x(\tau)$ as shown in Figure 2.22e.

If we let $r = t - \tau$, then substituting in (2.61) we obtain

$$y(t) = \int_{-\infty}^{\infty} x(t - r)h(r) \, dr$$

which implies that convolution is commutative: i.e.,

$$y(t) = h(t) * x(t) = x(t) * h(t)$$

EXAMPLE 2.12. Convolve a delta function $\delta(t)$ with an arbitrary function $f(t)$: i.e., find

$$y(t) = \delta(t) * f(t)$$

From the definition of convolution, we know that

$$y(t) = \int_{-\infty}^{\infty} f(\tau) \, \delta(t - \tau) \, d\tau$$

$$= f(t)$$

In words, convolution of a function $f(t)$ with $\delta(t)$ reproduces the function $f(t)$. This fact is shown schematically in Figure 2.23.

EXAMPLE 2.13. Let $f(t)$ and $g(t)$ be defined as

$$f(t) = \begin{cases} e^{-t}, & t \geq 0 \\ 0, & t < 0 \end{cases}$$

$$g(t) = \begin{cases} \alpha e^{-\alpha t}, & t \geq 0 \\ 0, & t < 0 \end{cases}$$

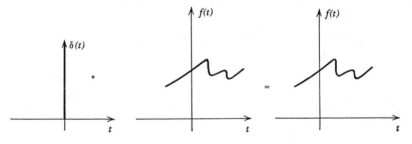

FIGURE 2.23

Find $y(t) = g(t) * f(t)$. By the definition of convolution, we know that

$$y(t) = \int_{-\infty}^{\infty} f(\tau)g(t - \tau)\, d\tau$$

$$= \begin{cases} \int_0^t e^{-\tau} \alpha e^{-\alpha(t-\tau)}\, d\tau, & t \geq 0 \\ 0, & t < 0 \end{cases}$$

$$= \alpha e^{-\alpha t} \int_0^t e^{\tau(\alpha - 1)}\, d\tau, \qquad t \geq 0$$

$$= \begin{cases} \dfrac{\alpha}{\alpha - 1}(e^{-t} - e^{-\alpha t}), & \alpha \neq 1 \\ te^{-t}, & \alpha = 1 \\ 0, & t < 0 \end{cases}, \qquad t \geq 0$$

In convolution problems, it is often easy to omit various cases by failing to calculate $y(t)$ for all t. A sketch of the various integrations which are implied by the convolution is often useful.

2.8   DISCUSSION AND SOME GENERALIZATIONS OF CONVOLUTION

The general form of the convolution of $h(t)$ and $x(t)$ given by (2.58) can often be simplified. In many applications, the impulse response is a *causal* function:[3] i.e., $h(t) = 0$, $t < 0$. In this case, $h(t - \tau)$ is zero for $\tau > t$, and so we can change the upper limit of integration in (2.58) from ∞ to t. If $x(t)$ is also causal, we can replace the lower limit of integration by 0. Thus we have four possible cases, summarized below.

$$y(t) = x(t) * h(t)$$

$$= \int_{-\infty}^{\infty} x(\tau)h(t - \tau)\, d\tau, \quad \text{general } x(t) \text{ and } h(t)$$

$$= \int_0^t x(\tau)h(t - \tau)\, d\tau, \quad \text{causal } x(t) \text{ and } h(t)$$

$$= \int_0^{\infty} x(\tau)h(t - \tau)\, d\tau, \quad \text{causal } x(t) \text{ and general } h(t)$$

$$= \int_{-\infty}^t x(\tau)h(t - \tau)\, d\tau, \quad \text{general } x(t) \text{ and causal } h(t)$$

[3] If $h(t)$ is noncausal, then one obtains an output before the input is applied.

The corresponding relationships for discrete-time convolutions are

$$y_k = x_k * h_k$$

$$= \sum_{n=-\infty}^{\infty} x_n h_{k-n}, \quad \text{general } x_n \text{ and } h_n$$

$$= \sum_{n=0}^{k} x_n h_{k-n}, \quad \text{causal } x_n \text{ and } h_n$$

$$= \sum_{n=0}^{\infty} x_n h_{k-n}, \quad \text{causal } x_n \text{ and general } h_n$$

$$= \sum_{n=-\infty}^{k} x_n h_{k-n}, \quad \text{general } x_n \text{ and causal } h_n$$

The derivation of the convolution sum and convolution integral for linear systems has been obtained by decomposing the input signal into a continuous sum of impulse functions. Implicit in this derivation is the assumption that the system is initially de-energized. If the system contains some initial stored energy, then an input which produces these initial stored energy sources must be included as an independent input function to the system. The total response is merely the sum of the responses that result from the input signal and the initial energy sources. (See Section 1.5.)

Relationships between the Step Response and Impulse Response

The step response of a linear system, denoted by $g(t)$, is the output that results from a step function input $u(t)$. In symbols, if $H(\cdot)$ represents the transformation performed by the linear system, then

$$g(t) = H(u(t))$$

The step response $g(t)$ can be obtained by convolving $u(t)$ and $h(t)$. That is,

$$g(t) = u(t) * h(t) = \int_{-\infty}^{\infty} u(\tau) h(t - \tau) \, d\tau \tag{2.62}$$

Because $u(t)$ is zero for $t < 0$, for a causal system (2.62) can be written as

$$g(t) = \int_{0}^{t} h(\tau) \, d\tau \tag{2.63}$$

Equation (2.63) states that the step response of a linear system is the integral of the impulse response. We can generalize (2.63) to arbitrary inputs. If $y(t)$ is the response resulting from an arbitrary input $x(t)$, then the response that results from an input $\int x(t') \, dt'$ is $\int y(t') \, dt'$. We can see this fact as follows. The output resulting from an input of $\int x(t')\, dt'$ is given by (2.64).

$$\left(\int x(t')\, dt'\right) * h(t) = \int_{-\infty}^{\infty} \left(\int x(t')\, dt'\right) h(t - \tau)\, d\tau \qquad (2.64)$$

Assuming that one can interchange the order of integration on t' and τ, we obtain

$$\left(\int x(t')\, dt'\right) * h(t) = \int y(t)\, dt$$

One can use the step response $g(t)$ to characterize the input-output relationship of a linear system. Consider the convolution of an arbitrary input $x(t)$ with the impulse response $h(t)$: i.e.,

$$y(t) = \int_{-\infty}^{\infty} x(\tau) h(t - \tau)\, d\tau \qquad (2.65)$$

Integrate (2.65) by parts and use (2.63) to obtain

$$y(t) = -x(\tau) g(t - \tau)\Big|_{-\infty}^{\infty} + \int_{-\infty}^{\infty} x'(\tau) g(t - \tau)\, d\tau$$

$$= \int_{-\infty}^{\infty} x'(\tau) g(t - \tau)\, d\tau \qquad (2.66)$$

provided that $x(t)$ and $g(t)$ are zero at $t = -\infty$. In words, the output response of a linear system is given by convolving the step response $g(t)$ with the derivative of the input $x'(t)$.

The function $x(t)$ may have a jump discontinuity at the origin. If it does and one is concerned with causal functions, then $x'(t)$ contains an impulse at the origin. This impulse can be handled by rewriting (2.66) as

$$y(t) = \int_{0^-}^{0^+} x'(\tau) g(t - \tau)\, d\tau + \int_{0^+}^{t} x'(\tau) g(t - \tau)\, d\tau$$

$$= \int_{0^-}^{0^+} x(0)\, \delta(\tau) g(t - \tau)\, d\tau + \int_{0^+}^{t} x'(\tau) g(t - \tau)\, d\tau$$

$$= x(0) g(t) + \int_{0^+}^{t} x'(\tau) g(t - \tau)\, d\tau \qquad (2.67)$$

where $x(0)$ is the value of the jump discontinuity at the origin.

In general, an arbitrary function $x(t)$ can be expressed as a continuous sum of the nth power function $t^n u(t)$ (for $n = 0$ we have step function, for $n = 1$ we have ramp functions, etc.). The output response can thus be expressed as a continuous sum of the responses resulting from the input power function.

Now $t^n u(t)$ is the $(n + 1)$st integral of $\delta(t)$. Thus, the response to $t^n u(t)$ is the $(n + 1)$st integral of $h(t)$, which we denote by $h^{-(n+1)}(t)$. The general result is that we can express the output $y(t)$ as

$$y(t) = x^{(n+1)}(t) * h^{-(n+1)}(t) \tag{2.68}$$

The proof involves repeated integration by parts, as in the case of the step response. This relationship can also be easily seen in terms of a transform domain formulation (see Chapters 5 and 6).

EXAMPLE 2.14. Find the response of the system in Figure 2.24 resulting from an input pulse of the form $x(t) = a[u(t) - u(t - T)]$. We find the impulse response of this system by applying an input $x(t) = \delta(t)$. Because of the simple nature of this system, we can calculate the output directly from the block diagram. Thus

$$h(t) = \int_{-\infty}^{t} (\delta(\tau) - \delta(\tau - T))\, d\tau$$

$$= \begin{cases} (u(t) - u(t - T)), & t \geq 0 \\ 0, & t < 0 \end{cases} \tag{2.69}$$

Graphically, $h(t)$ is shown in Figure 2.25. The response is therefore

$$y(t) = x(t) * h(t) = \int_{0}^{t} a[u(\tau) - u(\tau - T)][u(t - \tau) - u(t - \tau - T)]\, d\tau$$

$$= \begin{cases} a \displaystyle\int_{0}^{t} d\tau, & 0 \leq t < T \\ a \displaystyle\int_{t-T}^{T} d\tau, & T \leq t < 2T \\ 0, & \text{otherwise} \end{cases}$$

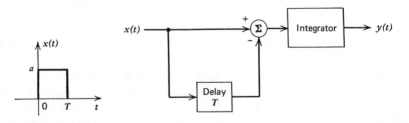

FIGURE 2.24

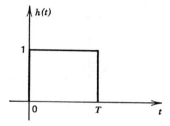

FIGURE 2.25

Performing the integrations, we have

$$y(t) = \begin{cases} at, & 0 \le t < T \\ a(2T - t), & T \le t < 2T \\ 0, & \text{otherwise} \end{cases}$$

The output response is shown in Figure 2.26 below.

This example points out the complexity involved in calculating convolutions analytically, even for very simple functions. A sketch of the convolved functions is often a great aid in setting up the various integrals involved in calculating a convolution.

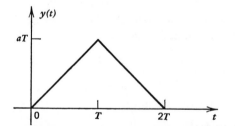

FIGURE 2.26

EXAMPLE 2.15.   Find the impulse response of the RC circuit shown in Figure 2.27. On the basis of (2.63), we can find the impulse response by first finding the step response $g(t)$ and then differentiating to obtain $h(t)$. Assuming that $RC = 1$, the differential equation relating $y(t)$ and $x(t)$ is

$$\frac{dy(t)}{dt} + y(t) = x(t)$$

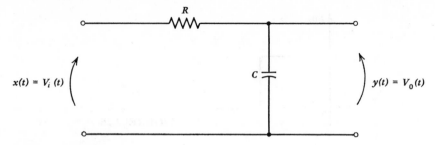

FIGURE 2.27

The step response is thus obtained by solving

$$\frac{dg(t)}{dt} + g(t) = \begin{cases} 1, & t \geq 0 \\ 0, & t < 0 \end{cases}$$

with

$$g(0) = 0.$$

The solution is easily obtained as

$$g(t) = (1 - e^{-t})u(t) \tag{2.70}$$

Thus, by (2.63), we have the impulse response

$$h(t) = \frac{d}{dt}(g(t)) = \frac{d}{dt}(1 - e^{-t})u(t)$$

$$= e^{-t}u(t) + (1 - e^{-t})\,\delta(t) = e^{-t}u(t) \tag{2.71}$$

EXAMPLE 2.16.   Find the output of the RC circuit of the previous example resulting from an input $x(t)$ given by

$$x(t) = \begin{cases} 0, & t < 0 \\ Ae^{-\beta t}, & t \geq 0 \end{cases}$$

Using the convolution integral, we have

$$y(t) = x(t) * h(t)$$

$$= \int_0^t Ae^{-\beta\tau}e^{-(t-\tau)}\,d\tau, \qquad t \geq 0$$

$$= Ae^{-t}\int_0^t e^{-\tau(\beta-1)}\,d\tau, \qquad t \geq 0$$

$$= \begin{cases} \dfrac{A(e^{-\beta t} - e^{-t})}{1 - \beta}\,u(t), & \beta \neq 1 \\ Ate^{-t}u(t), & \beta = 1 \\ 0, & t < 0 \end{cases}, \qquad \begin{matrix} t \geq 0 \\[4pt] \\ t < 0 \end{matrix}$$

EXAMPLE 2.17.   Find the output response for the *RC* circuit of the previous two examples to an input $x(t)$ given by

$$x(t) = \begin{cases} \sin t, & 0 < t < \dfrac{\pi}{2} \\ 0, & \text{otherwise} \end{cases}$$

The step response of this system is given by (2.70) as

$$g(t) = \begin{cases} 0, & t < 0 \\ 1 - e^{-t}, & t \geq 0 \end{cases}$$

Thus, the output $y(t)$ is

$$y(t) = x'(t) * g(t)$$

$$= \begin{cases} 0, & t < 0 \\ \displaystyle\int_0^t \cos \tau \,(1 - e^{-(t-\tau)})\, d\tau, & 0 \leq t < \dfrac{\pi}{2} \\ \displaystyle\int_0^{\pi/2} \cos \tau \,(1 - e^{-(t-\tau)})\, d\tau, & \dfrac{\pi}{2} \leq t \end{cases}$$

The limits on the integrals are most easily seen by drawing some sketches of $x'(\tau)$ and $g(t - \tau)$ for various values of t. Evaluating the above integrals, we obtain

$$y(t) = \begin{cases} 0, & t < 0 \\ \tfrac{1}{2}(\sin t - \cos t + e^{-t}), & 0 \leq t < \dfrac{\pi}{2} \\ \dfrac{e^{-t}}{2}(1 + e^{\pi/2}), & \dfrac{\pi}{2} \leq t \end{cases}$$

The input and output functions are plotted in Figure 2.28.

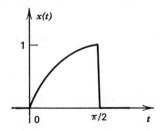
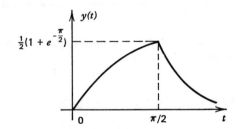

FIGURE 2.28

The step response $g(t)$ is useful from a practical point of view because a step input can easily be generated and applied to real systems. An impulse input may not be practical in many systems because a narrow pulse—our physical approximation of an impulse—contains little energy. The concept of convolution and the impulse or step response is also useful in another practical sense. If one must characterize a linear "black-box" system without knowledge of the internal structure of the system, then the step or impulse response becomes a natural characterization. It is probably not reasonable to develop a difference or differential equation model without some knowledge of the internal structure of a system.

2.9   FINDING THE IMPULSE RESPONSE— CONTINUOUS-TIME SYSTEMS

As in the case of discrete-time systems, we wish to develop techniques to find the impulse-response function of a continuous-time system. We have in Examples (2.14) and (2.15) demonstrated two methods. If the system is modeled or specified by a block diagram, it may be possible to obtain the impulse response directly from the block diagram, as in Example (2.14). Usually this method will work only for very simple systems.

Example (2.15) demonstrates another technique. Suppose we have a system defined by the differential equation

$$L[y(t)] = x(t) \tag{2.72}$$

We can find the step response of this system from

$$L[g(t)] = \begin{cases} 1, & t \geq 0 \\ 0, & t < 0 \end{cases} \tag{2.73}$$

with the appropriate initial conditions. The impulse response $h(t)$ can then be obtained from

$$h(t) = \frac{d}{dt}[g(t)] \tag{2.74}$$

A third and more powerful approach is based on knowledge of the homogeneous solutions to

$$L[y(t)] = x(t)$$

To develop this method, suppose we have a second-order system of the form

$$L[y(t)] = (D^2 + a_1 D + a_0)[y(t)] = x(t) \tag{2.75}$$

Assuming that the system is initially at rest, with no energy storage, implies that the initial conditions are

$$y(0) = 0$$
$$y'(0) = 0 \qquad (2.76)$$

Now if $h(t)$ is the impulse-response function, the output is given by

$$y(t) = \int_0^t x(\tau)h(t - \tau)\, d\tau \qquad (2.77)$$

Equations (2.75) and (2.77) represent two methods of finding the output response $y(t)$. Using (2.77) as a starting point, consider the conditions which (2.75) and (2.76) impose on the impulse-response function in (2.77).

We note that $y(0) = 0$ from (2.77), as required by (2.76). Differentiating (2.77) with respect to t, we have

$$y'(t) = h(t - \tau)x(\tau)\big|_{\tau=t} + \int_0^t h'(t - \tau)x(\tau)\, d\tau$$

$$= h(0)x(t) + \int_0^t h'(t - \tau)x(\tau)\, d\tau \qquad (2.78)$$

Equations (2.76) require that $y'(0) = 0$. Setting $y'(0) = 0$ in (2.78) implies that $h(0) = 0$. Differentiating again, we obtain

$$y''(t) = h'(0)x(t) + \int_0^t h''(t - \tau)x(\tau)\, d\tau \qquad (2.79)$$

Equations (2.77), (2.78), and (2.79) are expressions for $y(t)$, $y'(t)$, and $y''(t)$. Consider the result of forming the sum $y''(t) + a_1 y'(t) + a_0 y(t)$ using these expressions. The result is

$$h'(0)x(t) + \int_0^t h''(t - \tau)x(\tau)\, d\tau + a_1 \int_0^t h'(t - \tau)x(\tau)\, d\tau + a_0 \int_0^t h(t - \tau)x(\tau)\, d\tau$$

We observe that if

(a) $h'(0) = 1$ \qquad (2.80)

(b) $\displaystyle\int_0^t [h''(t - \tau) + a_1 h'(t - \tau) + a_0 h(t - \tau)]x(\tau)\, d\tau = 0$ \qquad (2.81)

then (2.77) will be a solution of (2.75)! Equation (2.81) implies that the integrand on the left-hand side is zero. If $x(t) \neq 0$, then the bracketed term is zero. Setting the bracketed term equal to zero yields

$$h''(t - \tau) + a_1 h'(t - \tau) + a_0 h(t - \tau) = 0 \qquad (2.82)$$

or

$$h''(t) + a_1 h'(t) + a_0 h(t) = 0 \qquad (2.83)$$

Equation (2.83) is, of course, the original homogeneous differential equation, (2.75). Thus the impulse response can be obtained by finding the homogeneous solutions to the original differential equation. Therefore $h(t)$ can be expressed as the sum of two linearly independent homogeneous solutions such as we found in Chapter 1.

$$h(t) = (c_1\phi_1(t) + c_2\phi_2(t))u(t)$$

where

$$L[\phi_i(t)] = 0, \qquad i = 1, 2 \tag{2.84}$$

The initial conditions which allow us to find c_1 and c_2 are given by (2.78) and (2.80) as

$$h(0) = 0$$
$$h'(0) = 1 \tag{2.85}$$

Thus we can evaluate c_1 and c_2 in (2.84) from

$$h(0) = 0 = c_1\phi_1(0) + c_2\phi_2(0)$$
$$h'(0) = 1 = c_1\phi_1'(0) + c_2\phi_2'(0)$$

EXAMPLE 2.18. Consider the system represented by the differential equation

$$L[y(t)] = y''(t) + y(t) = x(t) \tag{2.86}$$

The homogeneous solutions of (2.86) are easily found to be

$$\phi_1(t) = \sin t, \qquad \phi_2(t) = \cos t$$

Thus

$$h(t) = (c_1 \sin t + c_2 \cos t)u(t)$$

with

$$h(0) = 0, \qquad h'(0) = 1$$

The constants c_1 and c_2 therefore satisfy

$$h(0) = 0 = c_1 \sin 0 + c_2 \cos 0$$
$$h'(0) = 1 = c_1 \cos 0 - c_2 \sin 0$$

which imply that $c_1 = 1$ and $c_2 = 0$. The impulse response of the system modeled by (2.86) is thus

$$h(t) = \sin t \, u(t) \tag{2.87}$$

To verify this result, we substitute (2.87) into (2.86), with

$$h'(t) = \frac{d}{dt}(\sin t \, u(t)) = \cos t \, u(t) + \sin t \, \delta(t) = \cos t \, u(t)$$

$$h''(t) = \frac{d}{dt}(\cos t \, u(t)) = -\sin t \, u(t) + \cos t \, \delta(t) = -\sin t \, u(t) + \delta(t)$$

to obtain

$$h''(t) + h(t) = -\sin t\, u(t) + \delta(t) + \sin t\, u(t) = \delta(t).$$

Thus the output $y(t)$ for an arbitrary input $x(t)$ may be found as

$$y(t) = \int_0^t \sin(t - \tau)x(\tau)\, d\tau$$

We note that the linearly independent solutions

$$\phi_1(t) = e^{jt}$$
$$\phi_2(t) = e^{-jt}$$

could also have been chosen. The impulse-response function is

$$h(t) = (c_1 e^{jt} + c_2 e^{-jt})u(t)$$

with

$$h(0) = 0 = c_1 + c_2$$
$$h'(0) = 1 = jc_1 - jc_2$$

These equations give

$$c_1 = \frac{1}{2j} \qquad c_2 = -\frac{1}{2j}$$

from which

$$h(t) = \left(\frac{1}{2i}e^{jt} - \frac{1}{2i}e^{-jt}\right)u(t)$$

$$= \sin t\, u(t)$$

as before.

EXAMPLE 2.19.   Consider a system modeled by the differential equation

$$L[y(t)] = (D^2 + 2D + 2)[y(t)] = x(t) \tag{2.88}$$

The homogeneous solutions to $L[y(t)] = 0$ are $e^{-t}\sin t$ and $e^{-t}\cos t$. Thus, the impulse-response function is

$$h(t) = (c_1 e^{-t}\sin t + c_2 e^{-t}\cos t)u(t) \tag{2.89}$$

where the constants c_1 and c_2 are evaluated using

$$\left.\begin{array}{l} h(0) = 0 = c_2 \\ h'(0) = 1 = c_1 \end{array}\right\} \rightarrow c_1 = 1, \quad c_2 = 0$$

The impulse response of this system is

$$h(t) = e^{-t}\sin t\, u(t) \tag{2.90}$$

and the output response of the system for an arbitrary input $x(t)$ is

$$y(t) = \begin{cases} \int_0^t e^{-(t-\tau)} \sin (t - \tau) x(\tau) \, d\tau, & t \geq 0 \\ 0, & t < 0 \end{cases}$$

The reader should verify for himself that

$$(D^2 + 2D + 2)[h(t)] = \delta(t)$$

We can generalize this method to nth order systems in a straightforward manner. For the general case, we take

$$L[y(t)] = (D^n + b_{n-1}D^{n-1} + \cdots + b_1 D + b_0)[y(t)] = x(t) \quad (2.91)$$

with initial conditions

$$y(0) = y'(0) = \cdots = y^{(n-1)}(0) = 0$$

The output is again expressed by

$$y(t) = \int_0^t h(t - \tau) x(\tau) \, d\tau \quad (2.92)$$

Equating successive derivatives of $y(t)$ in (2.92) to zero yields

$$h(0) = h'(0) = \cdots = h^{(n-2)}(0) = 0 \quad (2.93)$$

The nth derivative of (2.92) yields

$$y^{(n)}(t) = h^{(n-1)}(0)x(t) + \int_0^t h^{(n)}(t - \tau) x(\tau) \, d\tau \quad (2.94)$$

By exactly the same argument as we use in the second-order case discussed previously, we find the impulse-response function for the system of (2.91) must satisfy the homogeneous equation

$$L[h(t)] = 0$$

with initial conditions $h(0) = h'(0) = \cdots = h^{(n-2)}(0) = 0$, $h^{(n-1)}(0) = 1$. Thus

$$h(t) = (c_1\phi_1(t) + c_2\phi_2(t) + \cdots + c_n\phi_n(t))u(t) \quad (2.95)$$

where these initial conditions are used to find the constants $c_1, c_2, \ldots, c_n$.

EXAMPLE 2.20. Consider a system modeled by the differential equation

$$L[y(t)] = (D^2 - 1)(D^2 - 1)[y(t)] = x(t)$$

The homogeneous solutions are e^t, e^{-t}, te^t, and te^{-t}. The impulse response is therefore

$$h(t) = (c_1 e^t + c_2 e^{-t} + c_3 te^t + c_4 te^{-t})u(t)$$

where the constants are evaluated by the use of

$$h(0) = 0 = c_1 + c_2$$
$$h'(0) = 0 = c_1 - c_2 + c_3 + c_4$$
$$h''(0) = 0 = c_1 + c_2 + 2c_3 - 2c_4$$
$$h'''(0) = 1 = c_1 - c_2 - 3c_3 + 3c_4$$

These equations yield $c_1 = \frac{1}{2}$, $c_2 = -\frac{1}{2}$, $c_3 = -\frac{1}{2}$, $c_4 = -\frac{1}{2}$. The impulse response is therefore

$$h(t) = \tfrac{1}{2}(e^t - e^{-t} - te^t - te^{-t})u(t)$$

The reader should verify that

$$L[h(t)] = (D^4 - 2D^2 + 1)[h(t)] = \delta(t)$$

To complete this section, we extend this method of finding impulse-response functions to systems which are forced by an input signal of the form $L_D[x(t)]$, rather than just $x(t)$. Consider a system of the form

$$L[y(t)] = L_D[x(t)] \tag{2.96}$$

Let the system $L[\hat{y}(t)] = x(t)$ have impulse response $\hat{h}(t)$. The response of the system modeled by $L[\hat{y}(t)] = x(t)$ is given by

$$\hat{y}(t) = \int_0^t \hat{h}(t - \tau)x(\tau)\,d\tau \tag{2.97}$$

This impulse response $\hat{h}(t)$ is found by the methods just described. However, we are forcing the system not with $x(t)$ but rather with $L_D[x(t)]$. Suppose that we apply the operator L_D to both sides of

$$L[\hat{y}(t)] = x(t) \tag{2.98}$$

We obtain

$$L_D[L[\hat{y}(t)]] = L_D[x(t)] \tag{2.99}$$

Making use of the commutative property for linear time-invariant differential operators, we write (2.99) as

$$L[L_D[\hat{y}(t)]] = L_D[x(t)] \tag{2.100}$$

Let $L_D[\hat{y}(t)] = y(t)$. We see that the output of the original system is merely

L_D operating on $\hat{y}(t)$. Thus the overall impulse response $h(t)$ for (2.96) must be

$$h(t) = L_D[\hat{h}(t)] \qquad (2.101)$$

We have assumed L_D to be of lower order than L. Otherwise, additional terms involving $\delta(t)$ and its derivatives will be generated in (2.101), because the $(n-1)$st and higher-order derivatives of $h(t)$ are not in general zero at $t = 0$.

To summarize this material, consider the problem given in Example 1.31. We solve this problem in the next example using the method of convolution.

EXAMPLE 2.21. We return to the circuit of Figure 1.18, redrawn below to show a general input $x(t)$. The relevant differential equation is

$$(D^2 + 2D + 2)[y(t)] = (D + 1)[x(t)] \qquad (2.102)$$

The first step is to determine the impulse response, $\hat{h}(t)$ of the system

$$(D^2 + 2D + 2)[y(t)] = x(t)$$

We have already solved this problem in Example 2.19. From (2.90) the impulse response $\hat{h}(t)$ is

$$\hat{h}(t) = e^{-t} \sin t \, u(t)$$

Thus, the impulse response $h(t)$ of (2.102) is given by

$$h(t) = (D + 1)[\hat{h}(t)] = (D + 1)[e^{-t} \sin t \, u(t)]$$
$$= -e^{-t} \sin t \, u(t) + e^{-t} \cos t \, u(t) + e^{-t} \sin t \, \delta(t) + e^{-t} \sin t \, u(t)$$
$$= e^{-t} \cos t \, u(t)$$

The output $y(t)$ is given by

$$y(t) = \int_0^t e^{-(t-\tau)} \cos (t - \tau) \, x(\tau) \, d\tau, \qquad t \geq 0$$

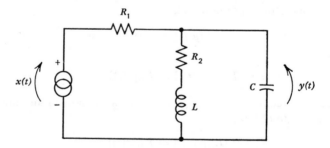

FIGURE 2.29

Let us take $x(t)$ to be a unit step input, as in Example 1.31. Then

$$y(t) = \int_0^t e^{-(t-\tau)} \cos (t - \tau) \, d\tau, \qquad t \geq 0$$

$$= \int_0^t e^{-\xi} \cos \xi \, d\xi$$

$$= \begin{cases} \frac{1}{2}(1 + e^{-t} \sin t - e^{-t} \cos t), & t \geq 0 \\ 0, & t < 0 \end{cases}$$

as before.

2.10   CONVOLUTION WITH TIME-VARYING SYSTEMS

The concept of an impulse response can be generalized to treat time-varying systems. The impulse response h in this case is a function of two parameters: t (or k for a discrete-time system), the time at which we observe the response; and τ (or m), the time at which the impulse $\delta(t - \tau)$ (or δ_{k-m}) is applied to the system. The output $y(t)$ is given by the integral

$$y(t) = \int_{-\infty}^{\infty} h(t, \tau) \, x(\tau) \, d\tau$$

or equivalently, for discrete-time systems, the sum

$$y_k = \sum_{m=-\infty}^{\infty} h_{k,m} \, x_m$$

(These formulations are valid for time-invariant systems also; in this case, the two parameters appear only in the form of a difference, $t - \tau$ or $k - m$—that is, the impulse response is a function only of the difference in time between the application of an impulse and the observation of its effect.)

EXAMPLE 2.22.   The modulator of Figure 2.30 is an example of a linear, time-varying, zero-memory system. An input impulse $\delta(t - \tau)$

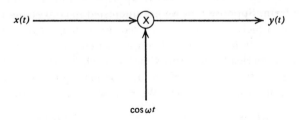

FIGURE 2.30

gives rise to an output $h(t, \tau) = \cos \omega \tau \, \delta(t - \tau)$. Hence a general input, $x(t)$, results in the output

$$y(t) = \int_{-\infty}^{\infty} \cos \omega \tau \, \delta(t - \tau) x(\tau) \, d\tau$$

$$= \cos \omega t \cdot x(t)$$

as is evident from the figure. If the multiplier is cascaded with an RC filter as in Figure 2.31 below, the impulse response is

$$h(t, \tau) = \cos \omega \tau \exp \left[-\frac{t - \tau}{RC} \right] u(t - \tau)$$

and the output is given by

$$y(t) = \int_{-\infty}^{t} \cos \omega \tau \exp \left[-\frac{t - \tau}{RC} \right] x(\tau) \, d\tau$$

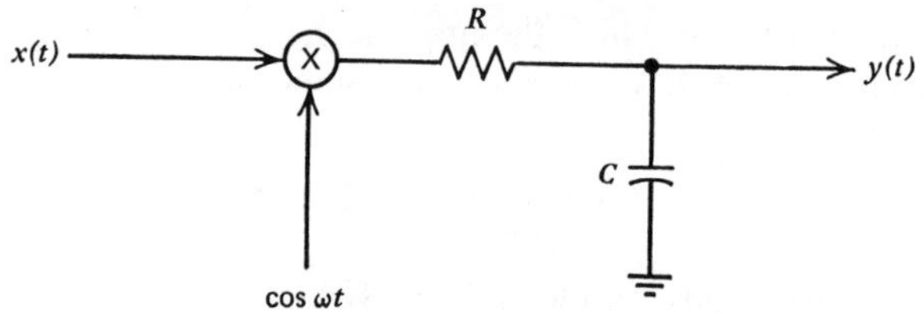

FIGURE 2.31

We shall not treat time-varying systems in any greater depth here. The reader should be aware, however, that techniques do exist for handling these systems by convolution in many situations where other approaches cannot be used.

2.11 NUMERICAL METHODS IN CONVOLUTION

We have seen previously that we can approximate an arbitrary function by an impulse train. Suppose that we are interested in convolving two arbitrary functions for which we have no closed-form analytical expression. For example, the functions may arise from measured data. We can perform the convolution numerically by approximating each of the functions with an impulse train and then convolving the approximating functions. Assume that the two functions are $x(t)$ and $f(t)$. We represent these functions as

$$f(t) \cong \Delta \cdot f(0) \, \delta(t) + \Delta \cdot f(\Delta) \, \delta(t - \Delta) + \cdots + \Delta \cdot f(n\Delta) \, \delta(t - n\Delta) + \cdots$$
$$x(t) \cong \Delta \cdot x(0) \, \delta(t) + \Delta \cdot x(\Delta) \, \delta(t - \Delta) + \cdots + \Delta \cdot x(n\Delta) \, \delta(t - n\Delta) + \cdots$$

$$(2.103)$$

The functions are sketched in Figure 2.32. The representation of (2.103) is really nothing more than a convenient method of representing the sampled functions, in this case sampled every Δ seconds. To convolve the sampled versions of $x(t)$ and $f(t)$ is nothing more than the discrete convolution we have previously discussed. The result is a sampled version of $y(t)$. That is, if we denote the sampled version of $y(t)$ by $y_s(t)$, we have

$$y_s(t) = f_s(t) * x_s(t)$$

$$= \Delta^2[f(0)\,\delta(t) + f(\Delta)\,\delta(t - \Delta) + \cdots] * [x(0)\,\delta(t) + x(\Delta)\,\delta(t - \Delta) + \cdots]$$

$$= \Delta^2 \Big\{ f(0)x(0)\,\delta(t) + [f(0)x(\Delta) + f(\Delta)x(0)]\,\delta(t - \Delta) + \cdots$$

$$+ \left[\sum_{m=0}^{n} f(m\Delta)x((n - m)\Delta) \right] \delta(t - n\Delta) + \cdots \Big\}$$

$$= \Delta^2 \sum_{n} \left[\sum_{m=0}^{n} f(m\Delta)x((n - m)\Delta) \right] \delta(t - n\Delta) \qquad (2.104)$$

The right-hand side is equal to the sampled version of $y(t)$ again sampled at intervals of Δ seconds. The sampled version of $y(t)$ can be represented as

$$y_s(t) = \Delta \cdot y(0)\,\delta(t) + \Delta y(\Delta)\,\delta(t - \Delta) + \cdots + \Delta y(n\Delta)\,\delta(t - n\Delta) + \cdots$$

$$(2.105)$$

Because the right-hand side of (2.105) is equal to the right-hand side of (2.104), we must have

$$y(0) = \Delta x(0)f(0)$$

$$y(\Delta) = \Delta[x(0)f(\Delta) + x(\Delta)f(0)]$$

$$\cdot$$
$$\cdot$$
$$\cdot$$

$$y(n\Delta) = \Delta \sum_{m=0}^{n} f(m\Delta)x((n - m)\Delta) \qquad (2.106)$$

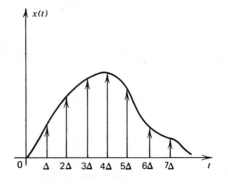
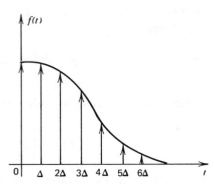

FIGURE 2.32

Equation (2.106) represents an approximation to the function $y(t)$ by the sampled version $y_s(t)$, sampled every Δ seconds.

2.12   DECONVOLUTION

We mentioned in the introduction a problem basic to any measurement process—the problem of determining the true state of nature based on a physical measurement. Because the measurement process must always smooth or filter the actual physical quantity being measured, we must undo this filtering effect to make the measurement more accurate. If we assume that the measuring process is a linear process, then the solution is to deconvolve the linear processing of the measuring instrument "out of" the resultant measurement. In symbols, we can state the problem as follows: Let $y(t) = h(t) * x(t)$. We are given $y(t)$, the result of a measuring process, and $h(t)$, the instrument function. The problem is to find $x(t)$, the actual physical process.

Consider the discrete version of this problem. That is, we are given $\{y(k)\}$ and $\{h(k)\}$ and we wish to find $\{x(k)\}$ from $\{y(k)\} = \{h(k)\} * \{x(k)\}$. If we convolve $\{h(0),\ h(1),\ h(2),\ \ldots\}$ and $\{x(0),\ x(1),\ x(2),\ \ldots\}$ we obtain the result,

$$y(k) = \sum_{n=0}^{k} x(n)\, h(k - n) \tag{2.107}$$

The first few terms of (2.107) yield

$$y(0) = x(0)h(0)$$
$$y(1) = x(0)h(1) + x(1)h(0)$$
$$y(2) = x(0)h(2) + x(1)h(1) + x(2)h(0)$$
$$\cdot$$
$$\cdot \tag{2.108}$$
$$\cdot$$

In the first equation, we can solve for $x(0)$, because we know both $y(0)$ and $h(0)$. Thus

$$x(0) = \frac{y(0)}{h(0)}, \quad \text{provided} \quad h(0) \neq 0 \tag{2.109}$$

Using (2.109), we can now solve the second equation of (2.108) for $x(1)$ as

$$x(1) = \left[y(1) - \frac{y(0)h(1)}{h(0)} \right] \cdot \frac{1}{h(0)}$$

We can continue this iterative process until we find the sequence $\{x(k)\}$.

We can also view this deconvolution from a somewhat different perspective. First notice the following. If we multiply the two polynomials

$$a(x) = a_0 + a_1 x + a_2 x^2 + a_3 x^3 + \cdots$$
$$b(x) = b_0 + b_1 x + b_2 x^2 + b_3 x^3 + \cdots$$

the result is another polynomial

$$c(x) = c_0 + c_1 x + c_2 x^2 + c_3 x^3 + \cdots \qquad (2.110)$$

where the c_i's are given by

$$c_0 = a_0 b_0$$
$$c_1 = a_0 b_1 + a_1 b_0$$
$$c_2 = a_0 b_2 + a_1 b_1 + a_2 b_0$$
$$\vdots$$

The nth coefficient of the product polynomial is given by

$$c_n = \sum_{i=0}^{n} a_i b_{n-i}$$

We recognize the immediate connection with convolution. Suppose now one were given the product polynomial $c(x)$ of (2.110) and the polynomial $a(x)$. To obtain the polynomial $b(x)$, we can divide $c(x)$ by $a(x)$. For example, let

$$c(x) = 12 + 10x + 14x^2 + 6x^3$$
$$a(x) = 4 + 2x$$

The polynomial $b(x)$ is

$$
\begin{array}{r}
3 + \quad x + \quad 3x^2 \\
\hline
4 + 2x\,\big)\,12 + 10x + 14x^2 + 6x^3 \\
\underline{12 + \quad 6x} \\
4x + 14x^2 \\
\underline{4x + \quad 2x^2} \\
12x^2 + 6x^3 \\
\underline{12x^2 + 6x^3}
\end{array}
$$

This is exactly the solution for the following problem. Find the sequence $\{b_0 \; b_1 \; b_2 \ldots\}$ which satisfies the equation

$$\{4 \quad 2\} * \{b_0 \quad b_1 \quad b_2 \quad \cdots\} = \{12 \quad 10 \quad 14 \quad 6\} \qquad (2.111)$$

The above long-division process determines that the sequence $\{b_n\}$ is

$$\{b_n\} = \{3 \quad 1 \quad 3\}$$

This is the process of solving for $\{b_n\}$ in (2.111).

Another method of deconvolving the sequence $\{b_n\}$ involves the following graphical approach. The situation is as depicted in Figure 2.33. We are given the left-hand column of numbers representing $\{a_n\}$ and the right-hand column of numbers representing $\{c_n\}$. The problem is to find the sequence $\{b_n\}$. Recall that in the convolution process, we reverse one of the convolving sequences. Thus, write the sequence $\{b_n\}$ on a movable strip of paper in reverse order, as shown in Figure 2.33b. Recalling the convolution process, we immediately recognize that b_0 must be 3: i.e., $4 \cdot b_0 = 12$. To obtain b_1, we slide the movable strip down one row and redo the calculation necessary to obtain 10 for c_1. Thus we have $4 \cdot b_1 + 3 \cdot 2 = 10$, or $b_1 = 1$. Repeating this process once more, we slide the sequence $\{b_n\}$ down one row and calculate b_2 from the equation $4b_2 + 2 \cdot 1 + 3 \cdot 0 = 14$, or $b_2 = 3$. The equations for b_0, b_1, etc. are precisely the iterative equations of (2.108). The algorithm depicted in Figure 2.33 is fairly simple to use. One merely forms the sum of products of the already calculated $\{b_n\}$ sequence in reverse order with the corresponding terms in the $\{a_n\}$ sequence. This sum is then subtracted from the sequence value of $\{c_n\}$ opposite b_0. Division of this number by a_0 yields the value for $\{b_n\}$.

EXAMPLE 2.23.   Suppose we have a sequence $\{a_n\} = \{1\ 2\ 3\ 4 \cdots\}$. What sequence convolved with $\{a_n\}$ yields the impulse sequencs $\{\delta_n\} = \{\ldots, 0, 1, 0, \ldots\}$? Using the graphical algorithm as shown in Figure 2.34a, we obtain $b_0 = 1$. To calculate b_1, we slide the strip with $\{b_n\}$ written in reverse order down one row as shown in Figure 2.34b. Thus

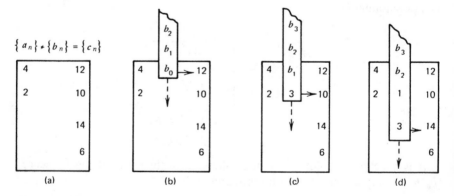

(a)   (b)   (c)   (d)

FIGURE 2.33

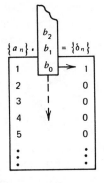 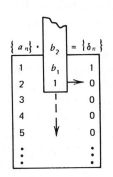 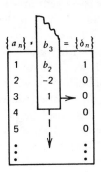

FIGURE 2.34

$1 \cdot b_1 + 2 \cdot 1 = 0$, or $b_1 = -2$. Continuing, we find that the sequence $\{b_n\}$ yields

$$\{b_n\} = \{1 \quad -2 \quad 1\}$$

In other words,

$$\{1 \quad 2 \quad 3 \quad 4 \quad \cdots\} * \{1 \quad -2 \quad 1\} = \{\cdots \quad 0 \quad 0 \quad 1 \quad 0 \quad 0 \quad \cdots\}$$

We call $\{1 \quad -2 \quad 1\}$ the reciprocal or inverse of $\{a_n\}$ and denote it by $\{a_n\}^{-1}$.

The reciprocal or inverse sequence can be used to solve the deconvolution problem. Suppose we wish to find $\{b_n\}$ given $\{a_n\}$ and $\{c_n\}$ from

$$\{a_n\} * \{b_n\} = \{c_n\} \qquad (2.112)$$

Convolving both sides of (2.112) with the inverse sequence $\{a_n\}^{-1}$, we obtain

$$\{a_n\}^{-1} * \{a_n\} * \{b_n\} = \{a_n\}^{-1} * \{c_n\} \qquad (2.113)$$

In (2.113), the left-hand side is $\{\delta_n\} * \{b_n\}$. However, any sequence $\{b_n\}$ convolved with the impulse sequence reproduces the sequence $\{b_n\}$. Thus, (2.113) yields

$$\{b_n\} = \{a_n\}^{-1} * \{c_n\} \qquad (2.114)$$

Thus, knowledge of the inverse sequence $\{a_n\}^{-1}$ can be used to find $\{b_n\}$ by means of (2.114).

The use of digital computers has become commonplace in engineering. A convenient way to view certain kinds of processing of numerical data is in terms of a discrete convolution of two sequences of numbers. For example, suppose $\{a_n\}$ represents a sequence of data points from some physical measurement process. If we wish to generate a running mean over k terms in order to smooth the sequence $\{a_n\}$, we can convolve $\{a_n\}$ and the sequence $\{1/k \; 1/k \cdots (k \text{ terms}) \cdots 1/k\}$. This procedure is the same as convolving $\{a_n\}$ with $1/k \{1 \quad 1 \quad \cdots \quad 1\}$ (k terms).

If we wanted to obtain an integrated version of the sequence $\{a_n\}$, we could convolve $\{a_n\}$ with the semi-infinite sequence $\{1 \quad 1 \quad 1 \quad \cdots\}$. This process again represents a type of smoothing or filtering of the data sequence $\{a_n\}$. In this case, $\{a_n\} * \{1 \quad 1 \quad 1 \quad \cdots\}$ generates a sequence $\{S_n\}$ in which S_n are the partial sums: that is, S_n is the sum of the first n terms of $\{a_n\}$.

These two examples point to the interpretation of the convolving of $\{a_n\}$ with a sequence $\{h_n\}$ as a filtering process on the sequence $\{a_n\}$. The choice of $\{h_n\}$ determines the particular processing involved. This view of linear filtering leads naturally to the idea of digital filters. The basic problem is to design a particular sequence $\{h_n\}$, representing some digital processing, to transform $\{a_n\}$ into some desired form. For example, we may desire to extract certain information contained in $\{a_n\}$. The use of a particular digital filter sequence $\{h_n\}$ can be used to enhance or extract this desired information.

2.13 SUMMARY

The convolution integral and sum are general ways to characterize linear time-invariant systems. They serve as an alternate formulation for solving the input-output problem vis-a-vis the differential or difference equation that describes the system. The characterization in terms of the impulse-response function can be generalized to systems which are time-varying, and so can be used in the analysis of systems where transform techniques are no longer valid. The convolution integral also provides an excellent conceptual way of viewing the operation of a time-invariant linear system.

PROBLEMS

1. Calculate the following discrete convolutions.
   a. $\{3 \quad 2 \quad 1 \quad -3\} * \{4 \quad 8 \quad -2\}$
   b. $\{3 \quad 2 \quad 1 \quad -3\} * \{4 \quad 8 \quad -2\}$
   $\qquad \uparrow \qquad\qquad\qquad \uparrow$
   c. $\{10 \quad -3 \quad 6 \quad 8 \quad 4 \quad 0 \quad 1\} * \{\tfrac{1}{2} \quad \tfrac{1}{2} \quad \tfrac{1}{2} \quad \tfrac{1}{2}\}$
   d. $\{1 \quad 1\} * \{1 \quad 1\} * \{2 \quad 2\}$
   e. $\{(\tfrac{1}{2})^k\}_{k=0}^{\infty} * \{u_k\}$, $\{u_k\}$ unit step sequence
   f. $\{(\tfrac{1}{3})^k\}_{k=0}^{\infty} * \{5 \quad 0 \quad 5 \quad 0 \quad 5\}$
2. Calculate the impulse response for the following discrete-time system.

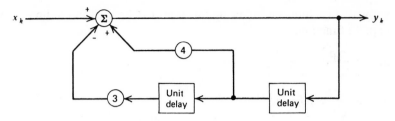

3. Calculate the impulse response for the following discrete-time system. Use the results of problem 2.

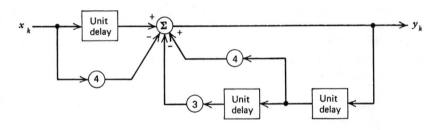

4. Find the impulse response for discrete-time systems defined by the following difference equations. Verify your solutions by substitution.
   a. $(S^2 - S + \frac{1}{4})[y_k] = x_k$
   b. $(S^2 - \frac{1}{4})[y_k] = x_k$
   c. $(S^3 - 3S^2 + 3S - 1)[y_k] = x_k$
   d. $(S^3 - 3S^2 + 3S - 1)[y_k] = S^3[x_k]$
   e. $(1 - 3S^{-1} + 3S^{-2} - S^{-3})[y_k] = x_k$

5. The output of a discrete-time system is $\{y_1(n)\}$, where

$$y_1(n) = \sum_{k=0}^{\infty} x(n - k)e^{-\alpha k}$$

The input to this system is $\{x(n)\}$. We wish to design another discrete-time system of the form shown below so that its output $\{y_2(n)\}$ is identical to $\{y_1(n)\}$. Can the gains a and b be chosen to make $\{y_2(n)\} = \{y_1(n)\}$? Explain.

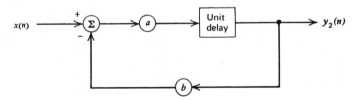

6. Show that each of the following functions can be used in the limit as an impulse function.

a. $\lim\limits_{x \to 0} \dfrac{\exp\left[-(t/x)^2\right]}{x\sqrt{\pi}}$

b. $\lim\limits_{x \to \infty} \dfrac{x}{\sqrt{\pi}} \exp\left[-(xt)^2\right]$

7. Calculate the following convolutions:

a. $u(t) * e^t u(t)$

b. $e^t u(t) * t^2 u(t)$

c. $\sin t\, u(t) * \sin t\, u(t)$

d. $e^t u(t) * e^{3t} u(t)$

e. $e^{\alpha t} u(t) * e^{\beta t} u(-t)$

8. The input voltage to the following system is $x(t)$ as shown. Find the output voltage for $0 < t < 1$.

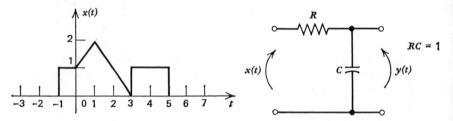

9. a. Consider two linear time-invariant systems connected in cascade as shown. If these two systems are interchanged, is the output response again $w(k)$? Show.

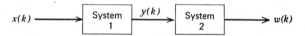

b. Repeat for continuous-time systems and signals.

10. a. Using the convolution integral, find the output signal for the system shown. This system is often used to smooth a sequence $\{x(nT)\}$ into a continuous-time function. It is called a zero-order hold circuit.

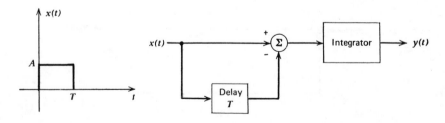

b. What is the output if $x(t)$ is applied to two of the above systems in cascade?

11. Consider a system which takes a sequence of sampled values $\{x(nT)\}$ and connects the sampled values with straight line segments as shown schematically below.

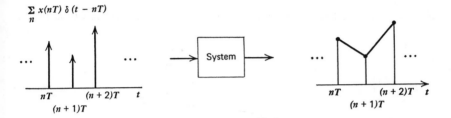

a. What is the impulse response of this system?
b. Sketch a block diagram realization of this system. This system, like that of Problem 10, can be used to smooth the sequence $\{x(nT)\}$ into a continuous-time function.

12. The step response of a system is $g(t)$ as shown. Find the output response to an input

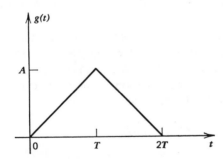

a. $u(t) - u(t - T)$
b. $tu(t) - tu(t - T)$

13. An amplifier has a step response $(1 - e^{-\alpha t})$. This amplifier is used to amplify rectangular pulses. What duration pulses can be accurately reproduced by such an amplifier?

*14. Given two independent random variables x and y and their associated probability density functions $p(x)$ and $p(y)$, show that the probability density function of the random variable $\omega(x, y) = x + y$ is $p(\omega) = p(x) * p(y)$.

15. Find the impulse response for the continuous-time systems defined by the following differential equations. Verify your solutions by substitution.

a. $(D^2 + 7D + 12)[y(t)] = x(t)$

b. $(D^2 + 6D + 9)[y(t)] = x(t)$

c. $(D^2 + 2D + 9)[y(t)] = x(t)$

d. $(D^3 + 6D^2 + 12D + 8)[y(t)] = x(t)$

e. $(D^3 + 6D^2 + 12D + 8)[y(t)] = (D - 1)[x(t)]$

16. A system has the following response to a step input. What would be the

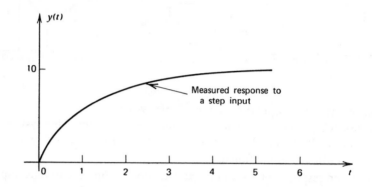

response of this system to a ramp input, $x(t) = t\,u(t)$? (Hint: assume that the system can be modeled by the use of a first-order differential equation).

17. In our treatment of convolution, we assumed that our systems were initially relaxed, with no initial energy storage. How can this method be extended to cover systems where there is some initial energy storage? Specifically, use the impulse-response function to express the output of a system described by the equation

$$(D^2 + 1)[y(t)] = x(t), \qquad t > 0$$

where $y(0) = 1$ and $y'(0) = 0$. Notice that initial conditions are non-zero, denoting some initial energy storage in the system.

$$Ans: \quad y(t) = \cos t + \int_0^t \sin(t - \tau)x(\tau)\,d\tau$$

18. Use the impulse-response function to express the output of a discrete-time system described by the equation

$$(S^2 - 2S + 1)[y_k] = x_k, \qquad k \geq 0$$

where $y_0 = 0$, $y_1 = 2$. Notice that the initial conditions are nonzero, indicating that there is initial energy storage in the system.

$$Ans: \quad y_k = 2k + \sum_{n=0}^{k-1}(k - n - 1)x_n$$

19. Use the convolution method to solve for the outputs in Problems 17 and 18 for the step inputs $u(t)$ and $\{u_k\}$, respectively. Use the direct method of the previous chapter to check your answers.

20. Solve the following equations using both the direct method and the convolution method.

   a. $(D^2 + 7D + 12)[y(t)] = e^t u(t)$

   b. $(S^2 - \frac{1}{9})[y_k] = (\frac{1}{3})^k u_k$

21. In transmitting a digital signal $\{x(k)\}$, $k = 0, 1, 2, \ldots, n$, a deterministic interference is introduced by the transmitter of the form $\alpha^k x(k - 1)$, $k = 1, 2, \ldots, n$. That is, the receiver receives the sequence $\{x(0), x(1) + \alpha x(0), x(2) + \alpha^2 x(1), \ldots, x(n) + \alpha^n x(n - 1)\}$ instead of $\{x(0), x(1), \ldots, x(n)\}$. Design a sequence $\{h(k)\}$ so that the convolution of $\{h(k)\}$ and the received sequence is $\{x(k)\}$.

SUGGESTED READINGS

Bracewell, R. N., *The Fourier Integral and Its Applications*, McGraw-Hill, New York, 1965.
   This text is somewhat more advanced than this work. It contains a very good discussion of both continuous-time and discrete-time convolutions.
Cooper, G. R., and C. D. McGillem, *Methods of Signal and System Analysis*, Holt, Rinehart and Winston, New York, 1967.
   Chapter 4 of this text contains a discussion of convolution for continuous-time systems written at the same level as that of this text.
Lathi, B. P., *Signals, Systems, and Communication*, Wiley, New York, 1967.
   Chapter 10 of this text contains a fairly complete discussion of convolution for continuous-time systems.

3

State Variables

3.1 INTRODUCTION

In the two preceding chapters we have investigated the input-output relationships for linear systems. We first solved linear difference or differential equations using classical techniques to find the system output, given the input. This solution process depends on modeling the physical system to obtain a relevant system equation.

We next used the superposition property of linear systems to obtain an input-output characterization in terms of a superposition integral or sum. For the case of time-invariant systems, this superposition integral or sum reduced to a convolution operation. This solution process depends on measuring the system output for a particular input (such as an impulse or step). The output of the system for a general input is then expressed in terms of this standard output.

We turn now to another time-domain characterization for linear systems. The state variable description we shall study is concerned with the system as a whole. Thus, we shall be concerned with the internal variables of the system in addition to the input-output variables. There are several reasons why this characterization is useful.

(1) It may be necessary to examine the behavior of all relevant signals in a system—not merely the input and output—in order to evaluate the model. For example, we must know the magnitude of internal variables to determine whether these variables remain within limits where linear approximations are valid. It may happen that an unobserved internal variable of the system is unstable. If we used only an input-output characterization for such a system, the system's instability would not be detected. Results obtained from such a model would likely be nonsense.

110

(2) We need a more general system description to treat multiple inputs and outputs. For example, with M inputs and N outputs, a solution by convolution methods requires the evaluation of MN convolution sums or integrals just to find the outputs, and we still have no knowledge of what is happening inside the system.

(3) In studying complex systems—such as a chemical plant, an ecological system, a power distribution system, and so on—we need a compact system description. The matrix description used in the state variable representation of linear systems is independent of system complexity. It can greatly facilitate the study of complex systems. It also furnishes us with a method of "computerizing" the solution process.

(4) Often in the study of systems, we require only a general qualitative description of system behavior. We need only general properties to answer such questions as these: Is the system stable? How can it be controlled? Which parameters are critical in achieving proper performance? What general types of output will the system exhibit? The state variable approach developed in this chapter provides direct answers to these questions.

We begin by considering discrete-time systems. We first develop a general method of specifying a system by its state variables and then treat methods for calculating the system behavior. This material is then extended to continuous-time systems. Along the way, we develop the ideas of observability and controllability for a linear system. There are some minor digressions to develop certain mathematical machinery that we will need.

3.2 STATE VARIABLE DESCRIPTION OF DISCRETE-TIME SYSTEMS

The concept of a state is very general and can be used to describe both linear and nonlinear systems. It can be used in a wide variety of systems and is not limited to those modeled by difference or differential equations. In general, the state of a system at time t is that (minimal) set of variables needed at time t so that, given the inputs to the system for $\tau > t$, one can exactly specify the future behavior of the system for $\tau > t$. For example, for a system modeled by a difference equation, the state at time $t = 0$ can be taken as the set of initial conditions for the difference equation.

Consider, for example, an nth order difference equation with input $u(k)$[1] and output $y(k)$.

$$y(k) + b_1 y(k - 1) + \cdots + b_n y(k - n) = a_0 u(k) \tag{3.1}$$

[1] In this chapter, we shall follow the common notation for state variable formulation. Here $u(k)$ is a general sequence, and not necessarily a unit step. Where confusion could arise, we shall be explicit.

Recall that the classical solution of this equation requires the knowledge of the initial conditions in order to obtain a unique solution. Because (3.1) is an nth order difference equation, n initial conditions must be specified. These n initial conditions and the input sequence $\{u(k)\}$, $k \geq 0$ determine the output sequence $\{y(k)\}$, $k \geq 0$. To exhibit the state variables explicitly, we first sketch a block diagram of the system represented by (3.1) as shown in Figure 3.1. The state variables can then be chosen as the output of the delay elements. The delay elements can be interpreted as memory states or cells. The contents of the memory cells must be specified at some time k_0 in order to determine the output $\{y(k)\}$, $k > k_0$ for a given input $\{u(k)\}$, $k > k_0$. Thus we define the state variables $x_i(k)$, $i = 1, 2, 3, \ldots, n$, as

$$x_1(k) = y(k - n)$$
$$x_2(k) = y(k - n + 1)$$
$$\cdot$$
$$\cdot \qquad\qquad\qquad (3.2)$$
$$\cdot$$
$$x_n(k) = y(k - 1)$$

Referring to Figure 3.1 or the equations of (3.2), we see that we can write each state variable at time $k + 1$ in terms of the state variables at time k.

$$x_1(k + 1) = x_2(k)$$
$$x_2(k + 1) = x_3(k)$$
$$\cdot$$
$$\cdot \qquad\qquad\qquad (3.3)$$
$$\cdot$$
$$x_{n-1}(k + 1) = x_n(k)$$
$$x_n(k + 1) = y(k) = a_0 u(k) - b_n x_1(k) - \cdots - b_1 x_n(k)$$

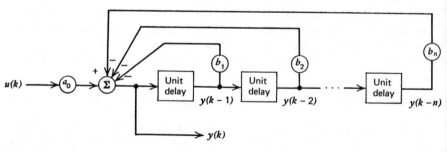

FIGURE 3.1

Using vector notation, we have

$$x(k + 1) = Ax(k) + Bu(k) \tag{3.4}$$

where

$$x(k) = \begin{bmatrix} x_1(k) \\ x_2(k) \\ \cdot \\ \cdot \\ \cdot \\ x_n(k) \end{bmatrix} \quad A = \begin{bmatrix} 0 & 1 & 0 & & 0 \\ 0 & 0 & 1 & & 0 \\ & & \cdots & & \\ 0 & 0 & 0 & & 1 \\ -b_n & -b_{n-1} & -b_{n-2} & \cdots & -b_1 \end{bmatrix} \quad B = \begin{bmatrix} 0 \\ 0 \\ \cdot \\ \cdot \\ \cdot \\ 0 \\ a_0 \end{bmatrix}$$

$$\tag{3.5}$$

The output can be written in terms of the state variables as

$$y(k) = Cx(k) + Du(k) \tag{3.6}$$

where

$$C = [-b_n \quad -b_{n-1} - \quad \cdots \quad -b_1] \quad \text{and} \quad D = a_0 \tag{3.7}$$

Equations (3.4) and (3.6), repeated below, constitute the *state variable representation* for a discrete-time system modeled by (3.1).

$$x(k + 1) = Ax(k) + Bu(k) \tag{3.8}$$

$$y(k) = Cx(k) + Du(k) \tag{3.9}$$

If we know the vector $x(k_0)$ for some k_0, then we can compute $x(k)$, and hence the output $y(k)$, for any $k \geq k_0$ in terms of the input sequence $u(k_0)$, $u(k_0 + 1), \ldots, u(k)$. Thus the vector x satisfies our definition for the state of a system. This n-dimensional vector is equivalent to the n initial conditions needed to solve the difference equation (3.1) in terms of the input u. Accordingly, x is called the *state vector* of the system and the $n \times n$ matrix A, relating the state at index $k + 1$ to the state at k, is called the *state* or *system matrix*. In their most general form (specifically, for a system with multiple inputs and multiple outputs as in Figure 1.1), B, C, and D are matrices. We shall treat this generalization shortly.

EXAMPLE 3.1. Consider a discrete-time system represented by the difference equation

$$y(k) + 2y(k - 1) + y(k - 2) = u(k) \tag{3.10}$$

A block diagram of a system described by (3.10) is shown in Figure 3.2. We have two delay elements. Therefore, two state variables should be

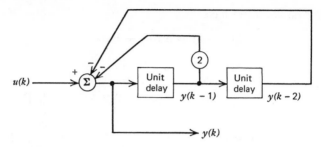

FIGURE 3.2

sufficient to describe the system. Defining the state variables as the outputs of delay elements, we obtain

$$x_1(k) = y(k - 2)$$
$$x_2(k) = y(k - 1)$$

(3.11)

Referring to Figure 3.2, we can write each state variable at time $k + 1$ in terms of those at time k as

$$x_1(k + 1) = x_2(k)$$
$$x_2(k + 1) = y(k) = u(k) - 2x_2(k) - x_1(k)$$

(3.12)

We can rewrite (3.12) in vector form to obtain a state variable representation as

$$\begin{bmatrix} x_1(k + 1) \\ x_2(k + 1) \end{bmatrix} = \begin{bmatrix} 0 & 1 \\ -1 & -2 \end{bmatrix} \begin{bmatrix} x_1(k) \\ x_2(k) \end{bmatrix} + \begin{bmatrix} 0 \\ 1 \end{bmatrix} u(k)$$

$$y(k) = [-1 \quad -2] \begin{bmatrix} x_1(k) \\ x_2(k) \end{bmatrix} + [1]u(k)$$

(3.13)

The appropriate matrices are therefore,

$$\mathbf{A} = \begin{bmatrix} 0 & 1 \\ -1 & -2 \end{bmatrix} \quad \mathbf{B} = \begin{bmatrix} 0 \\ 1 \end{bmatrix} \quad \mathbf{C} = [-1 \quad -2] \quad \mathbf{D} = [1] \quad (3.14)$$

Usually in system studies, one begins with a block diagram model of the system, as in the previous example. One advantage of the state variable method is that we can write down a state-variable description of our system by inspection without first determining the system difference equations. One can, of course, also write the state variables directly from the difference equation of the system. Without the block diagram, however, we lose our physical intuition for what the states represent in terms of the system.

The matrices **A**, **B**, and **C** for a given system are not unique. We can illustrate this fact by redefining the state variables in Example 3.1.

EXAMPLE 3.2.   The system of Example 3.1 is shown in Figure 3.3. Redefine the state variables by defining the output of the first delay as $x_1(k)$ instead of $x_2(k)$, thus:

$$x_1(k) = y(k - 1)$$
$$x_2(k) = y(k - 2) \tag{3.15}$$

The states at time $k + 1$ are

$$x_1(k + 1) = y(k) = -2x_1(k) - x_2(k) + u(k)$$
$$x_2(k + 1) = x_1(k)$$

Hence, the state variable representation is

$$\mathbf{x}(k + 1) = \begin{bmatrix} -2 & -1 \\ 1 & 0 \end{bmatrix} \mathbf{x}(k) + \begin{bmatrix} 1 \\ 0 \end{bmatrix} u(k)$$
$$y(k) = [-2 \quad -1]\mathbf{x}(k) + [1]u(k) \tag{3.16}$$

A comparison of equations (3.16) and (3.13) demonstrates that the matrices are not unique.

Before continuing, we present one more example of finding the state variable equations for a discrete time system.

EXAMPLE 3.3.   Consider the discrete-time system shown in Figure 3.4. Derive a state variable representation for this system. We identify the outputs of the delay elements as state variables. Writing equations for the inputs to the delay elements, we obtain

$$x_1(k + 1) = a_1 x_1(k) + u(k)$$
$$x_2(k + 1) = u(k) \tag{3.17}$$

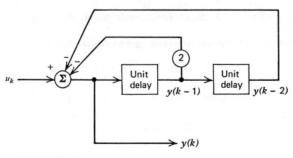

FIGURE 3.3

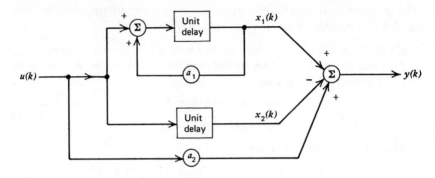

FIGURE 3.4

In vector form, we have

$$\mathbf{x}(k+1) = \begin{bmatrix} a_1 & 0 \\ 0 & 0 \end{bmatrix} \mathbf{x}(k) + \begin{bmatrix} 1 \\ 1 \end{bmatrix} u(k)$$

The output $y(k)$ is given by the equation

$$y(k) = [1 \quad -1]\, \mathbf{x}(k) + [a_2]\, u(k)$$

Thus one set of matrices for the system is

$$\mathbf{A} = \begin{bmatrix} a_1 & 0 \\ 0 & 0 \end{bmatrix} \qquad \mathbf{B} = \begin{bmatrix} 1 \\ 1 \end{bmatrix} \qquad \mathbf{C} = [1 \quad -1] \qquad \mathbf{D} = [a_2]$$

By renaming the unit delay outputs in the above example, we could again rewrite the state variable equations with another set of matrices as we did in Example 3.2. To demonstrate this nonuniqueness more generally, we can use the following argument. Choose any $n \times n$ matrix $\mathbf{W}$ whose inverse $\mathbf{W}^{-1}$ exists and define a new state vector $\mathbf{q}(k) = \mathbf{W}\mathbf{x}(k)$. The new state vector $\mathbf{q}(k)$ is a nonsingular linear transformation of $\mathbf{x}(k)$. We have

$$\begin{aligned} \mathbf{q}(k+1) = \mathbf{W}\mathbf{x}(k+1) &= \mathbf{W}[\mathbf{A}\mathbf{x}(k) + \mathbf{B}u(k)] \\ &= \mathbf{W}\mathbf{A}\mathbf{W}^{-1}\mathbf{q}(k) + \mathbf{W}\mathbf{B}u(k) \end{aligned} \tag{3.18}$$

and

$$y(k) = \mathbf{C}\mathbf{x}(k) + \mathbf{D}u(k) = \mathbf{C}\mathbf{W}^{-1}\mathbf{q}(k) + \mathbf{D}u(k) \tag{3.19}$$

Hence defining

$$\hat{\mathbf{A}} = \mathbf{W}\mathbf{A}\mathbf{W}^{-1}$$
$$\hat{\mathbf{B}} = \mathbf{W}\mathbf{B} \tag{3.20}$$
$$\hat{\mathbf{C}} = \mathbf{C}\mathbf{W}^{-1}$$

we have

$$q(k + 1) = \hat{A}q(k) + \hat{B}u(k) \qquad (3.21)$$

and

$$y(k) = \hat{C}q(k) + Du(k) \qquad (3.22)$$

Now q serves as our state vector, and $\hat{A} = WAW^{-1}$ is the new state matrix. In a later section, we shall investigate various choices for the state vector. We note here that all choices are equivalent in the sense that we shall always be able to define one choice in terms of another—for example, $q(k) = Wx(k)$ and $x(k) = W^{-1}q(k)$.

Our state variable formulation can readily be extended to systems with r inputs and s outputs. The relevant expressions follow directly from (3.8) and (3.9) as

$$x(k + 1) = Ax(k) + Bu(k) \qquad (3.23)$$

$$y(k) = Cx(k) + Du(k) \qquad (3.24)$$

Here $x(k)$ is an n-dimensional state vector, $u(k)$ is an r-dimensional input vector whose components are the r inputs $u_1(k), u_2(k), \ldots, u_r(k)$, and $y(k)$ is an s-dimensional output vector whose components are the s outputs $y_1(k), y_2(k), \ldots, y_s(k)$. The matrix A has dimensions $n \times n$, B has dimensions $n \times r$, C has dimensions $s \times n$, and D has dimensions $s \times r$, as shown in Figure 3.5.

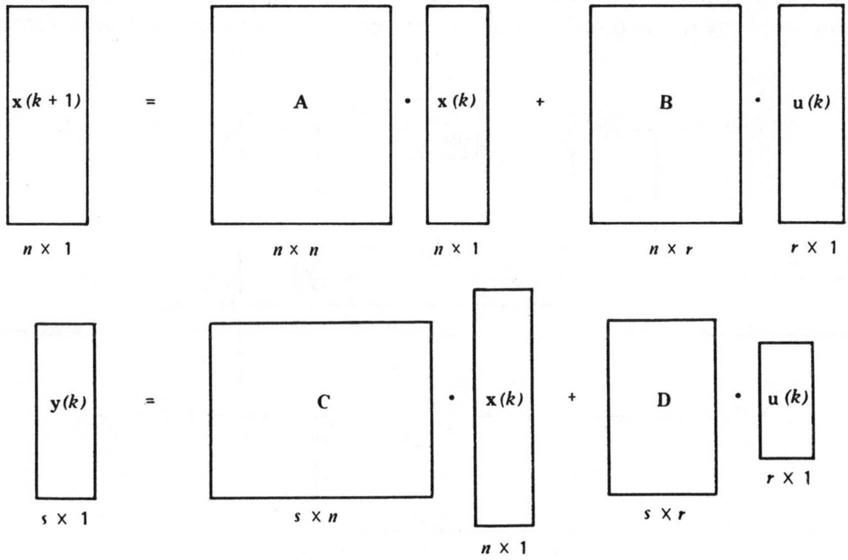

FIGURE 3.5

EXAMPLE 3.4. To illustrate the case of a multiple input-output system, consider the system shown in Figure 3.6. We again identify the outputs of the delay elements as state variables. If we write equations for the inputs to the delay elements, we obtain the equations

$$x_1(k + 1) = a_1 x_1(k) + u_1(k)$$

$$x_2(k + 1) = a_2 x_2(k) + u_2(k)$$

In vector form, these can be written

$$\mathbf{x}(k + 1) = \begin{bmatrix} a_1 & 0 \\ 0 & a_2 \end{bmatrix} \mathbf{x}(k) + \begin{bmatrix} 1 & 0 \\ 0 & 1 \end{bmatrix} \mathbf{u}(k)$$

The output $\mathbf{y}(k)$ is given by the vector equation

$$\mathbf{y}(k) = \begin{bmatrix} 1 & 1 \\ 0 & 1 \end{bmatrix} \mathbf{x}(k) + \begin{bmatrix} 0 & 0 \\ 1 & 0 \end{bmatrix} \mathbf{u}(k)$$

In this example, the matrices for the system are

$$\mathbf{A} = \begin{bmatrix} a_1 & 0 \\ 0 & a_2 \end{bmatrix} \qquad \mathbf{B} = \begin{bmatrix} 1 & 0 \\ 0 & 1 \end{bmatrix} \qquad \mathbf{C} = \begin{bmatrix} 1 & 1 \\ 0 & 1 \end{bmatrix} \qquad \mathbf{D} = \begin{bmatrix} 0 & 0 \\ 1 & 0 \end{bmatrix}$$

We have not stressed the technique of writing the state equations directly from the difference equation because it is generally easier to write the state variable equations directly from the system model in block diagram form.

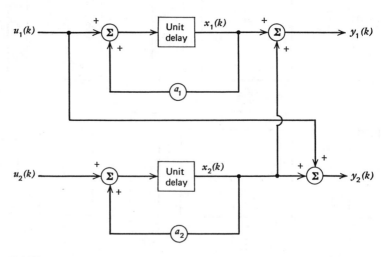

FIGURE 3.6

One can always sketch a block diagram of the difference equation as we have done in Chapters 2 and 3 and use the block diagram as an aid in obtaining the state variable equations.

3.3 THE SOLUTION OF STATE VARIABLE EQUATIONS—DISCRETE-TIME SYSTEMS

We have reformulated the solution to a general nth order difference equation in terms of a quantity we call the state vector of the system. The state vector $\mathbf{x}$ is a complete internal description of the system. Let us now consider the problem of determining the response $\mathbf{y}(k)$ of the system and the evolution of the state $\mathbf{x}(k)$ in terms of the input sequence $\{\mathbf{u}(k)\}$. To find the output $\mathbf{y}(k)$, we must know the state $\mathbf{x}(k)$, where $\mathbf{x}(k)$ evolves according to (3.23). One method of finding $\mathbf{x}(k)$ given $\mathbf{x}(0)$ and the sequence $\{\mathbf{u}(j)\}$, $j = 0, 1, \ldots, k$ is to solve (3.23) iteratively. Thus,

$$\mathbf{x}(1) = \mathbf{A}\mathbf{x}(0) + \mathbf{B}\mathbf{u}(0)$$

$$\mathbf{x}(2) = \mathbf{A}\mathbf{x}(1) + \mathbf{B}\mathbf{u}(1) = \mathbf{A}^2\mathbf{x}(0) + \mathbf{A}\mathbf{B}\mathbf{u}(0) + \mathbf{B}\mathbf{u}(1)$$

$$\cdot$$
$$\cdot$$
$$\cdot$$

and, in general,

$$\mathbf{x}(k) = \mathbf{A}^k\mathbf{x}(0) + \sum_{m=0}^{k-1} \mathbf{A}^{k-1-m}\mathbf{B}\mathbf{u}(m) \tag{3.25}$$

If the initial state is $\mathbf{x}(k_0)$ and we have $\mathbf{u}(k)$ for $k > k_0$, then (3.25) may be generalized to

$$\mathbf{x}(k) = \mathbf{A}^{k-k_0}\mathbf{x}(k_0) + \sum_{m=0}^{k-k_0-1} \mathbf{A}^{k-k_0-1-m}\mathbf{B}\mathbf{u}(k_0 + m), \qquad k > k_0 \tag{3.26}$$

The matrix $\mathbf{A}^k$ is the k-fold product[2] $\mathbf{A} \times \mathbf{A} \cdots \times \mathbf{A}$ and is often called the *fundamental* or *transition* matrix of the system. Referring to (3.25), we recognize two kinds of terms. The first term $\mathbf{A}^k\mathbf{x}(0)$ represents a source-free evolution of the state. The second term, which resembles a convolution sum, corresponds to a forced response of the state. We find the output $\mathbf{y}(k)$ from the state $\mathbf{x}(k)$ as

$$\mathbf{y}(k) = \mathbf{C}\mathbf{x}(k) + \mathbf{D}\mathbf{u}(k)$$

$$= \mathbf{C}\mathbf{A}^k\mathbf{x}(0) + \sum_{m=0}^{k-1} \mathbf{C}\mathbf{A}^{k-1-m}\mathbf{B}\mathbf{u}(m) + \mathbf{D}\mathbf{u}(k) \tag{3.27}$$

---

[2] $\mathbf{A}^0$ is defined as $\mathbf{I}$, the identity matrix.

Although this structure as represented by (3.25) and (3.27) may be cumbersome for small-order systems, it does furnish us with a general formulation for any size system. The solution process is also readily programed on a digital computer and yields a wealth of information about the system in addition to its output. The term $(\mathbf{CA}^{k-1}\mathbf{B})$ is exactly the impulse-response sequence we used in Chapter 2 to obtain the output from the input sequence, and $\mathbf{CA}^{k-1-m}\mathbf{B}$ corresponds to the shifted impulse-response sequence which appears in the convolution sum.

The basic calculation one must perform in solving (3.25) and (3.27) is finding $\mathbf{A}^n$. Thus, we need to develop a procedure, other than brute force, by which we can calculate a matrix multiplied by itself n times, where n is arbitrary. In the next section, we consider this problem. First, however, we present as examples the state and output calculations for two simple systems excited by an impulse sequence input. Let $k_0 = 0$ and the initial state be $\mathbf{x}(0) = \mathbf{0}$.

EXAMPLE 3.5. Suppose we have the following discrete-time system. What is the impulse response of this system? We identify the outputs of the delay elements as state variables and write equations for the input to the delay elements. Thus we have

$$x_1(k + 1) = \tfrac{1}{2}x_1(k) + u(k)$$
$$x_2(k + 1) = \tfrac{1}{4}x_1(k) + \tfrac{1}{4}x_2(k) + u(k)$$

In vector form, the above equations are

$$\mathbf{x}(k + 1) = \begin{bmatrix} \tfrac{1}{2} & 0 \\ \tfrac{1}{4} & \tfrac{1}{4} \end{bmatrix}\mathbf{x}(k) + \begin{bmatrix} 1 \\ 1 \end{bmatrix}u(k) \tag{3.28}$$

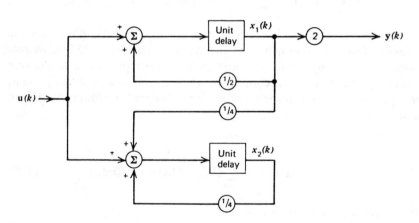

FIGURE 3.7

and the output is given by

$$y(k) = [2 \quad 0]\mathbf{x}(k) \tag{3.29}$$

Thus the appropriate matrices are

$$\mathbf{A} = \begin{bmatrix} \frac{1}{2} & 0 \\ \frac{1}{4} & \frac{1}{4} \end{bmatrix} \quad \mathbf{B} = \begin{bmatrix} 1 \\ 1 \end{bmatrix} \quad \mathbf{C} = [2 \quad 0] \quad \mathbf{D} = 0 \tag{3.30}$$

Using (3.25) with an impulse input, we obtain

$$\mathbf{x}(1) = \mathbf{B}\,1 = \begin{bmatrix} 1 \\ 1 \end{bmatrix}$$

$$\mathbf{x}(2) = \mathbf{A}\mathbf{x}(1) = \begin{bmatrix} \frac{1}{2} \\ \frac{1}{2} \end{bmatrix}$$

$$\mathbf{x}(3) = \mathbf{A}\mathbf{x}(2) = \begin{bmatrix} \frac{1}{4} \\ \frac{1}{4} \end{bmatrix}$$

and by induction

$$\mathbf{x}(k) = \begin{bmatrix} (\frac{1}{2})^{k-1} \\ (\frac{1}{2})^{k-1} \end{bmatrix}, \quad k > 0 \tag{3.31}$$

Thus

$$y(k) = h(k) = \mathbf{C}\mathbf{x}(k) = [2 \quad 0]\begin{bmatrix} (\frac{1}{2})^{k-1} \\ (\frac{1}{2})^{k-1} \end{bmatrix} = (\frac{1}{2})^{k-2}, \quad k > 0 \tag{3.32}$$

In this system, both the state variables and the impulse response decrease as 2^{-k} with increasing k.

EXAMPLE 3.6. Suppose in the discrete-time system of Example 3.5 we change the feedback gain in the lower delay loop from $\frac{1}{4}$ to 2, so that the matrices become

$$\mathbf{A} = \begin{bmatrix} \frac{1}{2} & 0 \\ \frac{1}{4} & 2 \end{bmatrix} \quad \mathbf{B} = \begin{bmatrix} 1 \\ 1 \end{bmatrix} \quad \mathbf{C} = [2 \quad 0] \quad \mathbf{D} = 0 \tag{3.33}$$

If we calculate the evolution of the state resulting from an impulse input, we obtain

$$\mathbf{x}(1) = \mathbf{B} \cdot 1 = \begin{bmatrix} 1 \\ 1 \end{bmatrix}$$

$$\mathbf{x}(2) = \mathbf{A}\mathbf{x}(1) = \begin{bmatrix} \frac{1}{2} \\ \frac{9}{4} \end{bmatrix}$$

$$\mathbf{x}(3) = \mathbf{A}\mathbf{x}(2) = \begin{bmatrix} \frac{1}{4} \\ \frac{37}{8} \end{bmatrix}$$

Continuing, we find the kth term as

$$\mathbf{x}(k) = \begin{bmatrix} (\frac{1}{2})^{k-1} \\ \dfrac{7(\frac{1}{2})^{1-k} - (\frac{1}{2})^{k-1}}{6} \end{bmatrix} \tag{3.34}$$

The output $y(k)$ is the same as in Example 3.4, namely,

$$y(k) = h(k) = [2 \quad 0]\mathbf{x}(k) = (\tfrac{1}{2})^{k-2}, \qquad k > 0 \tag{3.35}$$

Judged from its impulse response the system appears to be stable, because $h(k)$ dies away as k increases. However, one of the components of the state vector grows without bound as 2^k. Thus even though the output indicates that the system is stable, because one of the natural modes of the system is unobserved, this conclusion is incorrect. This example demonstrates one of the main reasons for the state variable approach to system analysis—there is more to a system than its output!

*3.4 THE CONCEPTS OF OBSERVABILITY AND CONTROLLABILITY

The last example demonstrates a problem that can arise when the state vector is not completely coupled into the output vector. This problem naturally leads to the idea of observability. A system is said to be *observable* if we can determine the initial state based on the observations $\mathbf{y}$ and the system matrices $\mathbf{A}$ and $\mathbf{C}$. Roughly speaking, the output vector $\mathbf{y}$ can be used to recover the system state vector $\mathbf{x}$. In the state variable formulation, the output is given by

$$\mathbf{y}(k) = \mathbf{C}\mathbf{x}(k) + \mathbf{D}\mathbf{u}(k) \tag{3.36}$$

If the matrix $\mathbf{C}$ contains a zero column, say the jth column, then x_j will not explicitly appear in the equations for $\mathbf{y}$. It is possible that the system may be interconnected in such a way that even though $\mathbf{C}$ has an all-zero column, the other columns furnish information about this natural mode. However, if the system is represented in "decoupled form" so that its natural modes do not interact, then the $\mathbf{C}$ matrix alone is sufficient to determine observability. (This discussion assumes that the eigenvalues of $\mathbf{A}$ are distinct.)

The idea of observability is concerned with determining the state vector from the output vector. The dual concept, called controllability, is concerned with controlling the state vector from the input vector.

$$\mathbf{x}(k + 1) = \mathbf{A}\mathbf{x}(k) + \mathbf{B}\mathbf{u}(k) \tag{3.37}$$

Roughly speaking, in controllable systems, the state vector components x_j are coupled to the input $\mathbf{u}$ in the same manner as the state components are coupled into the output $\mathbf{y}$ in observable systems. A system is said to be *controllable* if the initial state vector can be driven to the origin in a finite time by correct choice of the input $\{\mathbf{u}(k)\}$.

If a system is represented in so called *normal form*, then the natural modes of the systems are decoupled from each other.[3] In normal form, the matrix $\mathbf{A}$ is diagonal and (3.36) and (3.37) become

$$x_i(k+1) = \lambda_i x_i(k) + \sum_{j=1}^{r} b_{ij} u_j(k), \qquad i = 1, 2, 3, \ldots, n \qquad (3.38)$$

$$y_l(k) = \sum_{i=1}^{n} c_{li} x_i(k), \qquad l = 1, 2, \ldots, s \qquad (3.39)$$

In this decoupled form, if $b_{ij} = 0$ for all j and some i, then x_i cannot be affected by any input $\mathbf{u}(k)$. If this situation occurs, the system is not controllable. Likewise, if $c_{li} = 0$ for all i and some l, then the system is not observable, because x_i cannot be obtained from observation of $\mathbf{y}(k)$.

For a system in normal form, the absence of an all-zero column in $\mathbf{C}$ and an all-zero row in $\mathbf{B}$ guarantees that the system is observable and controllable, respectively. Again, this conclusion assumes that the system's eigenvalues are distinct. If the system's eigenvalues are not distinct, then a different characterization of controllability and observability is needed.[4] We shall return to the concepts of controllability and observability when we have developed some more machinery to handle the calculations needed to place a state variable representation in normal form.

EXAMPLE 3.7. We have thus far implicitly assumed that the state vector $\mathbf{x}$ evolves in time. This interpretation is actually too restrictive, as this example points out. In this problem, the state vector $\mathbf{x}$ evolves not in time but rather in a space dimension. Consider the problem of finding the currents in all the branches of the following network. All

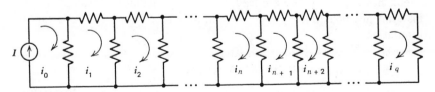

FIGURE 3.8

[3] The coordinate basis used to describe the system are in the directions of the system's eigenvectors. The procedure for obtaining this representation is discussed in Section 3.6.
[4] Athans, M., and P. L. Falb, *Optimal Control*, McGraw-Hill, New York, 1966, pp. 200–220.

the branch resistors are 1 ohm. Assume that there are q loops. If we write Kirchhoff's voltage for the nth loop, we obtain

$$3i(n + 1) - i(n) - i(n + 2) = 0 \qquad (3.40)$$

The initial or boundary conditions are

$$i(0) = I \text{ amp}$$
$$3i(q) = i(q - 1)$$

The second-order difference equation of (3.40) can be placed in state variable form by defining two state variables $x_1(n)$ and $x_2(n)$. Define the state variables as

$$x_1(n) = i(n)$$
$$x_2(n) = x_1(n + 1)$$

Then we see that (3.40) can be written

$$x_2(n + 1) = 3x_2(n) - x_1(n) \qquad (3.41)$$

We also have that

$$x_1(n + 1) = x_2(n) \qquad (3.42)$$

Combining (3.41) and (3.42) in vector notation we obtain

$$\mathbf{x}(n + 1) = \begin{bmatrix} 0 & 1 \\ -1 & 3 \end{bmatrix} \mathbf{x}(n)$$

Thus

$$\mathbf{x}(n + 1) = \mathbf{A}\mathbf{x}(n)$$

or alternately,

$$\mathbf{x}(n) = \mathbf{A}^n\mathbf{x}(0) \qquad (3.43)$$

If we can solve (3.43), we have the value for $i(n)$. Therefore we need a method for generating $\mathbf{A}^n$ easily. We can, of course, compute $\mathbf{A}^n$ by multiplying $\mathbf{A}$ by itself n times. This brute-force technique may not be feasible for large n, however. Thus we are lead to studying methods for calculating $\mathbf{A}^n$. This kind of calculation is basic for solutions based upon the state variable formulation.

3.5 FUNCTIONS OF MATRICES

We now see how to compute the state and output of a discrete-time system, given the initial state and the input sequence. Our solution, (3.25) and (3.27), is given in terms of various powers of the state matrix $\mathbf{A}$. As mentioned previously, the matrix $\mathbf{A}^k$ can certainly be calculated in a straightforward fashion by simply multiplying the matrix $\mathbf{A}$ by itself k times. However, for large k, the approach has two drawbacks: (1) the successive multiplications can involve excessive computation effort; (2) round-off errors soon reduce the significance of the result.

In this section, we digress briefly to examine a method for computing A^k and other functions of the matrix A. We begin with the *characteristic equation* of the $n \times n$ matrix A. The characteristic equation of A is defined to be

$$g(\lambda) = |A - \lambda I| = 0 \tag{3.44}$$

where $|\ |$ means "determinant of," and I is the identity matrix. For example, the characteristic equation of the matrix

$$A = \begin{bmatrix} 4 & 3 \\ 1 & 2 \end{bmatrix}$$

is

$$g(\lambda) = |A - \lambda I| = 0$$

that is,

$$g(\lambda) = \left| \begin{bmatrix} 4 & 3 \\ 1 & 2 \end{bmatrix} - \lambda \begin{bmatrix} 1 & 0 \\ 0 & 1 \end{bmatrix} \right| = 0$$

which becomes

$$g(\lambda) = \begin{vmatrix} 4 - \lambda & 3 \\ 1 & 2 - \lambda \end{vmatrix} = 0$$

or

$$g(\lambda) = (4 - \lambda)(2 - \lambda) - 3$$
$$= \lambda^2 - 6\lambda + 5$$
$$= (\lambda - 5)(\lambda - 1) = 0$$

Thus $\lambda^2 - 6\lambda + 5 = 0$ is the characteristic equation of the matrix $\begin{bmatrix} 4 & 3 \\ 1 & 2 \end{bmatrix}$.

The roots of the characteristic equation are called the *eigenvalues* of the matrix A. In this case, the eigenvalues are $\lambda_1 = 5$ and $\lambda_2 = 1$.

Now we use the *Caley-Hamilton theorem*, which states that every $n \times n$ matrix satisfies its characteristic equation. For example, if we substitute A^k for λ^k above, for $k = 0$, 1, and 2, we obtain the matrix equation

$$A^2 - 6A + 5A^0 = 0$$

that is,

$$\begin{bmatrix} 4 & 3 \\ 1 & 2 \end{bmatrix} \begin{bmatrix} 4 & 3 \\ 1 & 2 \end{bmatrix} - 6 \begin{bmatrix} 4 & 3 \\ 1 & 2 \end{bmatrix} + 5 \begin{bmatrix} 1 & 0 \\ 0 & 1 \end{bmatrix} \stackrel{?}{=} \begin{bmatrix} 0 & 0 \\ 0 & 0 \end{bmatrix}$$

$$\begin{bmatrix} 19 & 18 \\ 6 & 7 \end{bmatrix} - \begin{bmatrix} 24 & 18 \\ 6 & 12 \end{bmatrix} + \begin{bmatrix} 5 & 0 \\ 0 & 5 \end{bmatrix} = \begin{bmatrix} 0 & 0 \\ 0 & 0 \end{bmatrix}$$

The equation is certainly satisfied in this case. The theorem states that it is true for *any* square matrix. In general, the characteristic equation of a square $n \times n$ matrix $\mathbf{A}$ is

$$g(\lambda) \overset{\Delta}{=} |\mathbf{A} - \lambda\mathbf{I}|$$
$$= \lambda^n + a_{n-1}\lambda^{n-1} + \cdots + a_1\lambda + a_0 = 0 \qquad (3.45)$$

Now substituting $\mathbf{A}$ for λ as above, we have

$$g(\mathbf{A}) = \mathbf{A}^n + a_{n-1}\mathbf{A}^{n-1} + \cdots + a_1\mathbf{A} + a_0\mathbf{I} = \mathbf{0} \qquad (3.46)$$

where, again, $\mathbf{I}$ is the identity matrix, and $\mathbf{0}$ is a matrix whose elements are all zero. Equation (3.46) may be rewritten as

$$\mathbf{A}^n = -a_{n-1}\mathbf{A}^{n-1} - a_{n-2}\mathbf{A}^{n-2} - \cdots - a_1\mathbf{A} - a_0\mathbf{I} \qquad (3.47)$$

Thus $\mathbf{A}^n$ may be expressed in terms of the matrices $\mathbf{A}^{n-1}, \mathbf{A}^{n-2}, \ldots, \mathbf{A}$, and $\mathbf{I}$. To extend this result, we multiply (3.47) by $\mathbf{A}$ to obtain

$$\mathbf{A}^{n+1} = -a_{n-1}\mathbf{A}^n - a_{n-2}\mathbf{A}^{n-1} - \cdots - a_1\mathbf{A}^2 - a_0\mathbf{A} \qquad (3.48)$$

Substituting (3.47) for $\mathbf{A}^n$, we obtain

$$\mathbf{A}^{n+1} = -a_{n-1}[-a_{n-1}\mathbf{A}^{n-1} - a_{n-2}\mathbf{A}^{n-2} - \cdots - a_1\mathbf{A} - a_0\mathbf{I}]$$
$$- a_{n-2}\mathbf{A}^{n-1} - \cdots - a_1\mathbf{A}^2 - a_0\mathbf{A}$$
$$= (a_{n-1}^2 - a_{n-2})\mathbf{A}^{n-1} + (a_{n-1}a_{n-2} - a_{n-3})\mathbf{A}^{n-2} + \cdots + a_{n-1}a_0\mathbf{I}$$
$$(3.49)$$

This gives us an expression for $\mathbf{A}^{n+1}$, again in terms of $\mathbf{A}^{n-1}, \mathbf{A}^{n-2}, \ldots, \mathbf{A}$, and $\mathbf{I}$. Continuing this process, we see that any power of $\mathbf{A}$ can be represented as a weighted sum of matrices involving $\mathbf{A}$ to at most the $n-1$ power. Hence, functions of matrices which can be expressed as

$$f(\mathbf{A}) = \alpha_0\mathbf{I} + \alpha_1\mathbf{A} + \alpha_2\mathbf{A}^2 + \cdots + \alpha_k\mathbf{A}^k + \cdots$$
$$= \sum_{k=0}^{\infty} \alpha_k\mathbf{A}^k \qquad (3.50)$$

can be represented as

$$f(\mathbf{A}) = \beta_0\mathbf{I} + \beta_1\mathbf{A} + \beta_2\mathbf{A}^2 + \cdots + \beta_{n-1}\mathbf{A}^{n-1}$$
$$= \sum_{m=0}^{n-1} \beta_m\mathbf{A}^m \qquad (3.51)$$

where $\beta_0, \beta_1, \beta_2, \ldots, \beta_{n-1}$ are given terms of $\alpha_0, \alpha_1, \alpha_2, \ldots, \alpha_k, \ldots$. For example, the impulse response of a system with initial state $\mathbf{x}(0)$ may be found by use of (3.51).

$$\mathbf{x}(k) = \mathbf{A}^k\mathbf{x}(0) + \mathbf{A}^{k-1}\mathbf{B} \cdot \mathbf{1} = \sum_{m=0}^{n-1} \beta_m \mathbf{A}^m[\mathbf{A}\mathbf{x}(0) + \mathbf{B} \cdot \mathbf{1}]$$

and

$$\mathbf{h}(k) = \mathbf{y}(k) = \mathbf{C}\mathbf{x}(k) = \mathbf{C} \sum_{m=0}^{n-1} \beta_m \mathbf{A}^m[\mathbf{A}\mathbf{x}(0) + \mathbf{B} \cdot \mathbf{1}], \qquad k > 0$$

where we have written

$$\mathbf{A}^{k-1} = \sum_{m=0}^{n-1} \beta_m \mathbf{A}^m$$

The scalar calculation of $\beta_0, \beta_1, \ldots, \beta_{n-1}$ can be carried out by using the iterative method used in the calculation of $\mathbf{A}^n$ and $\mathbf{A}^{n+1}$ in (3.47) and (3.49). However, although straightforward and computationally easier than the brute force calculation of $\mathbf{A}^k$, this process can be lengthy.

To develop an easier method, let us return to the characteristic equation of the matrix $\mathbf{A}$.

$$g(\lambda) = |\mathbf{A} - \lambda\mathbf{I}| = \lambda^n + a_{n-1}\lambda^{n-1} + \cdots + a_1\lambda + a_0 = 0 \qquad (3.52)$$

Following the same steps as before, we can express the eigenvalues λ^n, λ^{n+1}, λ^{n+2} and so on in terms of $\lambda, \lambda^2, \ldots, \lambda^{n-1}$.

$$\lambda^n = -a_{n-1}\lambda^{n-1} - a_{n-2}\lambda^{n-2} - \cdots - a_1\lambda - a_0$$
$$\lambda^{n+1} = -a_{n-1}\lambda^n - a_{n-2}\lambda^{n-1} - \cdots - a_1\lambda^2 - a_0\lambda$$
$$= -a_{n-1}(-a_{n-1}\lambda^{n-1} - a_{n-2}\lambda^{n-2} - \cdots - a_1\lambda - a_0)$$
$$- a_{n-2}\lambda^{n-1} - \cdots - a_1\lambda^2 - a_0\lambda$$
$$= (a_{n-1}^2 - a_{n-2})\lambda^{n-1} + (a_{n-1}a_{n-2} - a_{n-3})\lambda^{n-2} + \cdots + a_{n-1}a_0$$

As before, we can write polynomials of λ in terms of $\lambda, \lambda^2, \ldots, \lambda^{n-1}$.

$$f(\lambda) = \alpha_0 + \alpha_1\lambda + \alpha_2\lambda^2 + \cdots + \alpha_k\lambda^k + \cdots = \sum_{k=0}^{\infty} \alpha_k\lambda^k$$

$$= \beta_0 + \beta_1\lambda + \cdots + \beta_{n-1}\lambda^{n-1} = \sum_{m=0}^{n-1} \beta_m\lambda^m \qquad (3.53)$$

Here we wish to find the n unknowns $\beta_0, \beta_1, \ldots, \beta_{n-1}$. We know that (3.53) holds for any λ which is a solution of the characteristic equation (3.52): that is, for any eigenvalue of the matrix $\mathbf{A}$. Assume first that the eigenvalues are distinct: that is, none are repeated. Substituting $\lambda_1, \lambda_2, \ldots, \lambda_n$ in (3.53) gives n equations in n unknowns

$$f(\lambda_1) = \beta_0 + \beta_1\lambda_1 + \beta_2\lambda_1^2 + \cdots + \beta_{n-1}\lambda_1^{n-1}$$
$$f(\lambda_2) = \beta_0 + \beta_1\lambda_2 + \beta_2\lambda_2^2 + \cdots + \beta_{n-1}\lambda_2^{n-1}$$

$$\tag{3.54}$$

$$f(\lambda_n) = \beta_0 + \beta_1\lambda_n + \beta_2\lambda_n^2 + \cdots + \beta_{n-1}\lambda_n^{n-1}$$

from which we can obtain $\beta_0, \beta_1, \ldots, \beta_{n-1}$. By comparing (3.53) and (3.51), we see that these coefficients are exactly the ones which appeared earlier in (3.51): i.e.,

$$f(\mathbf{A}) = \sum_{m=0}^{n-1} \beta_m \mathbf{A}^m$$

Hence, our problem is solved. The coefficients needed in the matrix expression for $f(\mathbf{A})$ are found as the solution to a linear system of scalar equations given by (3.54).

EXAMPLE 3.8. Let

$$f(\mathbf{A}) = \mathbf{A}^k, \quad \text{where} \quad \mathbf{A} = \begin{bmatrix} \frac{1}{2} & 0 \\ \frac{1}{4} & \frac{1}{4} \end{bmatrix} \tag{3.55}$$

The characteristic equation is

$$g(\lambda) = |\mathbf{A} - \lambda\mathbf{I}| = \left| \begin{bmatrix} \frac{1}{2} - \lambda & 0 \\ \frac{1}{4} & \frac{1}{4} - \lambda \end{bmatrix} \right| = 0$$

That is,

$$g(\lambda) = (\tfrac{1}{2} - \lambda)(\tfrac{1}{4} - \lambda) = 0$$

Thus the eigenvalues are

$$\lambda_1 = \tfrac{1}{2}, \quad \lambda_2 = \tfrac{1}{4}$$

Using (3.53), we obtain

$$f(\lambda) = \lambda^k = \beta_0 + \beta_1\lambda$$

and using (3.54), we have

$$(\tfrac{1}{2})^k = \beta_0 + \tfrac{1}{2}\beta_1$$
$$(\tfrac{1}{4})^k = \beta_0 + \tfrac{1}{4}\beta_1$$

Solving for the unknowns β_0 and β_1, gives

$$\beta_0 = (\tfrac{1}{4})^k(2 - 2^k)$$
$$\beta_1 = (\tfrac{1}{4})^k(4 \cdot 2^k - 4)$$

The solution for $\mathbf{A}^k$ is found by using (3.51)

$$\mathbf{A}^k = \begin{bmatrix} \frac{1}{2} & 0 \\ \frac{1}{4} & \frac{1}{4} \end{bmatrix}^k = \beta_0 \mathbf{I} + \beta_1 \mathbf{A}$$

$$= (\tfrac{1}{4})^k \begin{bmatrix} 2 - 2^k & 0 \\ 0 & 2 - 2^k \end{bmatrix} + (\tfrac{1}{4})^k \begin{bmatrix} 2 \cdot (2^k - 1) & 0 \\ 2^k - 1 & 2^k - 1 \end{bmatrix}$$

$$= (\tfrac{1}{4})^k \begin{bmatrix} 2^k & 0 \\ 2^k - 1 & 1 \end{bmatrix}$$

Thus we can return to Example 3.5 and write our state equation as

$$\mathbf{x}(k) = \mathbf{A}^{k-1}\mathbf{B} \cdot 1 = (\tfrac{1}{4})^{k-1} \begin{bmatrix} 2^{k-1} & 0 \\ 2^{k-1} - 1 & 1 \end{bmatrix} \begin{bmatrix} 1 \\ 1 \end{bmatrix} \cdot 1$$

$$= (\tfrac{1}{4})^{k-1} \begin{bmatrix} 2^{k-1} \\ 2^{k-1} \end{bmatrix} = \begin{bmatrix} 2^{1-k} \\ 2^{1-k} \end{bmatrix}$$

The output is

$$y(k) = \begin{bmatrix} 2 & 0 \end{bmatrix} \begin{bmatrix} 2^{1-k} \\ 2^{1-k} \end{bmatrix} = 2^{2-k}$$

EXAMPLE 3.9.   The reader is urged to apply this method to Example 3.6 with

$$\mathbf{A} = \begin{bmatrix} \frac{1}{2} & 0 \\ \frac{1}{4} & 2 \end{bmatrix}$$

and $\mathbf{B}$ and $\mathbf{C}$ as in Example 3.8. In this case,

$$\mathbf{A}^k = \begin{bmatrix} 2^{-k} & 0 \\ \dfrac{2^k - 2^{-k}}{6} & 2^k \end{bmatrix}$$

The state vector and output are thus

$$\mathbf{x}(k) = \begin{bmatrix} 2^{1-k} \\ \dfrac{7 \cdot 2^{k-1} - 2^{1-k}}{6} \end{bmatrix}, \qquad k > 0$$

$$y(k) = 2^{2-k}, \qquad\qquad k > 0$$

as we obtained previously.

Suppose now that the characteristic equation $g(\lambda) = 0$ has repeated roots[5] (for example, $\lambda_1 = \lambda_2$). In this case, we would be left with fewer than n linearly independent equations in (3.54). The following theorem[6] extends our results to the case of repeated eigenvalues.

THEOREM 3.1.   Let **A** be an $n \times n$ matrix with n_0 distinct eigenvalues, $\lambda_1, \lambda_2, \ldots, \lambda_{n_0}$ (if no eigenvalue is repeated, then $n_0 = n$, otherwise $n_0 < n$). Let the eigenvalue λ_i occur with multiplicity m_i, and define the polynomials

$$P(\mathbf{A}) = \sum_{m=0}^{n-1} \beta_m \mathbf{A}^m$$

$$P(\lambda) = \sum_{m=0}^{n-1} \beta_m \lambda^m$$

$$f(\mathbf{A}) = \sum_{k=0}^{\infty} \alpha_k \mathbf{A}^k$$

Then the matrix $f(\mathbf{A})$ is identical with the matrix $P(\mathbf{A})$ if and only if

a. $\qquad\qquad f(\lambda_i) = P(\lambda_i), \qquad i = 1, 2, \ldots, n_0 \qquad\qquad$ (3.56)

and

b. $\qquad \left.\dfrac{d^q}{d\lambda^q} f(\lambda)\right|_{\lambda=\lambda_i} = \left.\dfrac{d^q}{d\lambda^q} P(\lambda)\right|_{\lambda=\lambda_i}, \qquad \begin{matrix} i = 1, 2, \ldots, n_0 \\ q = 1, 2, \ldots, m_i - 1 \end{matrix} \qquad$ (3.57)

Note that (3.57), when rewritten in terms of the unknown coefficients $\beta_0, \beta_1, \ldots, \beta_{n-1}$, becomes

$$\left.\frac{d^q}{d\lambda^q} f(\lambda)\right|_{\lambda=\lambda_i} = \left.\frac{d^q}{d\lambda^q} \sum_{0}^{n-1} \beta_m \lambda_n^m\right|_{\lambda=\lambda_i} = \sum_{m=q}^{n-1} \underbrace{m(m-1)\cdots(m-q+1)\beta_m \lambda_i^{m-q}}_{q \text{ factors}}$$
(3.58)

Equation (3.58) yields the remaining equations needed to solve for $\beta_0, \beta_1, \ldots, \beta_{n-1}$.

EXAMPLE 3.10.   Let $\mathbf{A} = \begin{bmatrix} \frac{1}{2} & 2 \\ -\frac{1}{128} & \frac{1}{4} \end{bmatrix}$ and suppose that we wish to

compute a general form for the matrix $\mathbf{A}^k$. We find the characteristic equation

$$g(\lambda) = |\mathbf{A} - \lambda \mathbf{I}| = \lambda^2 - \frac{3\lambda}{4} + \frac{9}{64} = 0$$

---

[5] That is, the matrix **A** has repeated eigenvalues.
[6] See, for example, Zadeh, L. A., and C. A. Desoer, *Linear System Theory*, McGraw-Hill, New York, 1963, Appendix D and especially pp. 607–609.

with a double root $\lambda = \frac{3}{8}$. From (3.56), we have

$$f(\lambda) = \lambda^k = (\tfrac{3}{8})^k = \beta_0 + \tfrac{3}{8}\beta_1$$

and from (3.57),

$$f'(\lambda) = k\lambda^{k-1} = (\tfrac{3}{8})^{k-1} \cdot k = \beta_1$$

Hence

$$\beta_1 = k(\tfrac{3}{8})^{k-1}$$

$$\beta_0 = (\tfrac{3}{8})^k(1 - k)$$

Thus, by (3.51), we see that

$$\mathbf{A}^k = (\tfrac{3}{8})^k(1 - k)\mathbf{I} + k(\tfrac{3}{8})^{k-1}\mathbf{A}$$

$$= \left(\frac{3}{8}\right)^k \begin{bmatrix} 1 + \dfrac{k}{3} & \dfrac{16k}{3} \\[2ex] -\dfrac{k}{48} & 1 - \dfrac{k}{3} \end{bmatrix}$$

3.6   IMPORTANCE OF THE STATE MATRIX

In the previous discussion, we introduced the state variable representation of a system and saw how to write its response in terms of the state matrix $\mathbf{A}$. In the last section, we found a method for writing the relevant equations explicitly in terms of $\mathbf{A}$ and its powers up to $\mathbf{A}^{n-1}$. Let us digress briefly at this point to investigate those properties of our solution which can be deduced immediately from an inspection of $\mathbf{A}$.

Recall that the state vector and state matrix of a linear system are not unique. If $\mathbf{W}$ is an $n \times n$ nonsingular matrix, then $\mathbf{x}$ and $\mathbf{A}$ may be replaced by $\mathbf{q} = \mathbf{W}\mathbf{x}$ and $\hat{\mathbf{A}} = \mathbf{W}\mathbf{A}\mathbf{W}^{-1}$, respectively with no loss of information. This replacement is possible because the characteristic equations of the two matrices $\mathbf{A}$ and $\hat{\mathbf{A}}$ are identical, as is seen from

$$\hat{g}(\lambda) = |\hat{\mathbf{A}} - \lambda\mathbf{I}| = |\mathbf{W}\mathbf{A}\mathbf{W}^{-1} - \lambda\mathbf{W}\mathbf{W}^{-1}|$$
$$= |\mathbf{W}||\mathbf{A} - \lambda\mathbf{I}||\mathbf{W}^{-1}|$$
$$= |\mathbf{A} - \lambda\mathbf{I}| = g(\lambda)$$

where we have written $\mathbf{W}\mathbf{W}^{-1}$ for $\mathbf{I}$ and used the relations $|\mathbf{M}_1\mathbf{M}_2| = |\mathbf{M}_1||\mathbf{M}_2|$, and $|\mathbf{M}^{-1}| = |\mathbf{M}|^{-1}$. This result is an important one.

The eigenvalues of the state matrix of a system are independent of the choice of state vector.

This result suggests that the set of eigenvalues for a system somehow characterizes the system, which is indeed true. The eigenvalues are another representation of the system's natural modes. To exhibit the eigenvalues of the system in more explicit form, we can transform the system matrix **A** into a diagonal matrix with the eigenvalues of the system as the elements on the diagonal. To see how this is accomplished we need to digress somewhat and consider the problem of "change of basis."

Change of Basis

Consider the following problem. We have a set of linear equations given in vector form by

$$\mathbf{y} = \mathbf{A}\mathbf{x} \tag{3.59}$$

Assume for purposes of exposition that **y** and **x** are 2-dimensional and **A** is 2×2, so that (3.59) is equivalent to

$$\begin{bmatrix} y_1 \\ y_2 \end{bmatrix} = \begin{bmatrix} a_{11} & a_{12} \\ a_{21} & a_{22} \end{bmatrix} \begin{bmatrix} x_1 \\ x_2 \end{bmatrix} \tag{3.60}$$

Now to describe **y** and **x** we can use as a basis a coordinate system with two independent basis vectors. For example, we might use as basis vectors an orthogonal basis as shown in Figure 3.9a. Any vector **v** in the 2-dimensional space can now be represented by a pair v_1, v_2 which is the projection of **v** in the directions of $\mathbf{u}_1$ and $\mathbf{u}_2$, respectively. Equation (3.59) implicitly assumes some underlying basis set for a description of **y**, **x**, and **A**. Usually it is not important to state the basis set explicitly. The question arises as to what happens to (3.59) if we change the basis set. Suppose, for example, we wish to change from the **u** basis of Figure 3.9a to the **w** basis set shown in Figure 3.9b. We can write the **w** basis vectors in terms of the **u** basis vectors as

$$\mathbf{w}_1 = \alpha_1 \mathbf{u}_1 + \beta_1 \mathbf{u}_2$$
$$\mathbf{w}_2 = \alpha_2 \mathbf{u}_1 + \beta_2 \mathbf{u}_2 \tag{3.61}$$

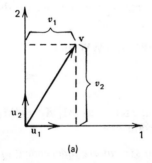

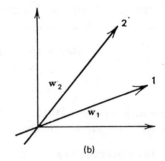

(a)   (b)

FIGURE 3.9

where the α's and β's are the projections of the $\mathbf{w}$ vectors on the $\mathbf{u}$ vectors. Now because $\mathbf{y}$ is the same vector in either basis, we have two representations for $\mathbf{y}$, namely,

$$\mathbf{y} = y_1^u\mathbf{u}_1 + y_2^u\mathbf{u}_2 = y_1^w\mathbf{w}_1 + y_2^w\mathbf{w}_2$$

$$= y_1^w[\alpha_1\mathbf{u}_1 + \beta_1\mathbf{u}_2] + y_2^w[\alpha_2\mathbf{u}_1 + \beta_2\mathbf{u}_2] \qquad (3.62)$$

$$= [\alpha_1 y_1^w + \alpha_2 y_2^w]\mathbf{u}_1 + [\beta_1 y_1^w + \beta_2 y_2^w]\mathbf{u}_2$$

Equating the coefficients of the first line to those of the last line in (3.62) yields

$$y_1^u = \alpha_1 y_1^w + \alpha_2 y_2^w$$

$$y_2^u = \beta_1 y_1^w + \beta_2 y_2^w$$

or, in vector notation,

$$\mathbf{y}^u = \boldsymbol{\Omega}\mathbf{y}^w \quad \text{where} \quad \boldsymbol{\Omega} = \begin{bmatrix} \alpha_1 & \alpha_2 \\ \beta_1 & \beta_2 \end{bmatrix} \qquad (3.63)$$

We can also express the vector $\mathbf{x}$ in the $\mathbf{u}$ basis in terms of the $\mathbf{w}$ basis as

$$\mathbf{x}^u = \boldsymbol{\Omega}\mathbf{x}^w \qquad (3.64)$$

Assume that the set of linear equations (3.59) is expressed in terms of the $\mathbf{u}$ basis: i.e.,

$$\mathbf{y}^u = \mathbf{A}\mathbf{x}^u \qquad (3.65)$$

In the $\mathbf{w}$ basis, (3.65) becomes

$$\boldsymbol{\Omega}\mathbf{y}^w = \mathbf{A}\boldsymbol{\Omega}\mathbf{x}^w$$

which means that

$$\mathbf{y}^w = \boldsymbol{\Omega}^{-1}\mathbf{A}\boldsymbol{\Omega}\mathbf{x}^w$$

$$= \hat{\mathbf{A}}\mathbf{x}^w$$

where

$$\hat{\mathbf{A}} = \boldsymbol{\Omega}^{-1}\mathbf{A}\boldsymbol{\Omega} \qquad (3.66)$$

In other words, under the change of basis, the $\mathbf{A}$ matrix is transformed into $\hat{\mathbf{A}} = \boldsymbol{\Omega}^{-1}\mathbf{A}\boldsymbol{\Omega}$.

Now consider the problem of transforming a matrix $\mathbf{A}$ into a diagonal matrix $\mathbf{D}$ such that the eigenvalues of $\mathbf{A}$ appear as the diagonal elements of $\mathbf{D}$. Assuming that the eigenvalues of $\mathbf{A}$ are distinct, we can accomplish this transformation by using a nonsingular matrix $\mathbf{P}$ whose columns consist of the eigenvectors of the matrix $\mathbf{A}$. The *eigenvectors* of $\mathbf{A}$ are those directions in the space that are not changed by the transformation $\mathbf{A}$. In symbols, the eigenvectors $\mathbf{v}_1, \mathbf{v}_2, \ldots, \mathbf{v}_n$ of $\mathbf{A}$ are defined by

$$\mathbf{A}\mathbf{v}_i = \lambda_i\mathbf{v}_i, \qquad i = 1, 2, \ldots, n \qquad (3.67)$$

where λ_i, $i = 1, 2, \ldots, n$ are the eigenvalues of $\mathbf{A}$. In this case, the diagonal matrix $\mathbf{D}$ is

$$\mathbf{D} = \mathbf{P}^{-1}\mathbf{A}\mathbf{P} \qquad (3.68)$$

EXAMPLE 3.11. Let $\mathbf{A}$ be the matrix $\begin{bmatrix} 3 & 4 \\ 2 & 1 \end{bmatrix}$. Find a diagonal matrix $\mathbf{D}$ that has the same eigenvalues as $\mathbf{A}$. The characteristic equation is

$$g(\lambda) = |\mathbf{A} - \lambda\mathbf{I}| = (\lambda - 5)(\lambda + 1) = 0$$

The eigenvalues of $\mathbf{A}$ are therefore $\lambda_1 = 5$ and $\lambda_2 = -1$. The corresponding eigenvectors are obtained by solving

$$(\mathbf{A} - \lambda_i\mathbf{I})\mathbf{v}_i = 0, \qquad i = 1, 2 \qquad (3.69)$$

Equation (3.69) leads to the two relations given below.

$$\begin{bmatrix} -2 & 4 \\ 2 & -4 \end{bmatrix}\mathbf{v}_1 = 0 \quad \text{and} \quad \begin{bmatrix} 4 & 4 \\ 2 & 2 \end{bmatrix}\mathbf{v}_2 = 0 \qquad (3.70)$$

From (3.70), we thus find $\mathbf{v}_1$ and $\mathbf{v}_2$ as

$$\mathbf{v}_1 = \begin{bmatrix} 2 \\ 1 \end{bmatrix}, \quad \mathbf{v}_2 = \begin{bmatrix} 1 \\ -1 \end{bmatrix}$$

(within an arbitrary multiplying constant). Therefore, the matrix $\mathbf{P}$ is

$$\mathbf{P} = \begin{bmatrix} 2 & 1 \\ 1 & -1 \end{bmatrix}$$

The inverse of $\mathbf{P}$ is

$$\mathbf{P}^{-1} = \frac{-1}{3}\begin{bmatrix} -1 & -1 \\ -1 & 2 \end{bmatrix}$$

This fact implies that $\mathbf{D}$ is

$$\mathbf{D} = \mathbf{P}^{-1}\mathbf{A}\mathbf{P} = \frac{-1}{3}\begin{bmatrix} -1 & -1 \\ -1 & 2 \end{bmatrix}\begin{bmatrix} 3 & 4 \\ 2 & 1 \end{bmatrix}\begin{bmatrix} 2 & 1 \\ 1 & -1 \end{bmatrix}$$

$$= \begin{bmatrix} 5 & 0 \\ 0 & -1 \end{bmatrix} = \begin{bmatrix} \lambda_1 & 0 \\ 0 & \lambda_2 \end{bmatrix} \qquad (3.71)$$

If one or more eigenvalues are repeated, then a form almost as simple can be obtained.

$$P^{-1}AP = J = \begin{bmatrix} J_1 & 0 & . & . & . & 0 \\ 0 & J_2 & . & . & . & 0 \\ & & . & & & \\ & & . & & & \\ & & . & & & \\ 0 & 0 & . & . & . & J_{n_0} \end{bmatrix} \tag{3.72}$$

where $J_1 \cdots J_{n_0}$ are elementary square matrices of the form

$$J_i = \begin{bmatrix} \lambda_i & 1 & 0 & & 0 \\ 0 & \lambda_i & . & . & . & 0 \\ & & . & . & & \\ & & & . & . & \\ & & & & . & 1 \\ 0 & 0 & . & . & . & \lambda_i \end{bmatrix} \tag{3.73}$$

and of dimension $m_i \times m_i$, where m_i is the multiplicity of λ_i. The important point here is not how to generate these matrices, but to note how they allow us to characterize our solution.

Assume for simplicity that the eigenvalues of A are distinct and that we have found the matrix P of (3.68). Our solution for the system state $x(k)$ and output $y(k)$ is expressed in terms of A^k. Now we can write A as

$$A = PDP^{-1} \tag{3.74}$$

and so

$$A^k = (PDP^{-1})^k = (PDP^{-1})(PDP^{-1}) \cdots (PDP^{-1}) = PD^kP^{-1} \tag{3.75}$$

where

$$D^k = \begin{bmatrix} \lambda_1^k & 0 & . & . & . & 0 \\ 0 & \lambda_2^k & . & . & . & 0 \\ & & . & & & \\ & & . & & & \\ & & . & & & \\ 0 & 0 & . & . & . & \lambda_n^k \end{bmatrix} \tag{3.76}$$

Thus, the elements of A^k, and hence of $x(k)$, are linear combinations of $\lambda_1^k, \lambda_2^k, \ldots, \lambda_n^k$. It follows that if $|\lambda_i| > 1$ for some i, then one or more of the state variables (and possibly the outputs also) will grow without bound with time. Hence, the system is unstable. Conversely, if $|\lambda_i| < 1$ for all $i = 1, \ldots, n$, and if the input is bounded, then all state variables and outputs will remain bounded. Finally, if $|\lambda_i| = 1$ for some i, then we can find a bounded input which will cause the sum in (3.25) to increase without bound as k increases. This result gives us a very easy method to determine system stability, and is one of the virtues of the state variable approach.

The eigenvalues of the state matrix characterize the system stability. A discrete-time system will be stable if, and only if, all eigenvalues have modulus less than 1.

The interested reader can see this conclusion reflected in the results of Examples 3.5 and 3.6.

EXAMPLE 3.12.   Suppose we have a discrete-time system whose state matrix is

$$A = \begin{bmatrix} 1 & a \\ 2 & \frac{1}{2} \end{bmatrix} \qquad (3.77)$$

For what values of a is the system stable?

We first find the eigenvalues of this matrix as a function of the parameter a. The characteristic equation is

$$g(\lambda) = (1 - \lambda)(\tfrac{1}{2} - \lambda) - 2a = 0$$

That is,

$$\lambda^2 - \tfrac{3}{2}\lambda + \tfrac{1}{2} - 2a = 0$$

with the roots

$$\lambda_1 = \tfrac{3}{4} + \sqrt{\tfrac{1}{16} + 2a}$$

$$\lambda_2 = \tfrac{3}{4} - \sqrt{\tfrac{1}{16} + 2a}$$

Two cases are possible.

Case 1:   Suppose $a \geq -\tfrac{1}{32}$. In this case, λ_1 and λ_2 are real valued, with $\lambda_1 > \lambda_2$. For the system to be stable, $|\lambda_i| < 1$, $i = 1, 2$. Thus set $\lambda_1 < 1$ to see what is implied about a.

$$\tfrac{3}{4} + \sqrt{\tfrac{1}{16} + 2a} < 1$$

That is,

$$\tfrac{1}{16} + 2a < \tfrac{1}{16}$$

which implies that $a < 0$. To insure stability, we must have $\lambda_2 > -1$. This restriction means that

$$\tfrac{3}{4} - \sqrt{\tfrac{1}{16} + 2a} > -1$$

$$-\sqrt{\tfrac{1}{16} + 2a} > -\tfrac{7}{4}$$

Thus

$$\sqrt{\tfrac{1}{16} + 2a} < \tfrac{7}{4}$$

which implies

$$\tfrac{1}{16} + 2a < \tfrac{49}{16}$$

$$a < \tfrac{3}{2}$$

which is assured by the preceding requirement that $a < 0$.

Case 2: Suppose now $a < -\tfrac{1}{32}$. In this case, λ_1 and λ_2 are complex valued. Setting $|\lambda_1| < 1$ or $|\lambda_2| < 1$ implies that

$$|\lambda_1|^2 = |\lambda_2|^2 = |\tfrac{3}{4} \pm j\sqrt{-\tfrac{1}{16} - 2a}|^2 < 1$$

$$\tfrac{9}{16} + (-\tfrac{1}{16} - 2a) < 1$$

or

$$a > -\tfrac{1}{4}$$

Combining these two results, we see that the system is stable for all values of a in the interval $(-\tfrac{1}{4}, 0)$.

*EXAMPLE 3.13. In our discussion on observability and controllability, we developed a method by which we could determine these system properties, provided that the system matrices were in normal form. That is, the system matrices are described in terms of a basis set that makes $\mathbf{A}$ a diagonal matrix. When the system matrices are in normal form, we can immediately determine whether a system is observable and controllable by examining the $\mathbf{B}$ and $\mathbf{C}$ matrices. For example, let us examine the observability and controllability of the following system. We identify as state variables the outputs of the two delay elements. Writing equations for the input variables to the delay elements yields

$$x_1(k + 1) = -2x_1(k) + 3x_2(k) + u(k)$$

$$x_2(k + 1) = x_1(k) + u(k)$$

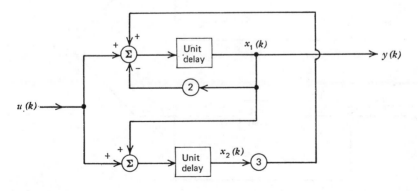

FIGURE 3.10

The output $y(k)$ is given by

$$y(k) = x_1(k)$$

The system matrices are therefore

$$A = \begin{bmatrix} -2 & 3 \\ 1 & 0 \end{bmatrix} \qquad B = \begin{bmatrix} 1 \\ 1 \end{bmatrix} \qquad C = [1 \quad 0] \qquad D = [0]$$

Solving $|A - \lambda I| = 0$, we find that the eigenvalues are 3, 1. For the diagonalizing transformation, we use

$$P = \begin{bmatrix} -3 & 1 \\ 1 & 1 \end{bmatrix} \quad \text{and} \quad P^{-1} = -\frac{1}{4}\begin{bmatrix} 1 & -1 \\ -1 & -3 \end{bmatrix}$$

Then

$$\hat{A} = P^{-1}AP = \begin{bmatrix} -3 & 0 \\ 0 & 1 \end{bmatrix} \qquad \hat{B} = P^{-1}B = \begin{bmatrix} 0 \\ 1 \end{bmatrix} \qquad \hat{C} = CP = [-3 \quad 1]$$

The system's eigenvalues are all distinct. Because **B** has a zero row (the first), the system is not controllable. Similarly, because **C** has no all-zero column, the system is observable. An equivalent representation of this system is obtained by use of the transformed state coordinates $x' = Px$. The system block diagram under these transformed state variables is shown in Figure 3.11.

We suggest that the reader apply a diagonalizing transformation to the system of Example 3.5 and transform the system to normal form, and thus determine whether the system is controllable and observable.

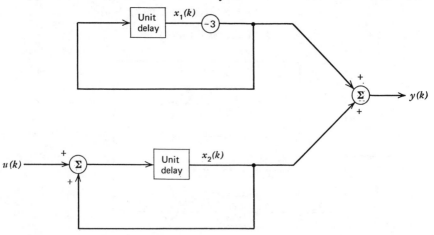

FIGURE 3.11

EXAMPLE 3.14. Let us return to an example previously solved in Chapters 1 and 2 (see Examples 1.32 and 2.6). To compare the various solution methods, consider a state variable formulation for this problem. The system is described by the difference equation

$$y(k + 2) - 5y(k + 1) + 6y(k) = u(k) \tag{3.78}$$

with zero initial conditions. We wish to obtain the output for an input step sequence. To obtain the state variables for (3.78), we first sketch a block diagram of the system as shown in Figure 3.12. We identify the outputs of the delay elements as the state variables. Thus

$$x_1(k) = y(k + 1)$$
$$x_2(k) = y(k)$$

the states at time $k + 1$ are

$$x_1(k + 1) = 5x_1(k) - 6x_2(k) + u(k)$$
$$x_2(k + 1) = x_1(k) \tag{3.79}$$

and the output is given by

$$y(k) = x_2(k)$$

Thus we have the state variable formulation from (3.79) as

$$\mathbf{x}(k + 1) = \begin{bmatrix} 5 & -6 \\ 1 & 0 \end{bmatrix} \mathbf{x}(k) + \begin{bmatrix} 1 \\ 0 \end{bmatrix} u(k)$$

$$y(k) = [0 \quad 1]\mathbf{x}(k) \tag{3.80}$$

The state vector $\mathbf{x}(k)$ is given by

$$\mathbf{x}_k = \mathbf{A}^k \mathbf{x}(0) + \sum_{m=0}^{k-1} \mathbf{A}^{k-1-m} \mathbf{B} u(m)$$

$$= \begin{cases} 0, & k \leq 0 \\ \sum_{m=0}^{k-1} \mathbf{A}^{k-1-m} \mathbf{B} \cdot 1 & k > 0 \end{cases} \quad \text{since} \quad \mathbf{x}(0) = 0 \tag{3.81}$$

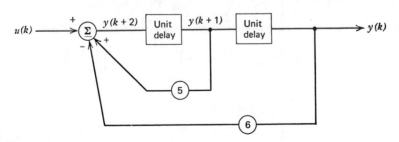

FIGURE 3.12

To calculate $\mathbf{A}^{k-1-m}$, we need an expression for $\mathbf{A}^n$. To find this general expression, we know by the Caley-Hamilton Theorem that

$$\mathbf{A}^n = \beta_0\mathbf{I} + \beta_1\mathbf{A} \tag{3.82}$$

To find the constants β_0 and β_1, we solve

$$\lambda_i^n = \beta_0 + \beta_1\lambda_i \qquad i = 1, 2 \tag{3.83}$$

where λ_1 and λ_2 are the eigenvalues of $\mathbf{A}$. Solving $|\mathbf{A} - \lambda\mathbf{I}| = 0$, we find that $\lambda_1 = 2$ and $\lambda_2 = 3$. Notice that these eigenvalues are precisely the roots of the characteristic equation of the original difference equation (3.78) which are used to form the unforced solution in Example 1.32. Thus, the equations we need to solve for β_0 and β_1 are

$$2^n = \beta_0 + 2\beta_1$$
$$3^n = \beta_0 + 3\beta_1$$

These equations yield

$$\beta_0 = 3 \cdot 2^n - 2 \cdot 3^n$$
$$\beta_1 = 3^n - 2^n$$

Hence

$$\mathbf{A}^n = \beta_0\mathbf{I} + \beta_1\mathbf{A}$$
$$= \begin{bmatrix} -2 \cdot 2^n + 3 \cdot 3^n & 6 \cdot 2^n - 6 \cdot 3^n \\ -2^n + 3^n & 3 \cdot 2^n - 2 \cdot 3^n \end{bmatrix} \tag{3.84}$$

Thus

$$\mathbf{A}^{k-1-m}\mathbf{B} = \mathbf{A}^{k-1-m}\begin{bmatrix} 1 \\ 0 \end{bmatrix} = \begin{bmatrix} -2 \cdot 2^{k-1-m} + 3 \cdot 3^{k-1-m} \\ -2^{k-1-m} + 3^{k-1-m} \end{bmatrix}$$
$$= 2^{-m}\begin{bmatrix} -2^k \\ -2^{k-1} \end{bmatrix} + 3^{-m}\begin{bmatrix} 3^k \\ 3^{k-1} \end{bmatrix} \tag{3.85}$$

To find $\mathbf{x}(k)$, we must sum (3.85) from $m = 0$ to $k - 1$. Thus

$$\mathbf{x}(k) = \sum_{m=0}^{k-1} \mathbf{A}^{k-1-m}\begin{bmatrix} 1 \\ 0 \end{bmatrix} \cdot 1$$
$$= \begin{bmatrix} -2^k \\ -2^{k-1} \end{bmatrix}\sum_{m=0}^{k-1} 2^{-m} + \begin{bmatrix} 3^k \\ 3^{k-1} \end{bmatrix}\sum_{m=0}^{k-1} 3^{-m}$$
$$= \begin{bmatrix} -2^k \\ -2^{k-1} \end{bmatrix}\left(\frac{1 - 2^{-k}}{1 - \frac{1}{2}}\right) + \begin{bmatrix} 3^k \\ 3^{k-1} \end{bmatrix}\left(\frac{1 - 3^{-k}}{1 - \frac{1}{3}}\right)$$
$$= 2\begin{bmatrix} -2^k + 1 \\ -2^{k-1} + \frac{1}{2} \end{bmatrix} + \frac{3}{2}\begin{bmatrix} 3^k - 1 \\ 3^{k-1} - \frac{1}{3} \end{bmatrix}$$
$$= \begin{bmatrix} \frac{1}{2} - 2 \cdot 2^k + \frac{3}{2} \cdot 3^k \\ \frac{1}{2} - 2^k + \frac{3}{2} \cdot 3^{k-1} \end{bmatrix} \tag{3.86}$$

where we have used the formula for the partial sum of a geometric series to obtain closed-form expressions for the two sums in (3.86). The output response to a step input is therefore

$$
y(k) = [0 \quad 1]\mathbf{x}(k) =
\begin{cases}
0, & k \le 0 \\
\frac{1}{2} + \frac{1}{2} \cdot 3^k - 2^k, & k > 0
\end{cases}
\tag{3.87}
$$

Comparing this formulation with the previous solution processes of Examples 1.32 and 2.6, it is clear that this formulation is more cumbersome to use. For small systems, the state variable approach is useful only if one is interested in the state evolution process, and the input-output characterization is best obtained through the methods of Chapters 1, 2, and 4. For large systems, however, the state variable approach may be the only tractable method of system analysis.

EXAMPLE 3.15. As a final example of a state-space formulation for discrete-time systems, consider an ecological system which has been proposed to model a trout fish hatchery. Four stages in the growth of a trout are identified: eggs, fry, young, and adults. The number of fry at year n is assumed to be proportional to the difference between the number of eggs released the previous year and the number of eggs eaten by the fry and young the previous year. The number of young at year n is assumed to be proportional to the number of fry at year $n - 1$ minus the number of young removed for stocking streams. The number of adults at year n is assumed to be proportional to the number of young at year $n - 1$ less the number of young removed at year $n - 1$ plus the number of adults at year $n - 1$. Finally, the number of eggs at year n will be assumed proportional to the number of adults at year $n - 1$ plus the number introduced artificially from other ponds. One possible sketch of this model is given in Figure 3.13. The variables in Figure 3.13 are

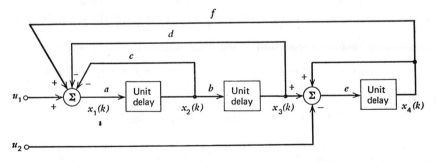

FIGURE 3.13

defined as

$x_1(n)$ = number of eggs at year n

$x_2(n)$ = number of fry at year n

$x_3(n)$ = number of young at year n

$x_4(n)$ = number of adults at year n

$u_1(n)$ = number of eggs added at year n

$u_2(n)$ = number of young removed at year n.

The equations relating these variables are

$$x_1(n) = u_1(n) - cx_2(n) - dx_3(n) + fx_4(n)$$
$$x_2(n) = ax_1(n-1)$$
$$x_3(n) = bx_2(n-1) \tag{3.88}$$
$$x_4(n) = ex_3(n-1) + ex_4(n-1) - eu_2(n-1)$$

Because only three delay elements are incorporated in our model, no more than three initial conditions are required to specify the initial state of our system. Thus, we choose a three-element state vector[7]

$$\mathbf{x}(n) = \begin{bmatrix} x_2(n) \\ x_3(n) \\ x_4(n) \end{bmatrix}$$

With the substitution from (3.88) for $x_1(n)$, the state equations are

$$x_2(n) = au_1(n-1) - acx_2(n-1) - adx_3(n-1) + afx_4(n-1)$$
$$x_3(n) = bx_2(n-1)$$
$$x_4(n) = ex_3(n-1) + ex_4(n-1) - eu_2(n-1) \tag{3.89}$$

or

$$\mathbf{x}(n+1) = \begin{bmatrix} -ac & -ad & af \\ b & 0 & 0 \\ 0 & e & e \end{bmatrix} \mathbf{x}(n) + \begin{bmatrix} a & 0 \\ 0 & 0 \\ 0 & -e \end{bmatrix} \mathbf{u}(n)$$

The characteristic equation of the state matrix is

$$g(\lambda) = (-ac - \lambda)(-\lambda e + \lambda^2) - b(-ade + ad\lambda - eaf) = 0$$

which reduces to

$$\lambda^3 + \lambda^2(e - ac) + \lambda(abd - ace) - abd(e + f) = 0$$

[7] For a particular system with n delay elements, it may happen that the n initial conditions are not linearly independent. In this case, a state vector of dimension less than n will suffice to describe the system. See Problem 2 at the end of this chapter.

The roots of this equation can be determined numerically from the parameters a, b, c, d, e, and f. From knowledge of these roots, we can predict whether the ecological system is stable, in which case eggs must be added to maintain the population, or unstable, in which case (with positive roots) young fish can be harvested yearly with no addition of eggs.

If we were interested in the number of eggs $x_1(n)$ and the number of adults $x_4(n)$ as a function of n, we could define the output vector

$$\mathbf{y}(n) = \begin{bmatrix} x_1(n) \\ x_4(n) \end{bmatrix} \tag{3.90}$$

with output equation

$$\mathbf{y}(n) = \begin{bmatrix} -c & -d & f \\ 0 & 0 & 1 \end{bmatrix} \mathbf{x}(n) + \begin{bmatrix} 1 & 0 \\ 0 & 0 \end{bmatrix} \mathbf{u}(n) \tag{3.91}$$

This example concludes our discussion of state variables for systems modeled by linear difference equations. This alternate formulation of linear systems models furnishes us with more information at the expense of more complexity. For complex systems, the state variable formulation furnishes us with a systematic and concise procedure for analyzing the system.

3.7   STATE VARIABLE FORMULATION FOR CONTINUOUS-TIME SYSTEMS

We proceed now to a discussion of continuous-time systems as modeled by linear differential equations with constant coefficients. The concepts of the state variable formulation discussed for discrete-time systems are essentially the same for continuous-time systems. We shall draw on these analogies to help formulate the state-space representation of continuous-time systems.

Consider a single-input, single-output system represented by the differential equation

$$\frac{d^n y(t)}{dt^n} + b_{n-1}\frac{d^{n-1}y(t)}{dt^{n-1}} + \cdots + b_0 y(t) = a_0 u(t) \tag{3.92}$$

To place this system in state variable form, we first sketch a block diagram of the system similar to the block diagrams we used to represent discrete-time systems. In the case of continuous-time systems, we use integrators as the basic element in the block diagram. Figure 3.14 depicts a block diagram representation of (3.92). In Figure 3.14, $y^{(n)}(t) \triangleq \dfrac{d^n y(t)}{dt^n}$. The integrators in Figure 3.14 correspond to the delay elements in the block diagrams for discrete-time systems. In order to calculate the output $y(t)$ for a given input

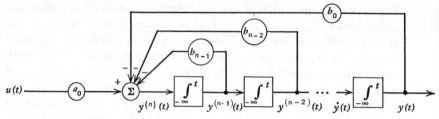

FIGURE 3.14

$u(t)$, we must know the initial values associated with each integrator, because these values can be specified independently of the input. In other words, the initial integrator outputs are like the memory of the system: thus, we define the state variables as the outputs from these memory units. We define the components of the state vector $\mathbf{x}(t)$ as

$$
\begin{aligned}
x_1(t) &= y(t) \\
x_2(t) &= \dot{y}(t) = \dot{x}_1(t) \\
x_3(t) &= \ddot{y}(t) = \dot{x}_2(t) \\
&\;\;\vdots \\
x_n(t) &= y^{(n-1)}(t) = \dot{x}_{n-1}(t)
\end{aligned} \tag{3.93}
$$

Rewriting (3.93), we can express $\dot{\mathbf{x}}(t)$ in terms of $\mathbf{x}(t)$ as

$$
\begin{aligned}
\dot{x}_1(t) &= x_2(t) \\
\dot{x}_2(t) &= x_3(t) \\
&\;\;\vdots \\
\dot{x}_{n-1}(t) &= x_n(t) \\
\dot{x}_n(t) &= y^{(n)}(t) = a_0 u(t) - b_0 x_1(t) - \cdots - b_{n-1} x_n(t)
\end{aligned} \tag{3.94}
$$

In vector form, (3.94) can be written as

$$
\dot{\mathbf{x}}(t) =
\begin{bmatrix}
\dot{x}_1(t) \\
\dot{x}_2(t) \\
\cdots \\
\dot{x}_{n-1}(t) \\
\dot{x}_n(t)
\end{bmatrix}
$$

$$= \begin{bmatrix} 0 & 1 & 0 & \cdots & 0 \\ 0 & 0 & 1 & \cdots & 0 \\ & & \vdots & & \vdots \\ & & \vdots & & \vdots \\ & & \vdots & & \vdots \\ 0 & 0 & 0 & \cdots & 1 \\ -b_0 & -b_1 & -b_2 & \cdots & -b_{n-1} \end{bmatrix} \begin{bmatrix} x_1(t) \\ x_2(t) \\ \vdots \\ \vdots \\ \vdots \\ x_{n-1}(t) \\ x_n(t) \end{bmatrix} + \begin{bmatrix} 0 \\ 0 \\ \vdots \\ \vdots \\ \vdots \\ 0 \\ a_0 \end{bmatrix} u(t) \qquad (3.95)$$

or

$$\dot{\mathbf{x}}(t) = \mathbf{A}\mathbf{x}(t) + \mathbf{B}u(t) \qquad (3.96)$$

Comparing (3.96) with (3.23), we observe one principal difference. Note that the state equation of a continuous-time system expresses the time derivative of the state vector in terms of the state vector and the input, whereas the state equation of a discrete-time system expresses the state at a given time in terms of the state at a previous time and the input.

We can, of course, express the output $y(t)$ in terms of the state $\mathbf{x}(t)$ and the input $u(t)$ as

$$y(t) = \mathbf{C}\mathbf{x}(t) + \mathbf{D}u(t) \qquad (3.97)$$

where, in the formulation above,

$$\mathbf{C} = [1 \quad 0 \quad 0 \quad \cdots \quad 0] \quad \text{and} \quad \mathbf{D} = 0$$

As before, we can handle multiple inputs and outputs with the same formulation

$$\dot{\mathbf{x}}(t) = \mathbf{A}\mathbf{x}(t) + \mathbf{B}u(t)$$
$$y(t) = \mathbf{C}\mathbf{x}(t) + \mathbf{D}u(t) \qquad (3.98)$$

where for n states, r inputs, and s outputs, the matrices $\mathbf{A}$, $\mathbf{B}$, $\mathbf{C}$, and $\mathbf{D}$ have dimensions as in (3.23) and (3.24).

As in the case of discrete-time systems, the state variables for a continuous-time system can be easily chosen by referring to a block diagram, such as Figure 3.14, or a schematic of the system. The state variables can again be interpreted as the "memory cells" of the system. For continuous-time systems, these memory cells can be chosen to correspond to the physical energy storing components within the system. For mechanical systems, the energy storing devices are springs and masses. For electrical systems, the energy storing devices are capacitors and inductors. The state variables can be chosen as the appropriate physical quantity associated with the energy storing device. For example, in springs we would choose as a state variable the contraction or expansion length. The next example demonstrates the choice in electrical systems.

EXAMPLE 3.16. Consider the electrical circuit shown in Figure 3.15. We choose as state variables the voltages across the capacitors as shown in Figure 3.15. If we now write Kirchhoff's current equations at the nodes between R_1 and R_2 and R_2 and R_3, we obtain the equations

$$\dot{x}_1 = \frac{1}{C_1}\left[\frac{u_1(t) - x_1(t)}{R_1} + \frac{x_2(t) - x_1(t)}{R_2}\right]$$

$$\dot{x}_2 = \frac{1}{C_2}\left[\frac{u_2(t) - x_2(t)}{R_3} + \frac{x_1(t) - x_2(t)}{R_2}\right] \tag{3.99}$$

The output $y(t)$ is clearly

$$y(t) = x_1(t) - x_2(t) \tag{3.100}$$

Equations (3.99) and (3.100) can be written in vector form as

$$\dot{\mathbf{x}}(t) = \begin{bmatrix} -\dfrac{1}{C_1}\left(\dfrac{1}{R_1} + \dfrac{1}{R_2}\right) & \dfrac{1}{R_2 C_1} \\ \dfrac{1}{R_2 C_2} & -\dfrac{1}{C_2}\left(\dfrac{1}{R_2} + \dfrac{1}{R_3}\right) \end{bmatrix}\mathbf{x}(t) + \begin{bmatrix} \dfrac{1}{R_1 C_1} & 0 \\ 0 & \dfrac{1}{R_3 C_2} \end{bmatrix}\mathbf{u}(t)$$

$$y(t) = \begin{bmatrix} 1 & -1 \end{bmatrix}\mathbf{x}(t)$$

The above equations constitute the state-space formulation of the circuit in Figure 3.15.

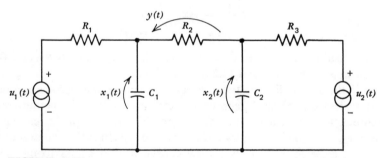

FIGURE 3.15

3.8 SOLUTION OF THE CONTINUOUS-TIME STATE VARIABLE EQUATION

We turn now to the solution of the state equation given by (3.96). We begin by considering the natural or unforced solution: i.e., $\mathbf{u}(t) = 0$.

Thus (3.96) simplifies to

$$\dot{\mathbf{x}}(t) = \mathbf{A}\mathbf{x}(t) \tag{3.101}$$

The scalar version of the vector problem is often a good indicator of the true solution. In this case, the scalar version of (3.101) is $\dot{x}(t) = ax(t)$ which has a solution $x(t) = e^{at}x(0)$. Suppose we try the same form for the solution of (3.101): that is,

$$\mathbf{x}(t) = e^{\mathbf{A}t}\mathbf{x}(0) \tag{3.102}$$

where $e^{\mathbf{A}t}$ is a function of the matrix $\mathbf{A}$ which we define as the $n \times n$ matrix

$$e^{\mathbf{A}t} \overset{\Delta}{=} \sum_{k=0}^{\infty} \frac{t^k}{k!} \mathbf{A}^k \tag{3.103}$$

The definition of $e^{\mathbf{A}t}$ given in (3.103) is a special case of a function of a matrix which we have previously solved in Section 3.5. If (3.102) is a solution of (3.101), then it must, of course, satisfy (3.101). Let us check:

$$\frac{d}{dt}[e^{\mathbf{A}t}\mathbf{x}(0)] \overset{?}{=} \mathbf{A}e^{\mathbf{A}t}\mathbf{x}(0)$$

Substituting for $e^{\mathbf{A}t}$ and differentiating term by term with respect to t yields

$$\left[\sum_{k=1}^{\infty} \frac{t^{k-1}}{(k-1)!} \mathbf{A}^k\right]\mathbf{x}(0) \overset{?}{=} \mathbf{A}\left[\sum_{k=0}^{\infty} \frac{t^k}{k!} \mathbf{A}^k\right]\mathbf{x}(0)$$

Letting $j = k - 1$ in the left-hand sum, we see that

$$\left[\sum_{j=0}^{\infty} \frac{t^j}{j!} \mathbf{A}^{j+1}\right]\mathbf{x}(0) \overset{?}{=} \mathbf{A}\left[\sum_{k=0}^{\infty} \frac{t^k}{k!} \mathbf{A}^k\right]\mathbf{x}(0) \tag{3.104}$$

In the left-hand side of (3.104), we can factor an $\mathbf{A}$ from $\mathbf{A}^{j+1}$ to obtain the right-hand side. Thus (3.104) is an equality and we have established that $\mathbf{x}(t) = e^{\mathbf{A}t}\mathbf{x}(0)$ is the solution to (3.101). Because the solution to (3.101) is unique, $e^{\mathbf{A}t}\mathbf{x}(0)$ is the unforced response of the system. To find the vector $\mathbf{x}(0)$, we evaluate (3.102) for some t_0 for which $\mathbf{x}(t)$ is known and obtain

$$\mathbf{x}(t_0) = e^{\mathbf{A}t_0}\mathbf{x}(0) \tag{3.105}$$

To find $\mathbf{x}(0)$ in (3.105), we multiply both sides of (3.105) on the right by the inverse of the matrix $e^{\mathbf{A}t_0}$; i.e., $(e^{\mathbf{A}t_0})^{-1}$. Thus

$$\mathbf{x}(0) = (e^{\mathbf{A}t_0})^{-1}\mathbf{x}(t_0)$$

One can show (see Problem 3.9) that the inverse of $e^{\mathbf{A}t}$ is

$$(e^{\mathbf{A}t_0})^{-1} = e^{-\mathbf{A}t_0} \tag{3.106}$$

Thus

$$\mathbf{x}(0) = e^{-\mathbf{A}t_0}\mathbf{x}(t_0) \tag{3.107}$$

Substituting the results of (3.107) in (3.102), our homogeneous solution is

$$\mathbf{x}(t) = e^{\mathbf{A}t}e^{-\mathbf{A}t_0}\mathbf{x}(t_0)$$
$$= e^{\mathbf{A}t - \mathbf{A}t_0}\mathbf{x}(t_0)$$
$$= e^{\mathbf{A}(t-t_0)}\mathbf{x}(t_0) \tag{3.108}$$

where we have used

$$e^{\mathbf{A}t}e^{-\mathbf{A}t_0} = e^{\mathbf{A}t - \mathbf{A}t_0} \tag{3.109}$$

Equation (3.109) is true because $\mathbf{A}$ and $-\mathbf{A}$ commute (see Problem 3.8).

To find the complete solution to (3.96), we must now find a particular solution to the differential equation. To find the particular solution, assume a solution of the form

$$\mathbf{x}_p(t) = e^{\mathbf{A}t}\mathbf{q}(t) \tag{3.110}$$

where $\mathbf{q}(t)$ is an unknown function to be determined.[8] We proceed to find the unknown vector $\mathbf{q}(t)$ by substitution in (3.96).

$$\dot{\mathbf{x}}_p(t) = \mathbf{A}\mathbf{x}_p(t) + \mathbf{B}u(t)$$

$$\mathbf{A}e^{\mathbf{A}t}\mathbf{q}(t) + e^{\mathbf{A}t}\dot{\mathbf{q}}(t) = \mathbf{A}e^{\mathbf{A}t}\mathbf{q}(t) + \mathbf{B}u(t)$$

Thus

$$\dot{\mathbf{q}}(t) = e^{-\mathbf{A}t}\mathbf{B}u(t)$$

Integrating, we obtain

$$\mathbf{q}(t) = \mathbf{q}(t_0) + \int_{t_0}^{t} e^{-\mathbf{A}\tau}\mathbf{B}u(\tau)\,d\tau$$

Thus, the particular or forced solution is

$$\mathbf{x}_p(t) = e^{\mathbf{A}t}\mathbf{q}(t) = e^{\mathbf{A}t}\mathbf{q}(t_0) + \int_{t_0}^{t} e^{\mathbf{A}(t-\tau)}\mathbf{B}u(\tau)\,d\tau$$

To evaluate $\mathbf{q}(t_0)$, we use the entire solution and set $\mathbf{x}(t)$ evaluated at t_0 equal to $\mathbf{x}(t_0)$. Thus, we obtain

$$\mathbf{x}(t)\bigg|_{t=t_0} = \left[e^{\mathbf{A}(t-t_0)}\mathbf{x}(t_0) + e^{\mathbf{A}t}\mathbf{q}(t_0) + \int_{t_0}^{t} e^{\mathbf{A}(t-\tau)}\mathbf{B}u(\tau)\,d\tau \right]_{t=t_0}$$

which implies that $\mathbf{q}(t_0) = 0$. Hence, the entire solution of the state equation (3.96) is

$$\mathbf{x}(t) = e^{\mathbf{A}(t-t_0)}\mathbf{x}(t_0) + \int_{t_0}^{t} e^{\mathbf{A}(t-\tau)}\mathbf{B}u(\tau)\,d\tau \tag{3.111}$$

The corresponding output is

$$\mathbf{y}(t) = \mathbf{C}\mathbf{x}(t) + \mathbf{D}u(t) \tag{3.112}$$

---

[8] This is called the method of variation of parameters.

If $\mathbf{D} = 0$, the more usual case, then the output is given by

$$\mathbf{y}(t) = \mathbf{C}e^{\mathbf{A}(t-t_0)}\mathbf{x}(t_0) + \int_{t_0}^{t} \mathbf{C}e^{\mathbf{A}(t-\tau)}\mathbf{B}\mathbf{u}(\tau)\,d\tau \qquad (3.113)$$

The quantity $\mathbf{C}e^{\mathbf{A}t}\mathbf{B}$ in the integrand of (3.113) is exactly the impulse-response function $h(t)$ defined in Chapter 2 for single input-output systems. For multiple-input, multiple-output systems, the quantity $\mathbf{C}e^{\mathbf{A}t}\mathbf{B}$ is an $r \times s$ matrix (r inputs and s outputs). The (i, j) element is the impulse response at the ith output resulting from an impulse at $t = 0$ applied to the jth input, with all other inputs zero.

To evaluate the state equation and the output in this formulation, the basic calculation is a determination of $e^{\mathbf{A}t}$. This function of a matrix can be evaluated in a manner discussed previously in Section 3.5.

EXAMPLE 3.17. Let us evaluate $e^{\mathbf{A}t}$ for the matrix

$$\mathbf{A} = \begin{bmatrix} 3 & 0 & 0 \\ 0 & -2 & 1 \\ 0 & 4 & 1 \end{bmatrix}$$

First we find the eigenvalues of $\mathbf{A}$ from the characteristic equation

$$g(\lambda) = |\mathbf{A} - \lambda\mathbf{I}| = \begin{vmatrix} 3 - \lambda & 0 & 0 \\ 0 & -2 - \lambda & 1 \\ 0 & 4 & 1 - \lambda \end{vmatrix}$$

$$= (3 - \lambda)(\lambda^2 + \lambda - 6) = 0$$

Thus, the eigenvalues of $\mathbf{A}$ are $\lambda_1 = 3$, $\lambda_2 = 2$, and $\lambda_3 = -3$. The function of a matrix we wish to find is $e^{\mathbf{A}t}$. Using the results of Section 3.5 we know by the Caley-Hamilton theorem that

$$e^{\mathbf{A}t} = \beta_0\mathbf{I} + \beta_1\mathbf{A} + \beta_2\mathbf{A}^2 \qquad (3.114)$$

We can evaluate the constants β_0, β_1, and β_2 from the equations

$$e^{3t} = \beta_0 + 3\beta_1 + 9\beta_2$$
$$e^{2t} = \beta_0 + 2\beta_1 + 4\beta_2 \qquad (3.115)$$
$$e^{-3t} = \beta_0 - 3\beta_1 + 9\beta_2$$

The β's are thus found to be

$$\beta_0 = -e^{-3t} + \tfrac{9}{5}e^{2t} + \tfrac{1}{5}e^{-3t}$$
$$\beta_1 = \tfrac{1}{6}e^{3t} - \tfrac{1}{6}e^{-3t} \qquad (3.116)$$
$$\beta_2 = \tfrac{1}{6}e^{3t} - \tfrac{1}{5}e^{2t} + \tfrac{1}{30}e^{-3t}$$

Thus the matrix $\mathbf{e}^{\mathbf{A}t}$ is

$$\mathbf{e}^{\mathbf{A}t} = \beta_0 \mathbf{I} + \beta_1 \mathbf{A} + \beta_2 \mathbf{A}^2$$

$$= \begin{bmatrix} \beta_0 & 0 & 0 \\ 0 & \beta_0 & 0 \\ 0 & 0 & \beta_0 \end{bmatrix} + \begin{bmatrix} 3\beta_1 & 0 & 0 \\ 0 & -2\beta_1 & \beta_1 \\ 0 & 4\beta_1 & \beta_1 \end{bmatrix} + \begin{bmatrix} 9\beta_2 & 0 & 0 \\ 0 & 8\beta_2 & -\beta_2 \\ 0 & 4\beta_2 & 5\beta_2 \end{bmatrix}$$

$$= \frac{1}{5} \begin{bmatrix} 5e^{3t} & 0 & 0 \\ 0 & e^{2t} + 4e^{-3t} & e^{2t} - e^{-3t} \\ 0 & 4(e^{2t} - e^{-3t}) & 4e^{2t} + e^{-3t} \end{bmatrix} \tag{3.117}$$

If, for example, the state at $t = 0$ is $\mathbf{x}(0) = \begin{bmatrix} 1 \\ 1 \\ 1 \end{bmatrix}$, then for $\mathbf{u}(t) \equiv 0$, the state at time t is

$$\mathbf{x}(t) = \mathbf{e}^{\mathbf{A}t}\mathbf{x}(0) = \frac{1}{5} \begin{bmatrix} 5e^{3t} \\ 2e^{2t} + 3e^{-3t} \\ 8e^{2t} - 3e^{-3t} \end{bmatrix}$$

In the above example, we found a general expression for the state of a system with state matrix $\mathbf{A}$ and initial state $\mathbf{x}(0)$ as given. One of the first things we should do is verify that all signals internal to system remain within bounds wherein the system is stable. We see that all states above grow without bound as e^{3t} or e^{2t}. Hence, this particular system is unstable, and we must check the suitability of our model. After a few calculations like this, we shall begin to ask whether there is not some easier way to examine system stability, given the system matrix $\mathbf{A}$. Just as with the discrete-time case, the answer is yes. We can verify that our system is either stable or unstable without solving for the state and output behavior in detail. Again, it is the set of eigenvalues of $\mathbf{A}$ which determine system behavior.

We proved previously that the eigenvalues of $\mathbf{A}$ are the same as those of the matrix $\mathbf{WAW}^{-1}$ for any $n \times n$ matrix $\mathbf{W}$ whose inverse exists. Let us assume for convenience that the eigenvalues of the state matrix $\mathbf{A}$ are distinct, and choose for $\mathbf{W}$ the diagonalizing matrix $\mathbf{P}$ whose columns are the eigenvectors of $\mathbf{A}$. Then, as before,

$$\mathbf{P}^{-1}\mathbf{A}\mathbf{P} \triangleq \mathbf{D} = \begin{bmatrix} \lambda_1 & 0 & \cdots & 0 \\ 0 & \lambda_2 & \cdots & 0 \\ & & \cdot & \\ & & \cdot & \\ & & \cdot & \\ 0 & 0 & \cdots & \lambda_2 \end{bmatrix}$$

and

$$\mathbf{A} = \mathbf{PDP}^{-1}$$

Substituting in the definition for $\mathbf{e}^{\mathbf{A}t}$, we find

$$\mathbf{e}^{\mathbf{A}t} = \sum_{k=0}^{\infty} \frac{t^k}{k!} \mathbf{A}^k$$

$$= \sum_{k=0}^{\infty} \frac{t^k}{k!} (\mathbf{PDP}^{-1})^k$$

$$= \mathbf{P} \left[\sum_{k=0}^{\infty} \frac{t^k}{k!} \mathbf{D}^k \right] \mathbf{P}^{-1}$$

However, the matrix $\sum_{k=0}^{\infty} \frac{t^k}{k!} \mathbf{D}^k$ is merely

$$\sum_{k=0}^{\infty} \frac{t^k}{k!} \begin{bmatrix} \lambda_1^k & 0 & \cdots & 0 \\ 0 & \lambda_2^k & \cdots & 0 \\ & & \ddots & \\ 0 & 0 & \cdots & \lambda_n^k \end{bmatrix} = \begin{bmatrix} e^{\lambda_1 t} & 0 & \cdots & 0 \\ 0 & e^{\lambda_2 t} & \cdots & 0 \\ & & \ddots & \\ 0 & 0 & \cdots & e^{\lambda_n t} \end{bmatrix}$$

Hence,

$$\mathbf{e}^{\mathbf{A}t} = \mathbf{P} \begin{bmatrix} e^{\lambda_1 t} & 0 & \cdots & 0 \\ 0 & e^{\lambda_2 t} & \cdots & 0 \\ & & \ddots & \\ 0 & 0 & \cdots & e^{\lambda_n t} \end{bmatrix} \mathbf{P}^{-1} \tag{3.118}$$

Therefore the elements of $\mathbf{e}^{\mathbf{A}t}$, and hence the components of $\mathbf{x}(t)$, are composed of sums of the functions $e^{\lambda_1 t}$, $e^{\lambda_2 t}$, $\ldots$, $e^{\lambda_n t}$. If any of these funtions grows without bound, the system will be unstable; this growth will happen if the real part of any eigenvalue is greater than zero. Also, if the real part of any eigenvalue equals zero, we can find a bounded input which will cause the integral in (3.111) to become unbounded. Conversely, if the real parts of all eigenvalues are less than zero, our solution will remain bounded for all bounded inputs. Thus our conclusion:

A continuous-time system is stable if, and only if, all eigenvalues of the state matrix have real parts less than zero.

EXAMPLE 3.18. We are given a continuous-time system whose state matrix is

$$A = \begin{bmatrix} \alpha & 0 & 0 \\ 0 & \beta & -1 \\ 0 & 1 & -2 \end{bmatrix}$$

For what range of values of α and β is the system stable (assume that α and β are real)?

The characteristic equation for **A** is

$$g(\lambda) = (\alpha - \lambda)[(\beta - \lambda)(-2 - \lambda) + 1] = 0$$

that is,

$$(\alpha - \lambda)[\lambda^2 + \lambda(2 - \beta) + 1 - 2\beta] = 0 \qquad (3.119)$$

The roots of (3.119) are

$$\lambda_1 = \alpha, \quad \lambda_2 = \frac{\beta - 2}{2} + \frac{\sqrt{\beta(\beta + 4)}}{2}, \quad \lambda_3 = \frac{\beta - 2}{2} - \frac{\sqrt{\beta(\beta + 4)}}{2}$$

For stability, we clearly must have $\alpha < 0$. Investigating the range of β for stability involves two cases.

Case 1: λ_2 and λ_3 are complex for $-4 < \beta < 0$. In this case we must have $(\beta - 2)/2 < 0$, which means $\beta < 2$. Thus, the system is stable for all β in the range $-4 < \beta < 0$.

Case 2: λ_2 and λ_3 are real for $\beta \le -4$ or $\beta \ge 0$. Because $\lambda_2 \ge \lambda_3$, we set

$$0 > \lambda_2 = \frac{\beta - 2}{2} + \frac{\sqrt{\beta^2 + 4\beta}}{2}$$

Thus we have

$$2 - \beta > \sqrt{\beta^2 + 4\beta}$$

from which we find

$$\beta < \tfrac{1}{2}$$

The conclusion is that the system is stable for all $\alpha < 0$ and $\beta < \tfrac{1}{2}$.

EXAMPLE 3.19. For the circuit shown in Figure 3.16, find the step response using state variable methods. Also calculate the impulse response using two different methods. Because we wish to find the impulse response by two methods, let us first derive the differential equation relating the input $u(t)$ and the output $y(t)$. If we write Kirchhoff's current equation at node A, we have

$$i_1(t) + i_2(t) + i_3(t) = 0 \qquad (3.120)$$

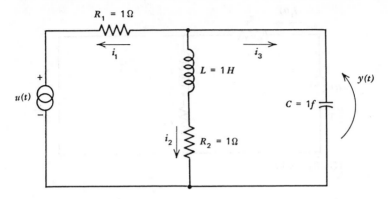

FIGURE 3.16

The various currents are given by

(1) $\quad i_1(t) = \dfrac{y(t) - u(t)}{R_1}$

(2) $\quad i_2(t) = \dfrac{1}{L} \displaystyle\int_{-\infty}^{t} v_L(\tau)\, d\tau = \dfrac{1}{L} \displaystyle\int_{-\infty}^{t} (y(\tau) - i_2(\tau)R_2)\, d\tau$

or $\quad i_2'(t) = \dfrac{1}{L}(y(t) - i_2(t)R_2)$

where $v_L(t)$ is the voltage drop across the inductor L.

(3) $\quad i_3(t) = C\dot{y}(t)$

If we differentiate (3.120), we have that

$$i_1'(t) + i_2'(t) + i_3'(t) = 0 \qquad (3.121)$$

Now use the expressions for the currents obtained above and substitute into (3.121). We have

$$\frac{\dot{y}(t) - \dot{u}(t)}{R_1} + \frac{1}{L}(y(t) - i_2(t)R_2) + C\ddot{y}(t) = 0$$

Now substitute for $i_2(t) = -i_1(t) - i_3(t)$ to obtain

$$\ddot{y}(t) + 2\dot{y}(t) + 2y(t) = \dot{u}(t) + u(t) \qquad (3.122)$$

Equation (3.122) is the differential equation that relates the input $\mathbf{u}(t)$ to the output $\mathbf{y}(t)$. We can use (3.122) to find the impulse response of the system using the methods of Chapter 2. However, before we do, we shall calculate the impulse response using a state-space formulation. To obtain a state-space representation, define the state variables as the current through the inductor and the voltage across the capacitor as shown in

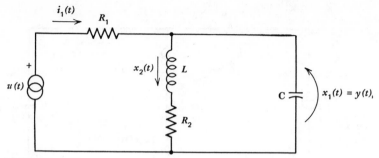

FIGURE 3.17

Figure 3.17. Writing Kirchhoff's voltage equation around the loop containing $u(t)$, we obtain

$$u(t) = x_1(t) + i_1(t)R_1$$

The current $i_1(t)$ is

$$i_1(t) = x_2(t) + C\dot{x}_1(t)$$

Thus

$$u(t) = x_1(t) + R_1[x_2(t) + C\dot{x}_1(t)]$$

and so

$$\dot{x}_1(t) = u(t) - x_1(t) - x_2(t) \qquad (3.123)$$

A second equation can be obtained by writing a voltage equation around the second loop.

$$x_1(t) = L\dot{x}_2(t) + Rx_2(t)$$

and so

$$\dot{x}_2(t) = x_1(t) - x_2(t) \qquad (3.124)$$

Combining (3.123) and (3.124), we obtain

$$\dot{\mathbf{x}}(t) = \begin{bmatrix} -1 & -1 \\ 1 & -1 \end{bmatrix} \mathbf{x}(t) + \begin{bmatrix} 1 \\ 0 \end{bmatrix} u(t) \qquad (3.125)$$

The output is clearly

$$y(t) = [1 \quad 0]\mathbf{x}(t) \qquad (3.126)$$

In general, for zero initial conditions, the output $y(t)$ for an input $u(t)$ is

$$y(t) = \int_0^t \mathbf{C}e^{\mathbf{A}(t-\tau)}\mathbf{B}u(\tau)\,d\tau \qquad (3.127)$$

From (3.127) we conclude that the impulse response is identically $\mathbf{C}e^{\mathbf{A}t}\mathbf{B}$: i.e.,

$$h(t) = \mathbf{C}e^{\mathbf{A}t}\mathbf{B} \qquad (3.128)$$

To calculate $h(t)$ from (3.128), we need an expression for $e^{\mathbf{A}t}$. Thus, we first obtain the eigenvalues of $\mathbf{A}$ from

$$g(\lambda) = |\mathbf{A} - \lambda\mathbf{I}| = \lambda^2 + 2\lambda + 2 = 0$$

The roots of $g(\lambda) = 0$ are $\lambda_1 = -1 + j$ and $\lambda_2 = -1 - j$. Now

$$e^{\lambda t} = \beta_0 + \beta_1\lambda$$

where β_0 and β_1 are found from the equations

$$e^{(-1+j)t} = \beta_0 + \beta_1(-1 + j)$$
$$e^{(-1-j)t} = \beta_0 + \beta_1(-1 - j)$$

These equations yield

$$\beta_0 = e^{-t}\sin t$$
$$\beta_1 = e^{-t}(\sin t + \cos t)$$

Thus $e^{\mathbf{A}t}$ is the matrix

$$e^{\mathbf{A}t} = \beta_0\mathbf{I} + \beta_1\mathbf{A} = e^{-t}\begin{bmatrix} \cos t & -\sin t \\ \sin t & \cos t \end{bmatrix}$$

The impulse response of the system is therefore

$$h(t) = \mathbf{C}e^{\mathbf{A}t}\mathbf{B} = [1 \quad 0]e^{-t}\begin{bmatrix} \cos t & -\sin t \\ \sin t & \cos t \end{bmatrix}\begin{bmatrix} 1 \\ 0 \end{bmatrix}, \qquad t \geq 0$$

$$= e^{-t}\cos t\, u(t) \tag{3.129}$$

Let's return to the original differential equation (3.122) that relates the input and output. Using the methods of Chapter 2, we first find the homogeneous solution to (3.122). The characteristic equation of (3.122) is

$$r^2 + 2r + 2 = 0$$

which has roots $r_1 = -1 + j$ and $r_2 = -1 - j$. Thus, the impulse-response function $\hat{h}(t)$ for the system represented by $\ddot{y}(t) + 2\dot{y}(t) + 2y(t) = u(t)$ is given by

$$\hat{h}(t) = e^{-t}(c_1\cos t + c_2\sin t)u(t)$$

We evaluate the constants c_1 and c_2 using the initial conditions $\hat{h}(0) = 0$ and $\hat{h}'(0) = 1$. Thus we have

$$\hat{h}(0) = 0 = c_1$$
$$\hat{h}'(0) = 1 = e^{-t}(-c_2\sin t + c_2\cos t)\big|_{t=0}$$
$$= c_2$$

Thus,

$$\hat{h}(t) = e^{-t} \sin t \, u(t) \tag{3.130}$$

The impulse-response function corresponding to (3.122) is found by applying the differential operator $(D + 1)$ to $\hat{h}(t)$. Thus,

$$
\begin{aligned}
h(t) &= (D + 1)[e^{-t} \sin t \, u(t)] \\
&= -e^{-t} \sin t \, u(t) + e^{-t} \cos t \, u(t) \\
&\quad + e^{-t} \sin t \, \delta(t) + e^{-t} \sin t \, u(t) \\
&= e^{-t} \cos t \, u(t)
\end{aligned} \tag{3.131}
$$

Comparing (3.131) and (3.129), we see that they are identical.

To find the step response of this system, we can proceed in various ways. Probably the simplest method is to convolve the step input with the impulse-response function. Thus

$$
\begin{aligned}
y(t) &= \int_0^t h(\tau)u(t - \tau)\, d\tau \\
&= \int_0^t e^{-\tau} \cos \tau \, d\tau \\
&= \begin{cases} \frac{1}{2}[1 + e^{-t} \sin t - e^{-t} \cos t], & t \geq 0 \\ 0, & t < 0 \end{cases}
\end{aligned} \tag{3.132}
$$

We can also calculate the step response from the state-space formulation. In this case,

$$
\begin{aligned}
y(t) &= \int_0^t \mathbf{C}e^{\mathbf{A}(t-\tau)}\mathbf{B}u(\tau)\, d\tau \\
&= \int_0^t \mathbf{C}e^{\mathbf{A}\tau}\mathbf{B} \, d\tau \\
&= \int_0^t \mathbf{C}\begin{bmatrix} e^{-\tau} \cos \tau \\ e^{-\tau} \sin \tau \end{bmatrix} d\tau \\
&= \mathbf{C}\begin{bmatrix} \frac{1}{2}(1 + e^{-t} \sin t - e^{-t} \cos t) \\ \frac{1}{2}(1 - e^{-t} \sin t - e^{-t} \cos t) \end{bmatrix}, \quad t \geq 0
\end{aligned}
$$

Therefore, as before

$$
y(t) = \begin{cases} \frac{1}{2}(1 + e^{-t} \sin t - e^{-t} \cos t), & t \geq 0 \\ 0, & t < 0 \end{cases} \tag{3.133}
$$

As a bonus, we obtain that $x_2(t)$, the current through the inductor, is

$$i_L(t) = x_2(t) = \tfrac{1}{2}(1 - e^{-t}\sin t - e^{-t}\cos t)u(t).$$

3.9 OBSERVABILITY AND CONTROLLABILITY FOR CONTINUOUS-TIME SYSTEMS

The ideas of observability and controllability for continuous-time systems parallel the concepts introduced in our discussion of discrete-time systems. Under the assumptions that the system matrix **A** has distinct eigenvalues and that the coordinate basis is chosen so that the state variable representation is in normal form, one can again determine observability and controllability by simply examining the **B** and **C** matrices. If **B** and **C** have only nonzero rows and columns, respectively, then the system is controllable and observable, respectively. The next example illustrates these ideas for the case of continuous-time systems.

EXAMPLE 3.20. Consider the system shown in Figure 3.18. We assume that the RC circuit in the feedback path is not loaded by either the input or output circuits. We define state variables as shown in Figure 3.18. The differential equations describing the two systems enscribed by the dotted lines are

$$\dot{x}_1(t) = -2x_2(t) + 2u(t)$$
$$\dot{x}_2(t) = x_1(t) - 3x_2(t)$$

(3.134)

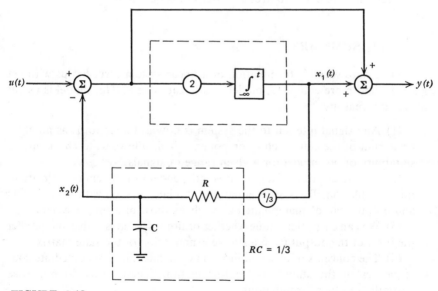

FIGURE 3.18

In state-space form we have

$$\dot{\mathbf{x}}(t) = \begin{bmatrix} 0 & -2 \\ 1 & -3 \end{bmatrix} \mathbf{x}(t) + \begin{bmatrix} 2 \\ 0 \end{bmatrix} u(t)$$

$$y(t) = \begin{bmatrix} 1 & -1 \end{bmatrix} \mathbf{x}(t) + u(t)$$

The eigenvalues of the **A** matrix are $\lambda_1 = -1$ and $\lambda_2 = -2$. The corresponding eigenvectors are $\begin{bmatrix} 2 \\ 1 \end{bmatrix}$ and $\begin{bmatrix} 1 \\ 1 \end{bmatrix}$, respectively. The diagonalizing transformation is

$$\mathbf{P} = \begin{bmatrix} 2 & 1 \\ 1 & 1 \end{bmatrix} \quad \text{and} \quad \mathbf{P}^{-1} = \begin{bmatrix} 1 & -1 \\ -1 & 2 \end{bmatrix}$$

Hence, the system matrices in normal form are

$$\hat{\mathbf{A}} = \mathbf{P}^{-1}\mathbf{A}\mathbf{P} = \begin{bmatrix} -1 & 0 \\ 0 & -2 \end{bmatrix}$$

$$\hat{\mathbf{B}} = \mathbf{P}^{-1}\mathbf{B} = \begin{bmatrix} 2 \\ -2 \end{bmatrix}$$

$$\hat{\mathbf{C}} = \mathbf{C}\mathbf{P} = \begin{bmatrix} 1 & 0 \end{bmatrix}$$

Thus, this system is controllable but not observable.

3.10   SUMMARY

The state variable formulation of a linear system furnishes us with yet another, and more complete, model of a linear system. This model is useful for several reasons.

(1) Any signal internal to the system is now available to us as an algebraic sum of the state vector components. This allows us to check on the suitability of our model for a given range of signals.

(2) The description of a system is independent of system complexity in this formulation. The system equations of a large system with many inputs and outputs can be manipulated as easily as those of a simple system.

(3) We can easily determine whether or not a system is stable and predict the form of the output from a simple examination of the state matrix.

(4) The concepts of observability and controllability are immediate consequences of the state variable formulation. These ideas do not arise naturally in other formulations.

PROBLEMS

1. Write state variable equations for the following systems:

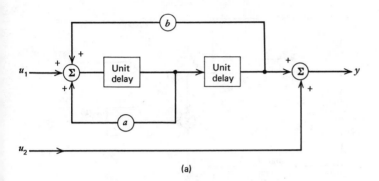

(a)

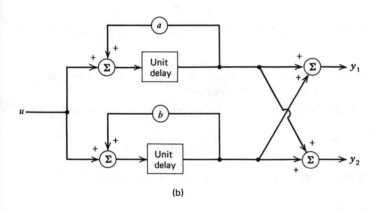

(b)

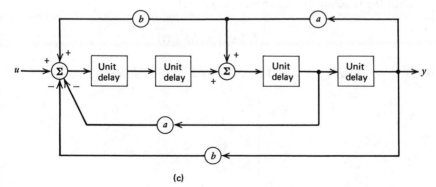

(c)

2. Write state variable equations for the following system, using a three-component state vector with x_1, x_2, and x_3 as shown.
   a. Evaluate the determinant, $|A|$, of the state matrix.
   b. What are the roots of the characteristic equation of the state matrix?
   c. Can you generalize the results of a and b?

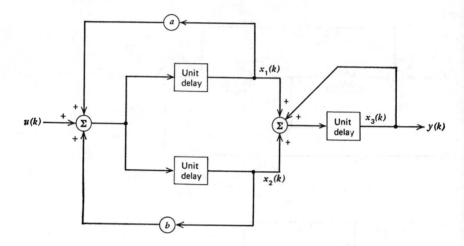

   d. Can you write state variable equations for this system using only a two-component state vector? Comment.
3. Assume that the function $f(\lambda)$ can be represented by an infinite series. Prove that

$$A \cdot f(A) = f(A) \cdot A$$

   for any square matrix A: i.e., prove that the matrices A and $f(A)$ commute.

4. With the function $f(\cdot)$ as above, prove that

$$\frac{d}{dt}[f(At)] = Af'(At)$$

5. Find a general expression for A^k for each of the following matrices.

   a. $$A = \begin{bmatrix} \frac{3}{4} & 0 \\ \frac{1}{2} & \frac{1}{2} \end{bmatrix}$$

   b. $$A = \begin{bmatrix} \frac{1}{2} & \frac{1}{4} \\ \frac{1}{16} & \frac{1}{2} \end{bmatrix}$$

c.
$$A = \begin{bmatrix} \frac{1}{2} & \frac{1}{4} \\ 1 & \frac{1}{2} \end{bmatrix}$$

d.
$$A = \begin{bmatrix} \frac{1}{2} & 0 \\ \frac{1}{2} & \frac{1}{2} \end{bmatrix}$$

e.
$$A = \begin{bmatrix} \frac{3}{4} & -\frac{1}{2} \\ -\frac{15}{32} & \frac{1}{2} \end{bmatrix}$$

6. Find a general expression for the state, $x(k)$, of the systems shown on p. 162, for $k \geqslant 0$. Let $x(0) = \begin{bmatrix} 1 \\ 1 \end{bmatrix}$ and assume that $u(k) = 0$ for all $k \geqslant 0$.

7. For each of the systems in Problem 1, what constraints must be placed on a and b to assure system stability?

8. Under what conditions on A and B (both of dimension $n \times n$) is it true that

$$e^{A+B} = e^A e^B$$

[Hint: Compare e^{A+B} and $e^A e^B$ term by term using (3.103).]

9. Using the results of Problem 8, prove that $[e^A]^{-1} = e^{-A}$.

10. For the matrices of Problem 5, find an expression for the matrix e^{At}.

11. Write state variable equations for each of the systems shown on p. 163.

12. For what values of the parameters involved is each of the systems above stable?

13. Show that the following system is not controllable by exhibiting explicitly a region which cannot be reached by the state vector x for any input $u(t)$.

$$\dot{x}(t) = \begin{bmatrix} 1 & 0 \\ 0 & 1 \end{bmatrix} x(t) + \begin{bmatrix} 1 \\ 1 \end{bmatrix} u(t), \qquad x(0) = 0$$
$$y(t) = [1 \quad 1]x(t)$$

Change the A matrix so that the system is controllable and give an input which can be used to reach the previously unreachable region.

14. Use the state variable representation for the impulse response function $Ce^{At}B$ to find the impulse response of the system represented by the differential equation

$$\ddot{y}(t) + y(t) = u(t)$$

15. Consider the system of three difference equations

$$x(n + 1) = 3x(n) + 5y(n) + 2z(n)$$
$$y(n + 1) = x(n) - y(n) + z(n) \qquad n = 0, 1, 2, \ldots$$
$$z(n + 1) = 2x(n) + y(n) + 3z(n)$$

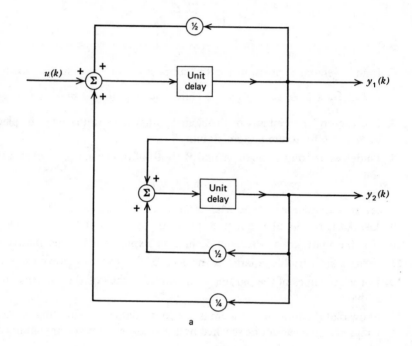

a

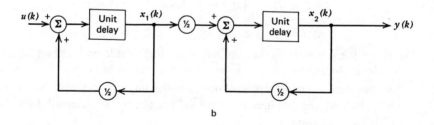

b

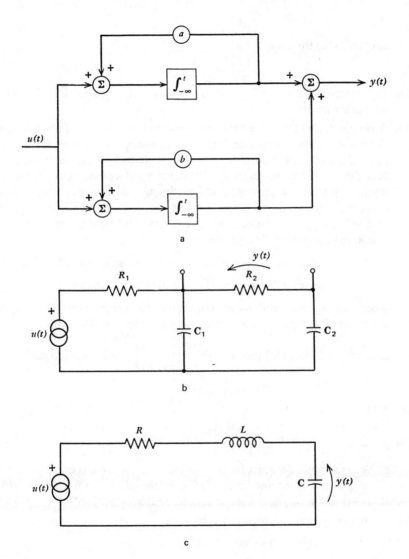

a

b

c

Write this system in the form

$$\mathbf{v}(n + 1) = \mathbf{A}\mathbf{v}(n), \qquad n = 1, 2, \ldots$$

and find $\mathbf{v}(n)$ for $\mathbf{v}(0) = \begin{bmatrix} 1 \\ 0 \\ 0 \end{bmatrix}$.

16. In Examples 3.1 and 3.2, determine whether the systems are observable and controllable.

17. A simple model for trade between two countries can be set up as follows. We assume national income (y) equals consumption outlays (c) plus net investment (i) plus exports (e) minus imports (m). Also, we assume outlays for domestic consumption (d) equal total consumption (c) minus imports (m). Let subscripts 1 and 2 distinguish the two countries. Assume domestic consumption (d) and imports (m) of each country in any time unit are constant multipliers of the country's national income (y) in the previous time period. In symbols,

$$d_1(n) = \alpha_{11}y_1(n - 1), \qquad m_1(n) = \alpha_{21}y_1(n - 1)$$
$$d_2(n) = \alpha_{22}y_2(n - 1), \qquad m_2(n) = \alpha_{12}y_2(n - 1)$$

where α_{11}, α_{12}, α_{21}, and α_{22} are constants. Because we consider only two countries, the exports of one must be the imports of the other.

a. Show that national income $\mathbf{y}(n) = \begin{bmatrix} y_1(n) \\ y_2(n) \end{bmatrix}$ can be written as

$$\mathbf{y}(n) = \mathbf{A}\mathbf{y}(n - 1) + \mathbf{I}(n)$$

where

$$\mathbf{A} = \begin{bmatrix} \alpha_{11} & \alpha_{12} \\ \alpha_{21} & \alpha_{22} \end{bmatrix} \quad \text{and} \quad \mathbf{I}(n) = \begin{bmatrix} i_1(n) \\ i_2(n) \end{bmatrix}$$

b. What conditions must be true so that the national income $\mathbf{y}(n)$ converges to a constant vector independent of the initial vector $\mathbf{y}(0)$? What is the physical interpretation for this condition?

18. Write the system of difference equations

$$x(n + 1) = x(n) + 2y(n)$$
$$y(n + 1) = 3x(n) + 2y(n) \qquad n = 0, 1, 2, \ldots$$

in matrix form and find the general solution for $x(0) = 1$ and $y(0) = 0$.

$$\text{Ans.} \quad \begin{cases} x(n) = \frac{1}{5}[2 \cdot 4^n + 3 \cdot (-1)^n] \\ y(n) = \frac{3}{5}[4^n - (-1)^n] \end{cases}$$

*19. Consider a model called a Markov chain that can be used in many applications. The system under consideration has two states, say 1 and 2, and at discrete instants of time, the system can pass from any state to any other state with the transition probabilities

$$\Pr(1 \to 1) = 1 - \alpha, \qquad \Pr(1 \to 2) = \alpha$$
$$\Pr(2 \to 1) = \beta, \qquad \Pr(2 \to 2) = 1 - \beta$$

where $0 < \alpha, \beta < 1$ and $\Pr(i \to j)$ denotes the probability of going from state i to state j at any time n. Let $p(k)$ and $q(k)$ be the probabilities of the system being in state 1 or 2, respectively, at time k. Derive a system of two first-order difference equations for $p(n)$ and $q(n)$. Given $p(0)$ and $q(0)$, indicate how one would determine $p(n)$ and $q(n)$ using the techniques of this chapter.

20. Find the inverse system for the system shown below. The inverse system is defined as that system whose response to the output of the given system is the input to the given system.

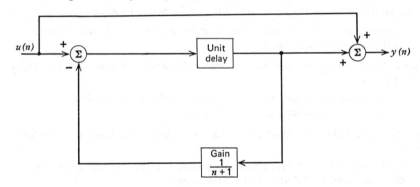

21. In Sections 3.2–3.6, we saw how to write and solve the state variable equations corresponding to the difference equation

$$y(k) + b_1 y(k-1) + \cdots + b_n y(k-n) = a_0 u(k)$$

How would you approach the problem of solving the equation

$$y(k) + b_1 y(k-1) + \cdots + b_n y(k-n)$$
$$= a_0 u(k) + a_1 u(k-1) + \cdots + a_m u(k-m)$$

using state variable methods?

22. In Sections 3.7–3.8, we discussed the problem of writing and solving the state variable equations corresponding to the differential equation

$$b_n \frac{d^n y(t)}{dt^n} + \cdots + b_1 \frac{dy(t)}{dt} + b_0 y(t) = u(t)$$

How would you approach the problem of solving the equation

$$b_n \frac{d^n y(t)}{dt^n} + \cdots + b_1 \frac{dy(t)}{dt} + b_0 y(t)$$

$$= a_0 u(t) + a_1 \frac{du(t)}{dt} + \cdots + a_m \frac{d^m u(t)}{dt^m}$$

using state variable methods?

SUGGESTED READINGS

Director, S. W., and R. A. Rohrer, *Introduction to System Theory*, McGraw-Hill, New York, 1972.

This text presents state variable methods for both continuous-time and discrete-time systems at a level comparable to the present text. Their discussion is much more extensive than this text.

Freeman, H., *Discrete-Time Systems*, Wiley, New York, 1965.

A good discussion of state variable methods for discrete-time systems at the senior-graduate level.

Gupta, S. C., *Transform and State Variable Methods in Linear Systems*, Wiley, New York, 1966.

A complete discussion of state variable methods for continuous-time systems at a level comparable with this text.

Zadeh, L. A., and C. A. Desoer, *Linear System Theory*, McGraw-Hill, New York, 1963.

A graduate-level text which contains a complete and rigorous discussion of linear systems, including state variable methods.

PART TWO

Transform-Domain Techniques

4

The Z-Transform

4.1 INTRODUCTION

In the preceding chapters, we have investigated various methods of formulating models for linear systems and of analyzing their behavior. In Chapter 1, we described systems by means of linear difference or linear differential equations. We reviewed classical solution methods for obtaining the output signal for a given system input. In Chapter 2, we used the impulse-response function, $h(t)$ or $\{h(n)\}$, to characterize linear systems. For linear time-invariant systems, we showed that the output could always be obtained by convolving the impulse-response function with the input signal. In Chapter 3, we considered yet another view of linear systems. The so-called state variable formulation takes the nth order equation of Chapter 1 and represents this equation as n first-order equations. The system output and system state are obtained by solving matrix equations. The system matrices **A**, **B**, **C**, and **D** furnish information about the system that is not easily obtained in other formulations. These three approaches to the analysis of linear systems have one common factor. They are all *time-domain* descriptions of a linear system. The independent parameter is either a continuous variable t or a discrete variable k.

We turn now to a transform-domain description of linear systems. We shall find that if we transform the time signals within our system to another form, we can often express relationships more simply than we have previously. In the following chapters, we develop a transform calculus for both continuous-time and discrete-time systems. One important consequence of the transform-domain description of linear systems is that the convolution operation in the time domain is converted to a multiplication operation in the transform domain. The procedure is analogous to replacing the multiplication of two numbers by the addition of their logarithms. The situation

169

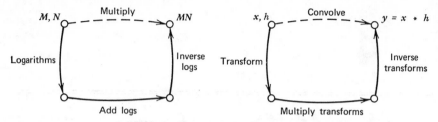

FIGURE 4.1

is shown schematically in Figure 4.1. The familiar Laplace or Fourier transforms are used in the continuous-time domain. The Z-transform is the appropriate transformation for discrete-time systems.

Transformation methods are of most value in studying linear time-invariant systems. The transform calculus simplifies the study of these systems by

(1) providing intuition that is not evident in the time domain solution;
(2) including initial conditions in the solution process automatically;
(3) reducing the solution process of many problems to a simple table look-up, much as one does for logarithms.

However, one disadvantage of transform-domain solutions, at least pedagogically, seems to be that we often forget the physical nature of the problem in the transform domain. The solution process becomes pretty much a "turn the crank" process. This tendency is unfortunate, because we then often do not use physical attributes of the problem that are useful.

An important aim of this study is the development of an awareness in the reader of what is involved, in a general sense, in applying the transform method. For this reason, we first examine the Z-transform, about which the reader most likely has fewer preconceived ideas. This discrete-time transformation calculus has been known in probability theory as the "moment-generating function" method. Although we refer to the discrete-time domain, the index does not have to be interpreted as time. In physical applications, other interpretations may be appropriate. Most of the sequences considered here are zero for the index less than zero. However, the development presented is not restricted to one-sided sequences. We have chosen to define the Z-transform in positive powers of z (rather than negative powers) even though this convention does not conform to many engineering texts on the subject. This definition is consistent with work in mathematics and probability theory and is generally easier to use. The conversion between the two definitions involves a simple substitution of z for z^{-1}.

4.2 THE Z-TRANSFORM

Given a finite sequence $\{x_k\}$, we define a function $X(z)$ of the complex variable z by forming a polynomial

$$X(z) = x_l z^l + x_{l+1} z^{l+1} + \cdots + x_m z^m = \sum_{k=l}^{m} x_k z^k \tag{4.1}$$

The function $X(z)$ is called the *generating function* or *Z-transform* of the sequence $\{x_k\}$. (We should distinguish between the function $X(\cdot)$ and the value the function assumes for a given z. However, in agreement with standard notation, we shall follow the accepted practice of using $X(z)$ to mean either.) In (4.1), the beginning and ending indices of the sequence, l and m, can be any integers including $-\infty$ and ∞. The next examples illustrate the calculations used in obtaining the Z-transform for some simple sequences.

EXAMPLE 4.1. Suppose the sequence $\{x_k\}$ is the finite sequence $\{x_k\} = \{8 \ 3 \ -2 \ 0 \ 4 \ -6\}$. The Z-transform using (4.1) is

$$X(z) = \sum_{k=-2}^{3} x_k z^k = 8z^{-2} + 3z^{-1} - 2 + 4z^2 - 6z^3$$

EXAMPLE 4.2. Consider the Z-transform of the sequence $\{x_k\}$ defined by

$$x_k = \begin{cases} 0, & k < 0 \\ 2^k, & k \geq 0 \end{cases}$$

Again using the definition (4.1), we see that

$$X(z) = \sum_{k=-\infty}^{\infty} x_k z^k = \sum_{k=0}^{\infty} 2^k z^k = \sum_{k=0}^{\infty} (2z)^k = \frac{1}{1 - 2z}$$

EXAMPLE 4.3. Consider the Z-transform for the sequence $\{x_k\}$ defined by

$$x_k = \begin{cases} 2^k, & k < 0 \\ 0, & k \geq 0 \end{cases}$$

In this case, the Z-transform is given by

$$X(z) = \sum_{k=-\infty}^{-1} x_k z^k = \sum_{k=-\infty}^{-1} (2z)^k = \sum_{k=1}^{\infty} (2z)^{-k}$$

$$= \frac{\dfrac{1}{2z}}{1 - \dfrac{1}{2z}} = \frac{1}{2z - 1}$$

where we have used the results of Appendix B to evaluate the geometric series.

Given $X(z)$, we can expand $X(z)$ in a power series in z (by any of several methods) to recover the sequence $\{x_k\}$. However, in cases where $\{x_k\}$ is nonzero for both positive and negative indices, it is imperative to realize that $X(z)$ itself cannot uniquely determine the sequence $\{x_k\}$. The reason is that we can expand $X(z)$ in a power series in more than one way if we have the option of considering sequences $\{x_k\}$ that can be nonzero for either positive or negative k. For example, consider the sequences $\{x_k\}$ and $\{y_k\}$ defined in (4.2) and (4.3).

$$x_k = \begin{cases} -a^k, & k < 0 \\ 0, & k \geq 0 \end{cases} \qquad (4.2)$$

$$y_k = \begin{cases} 0, & k < 0 \\ a^k, & k \geq 0 \end{cases} \qquad (4.3)$$

The Z-transform of $\{x_k\}$ is

$$X(z) = Z\{\{-a^k\}\} = \sum_{k=-\infty}^{-1} -a^k z^k = -\sum_{k=1}^{\infty} (az)^{-k}$$

$$= \frac{-(az)^{-1}}{1 - (az)^{-1}}$$

$$= \frac{1}{1 - az} \qquad (4.4)$$

The Z-transform of $\{y_k\}$ is

$$Y(z) = Z\{\{a^k\}\} = \sum_{k=0}^{\infty} (az)^k = \frac{1}{1 - az} \qquad (4.5)$$

Comparing (4.4) and (4.5), we see that the sequences of (4.2) and (4.3) have the same Z-transforms. This result is disconcerting at first because it implies that Z-transforms and the corresponding sequences are not unique. We can correct this situation by specifying the region of convergence for which the sums in (4.4) and (4.5) converge absolutely. We shall see that a given $X(z)$ along with its radius of convergence uniquely specifies the sequence $\{x_k\}$ which generated the polynomial $X(z)$.

The region of convergence of $X(z)$ is the set of complex valued z for which the sum $\sum_k |x_k z^k|$ exists: i.e., the set of z for which $X(z)$ has finite value. For example, in (4.4) $X(z)$ converges absolutely for $|1/az| < 1$ or $|z| > 1/a$. Similarly, in (4.5) $X(z)$ converges absolutely for $|az| < 1$ or $|z| < 1/a$. In Section 4.3, we show that knowledge of $X(z)$ and its region of convergence uniquely determines the corresponding sequence $\{x_k\}$. If we were concerned only with one-sided sequences, sequences nonzero for either positive or negative indices only, then the region of convergence is not needed to specify $\{x_k\}$ uniquely.

Certain authors define the Z-transform of a sequence $\{x_k\}$ as

$$\tilde{X}(z) = \sum_{k=-\infty}^{\infty} x_k z^{-k} \tag{4.6}$$

We can obtain the Z-transform $\tilde{X}(z)$ of (4.6) from $X(z)$ as defined by (4.1) by substituting z for z^{-1} and vice versa. The corresponding region of convergence is obtained by again substituting z for z^{-1} in the expression defining the region of convergence. For example, if $X(z) = a/(a - z)$, $|z| < a$, then $\tilde{X}(z) = a/(a - 1/z)$, $|z| > 1/a$.

4.3 CONVERGENCE OF THE Z-TRANSFORM

Consider the Z-transform of a sequence $\{x_k\}$, $-\infty \leq k \leq +\infty$.

$$X(z) = \sum_{k=-\infty}^{\infty} x_k z^k$$

We ask, for what region of the complex z plane does this sum converge absolutely?

Given the sequence $\{x_k\} = \{\ldots, x_{-2}, x_{-1}, x_0, x_1, \ldots\}$, assume[1] that there exist four positive numbers M_1, M_2, R_1, R_2 such that

$$\begin{aligned}|x_k| &\leq M_1 R_1^k, \quad k < 0 \\ |x_k| &\leq M_2 R_2^k, \quad k \geq 0\end{aligned} \tag{4.7}$$

Denote the upper bound of R_1 by R_- and the lower bound of R_2 by R_+. With $z = re^{j\theta}$, we have

$$\begin{aligned}\sum_{k=-\infty}^{\infty} |x_k z^k| &= \sum_{k=-\infty}^{\infty} |x_k (re^{j\theta})^k| \\ &= \sum_{k=-\infty}^{-1} |x_k r^k e^{j\theta k}| + \sum_{k=0}^{\infty} |x_k r^k e^{j\theta k}| \\ &= \sum_{k=1}^{\infty} |x_{-k}| r^{-k} + \sum_{k=0}^{\infty} |x_k| r^k \\ &\leq M_1 \sum_{k=1}^{\infty} (R_- r)^{-k} + M_2 \sum_{k=0}^{\infty} (R_+ r)^k \end{aligned} \tag{4.8}$$

The sums in Equation (4.8) are finite if and only if $R_- r > 1$ and $R_+ r < 1$. That is, the sum converges absolutely for all z in the annulus $1/R_- < |z| < 1/R_+$. Of course, if $R_- < R_+$, the sum does not converge absolutely for any

---

[1] If no such numbers exist, then the sum does not converge absolutely for any z.

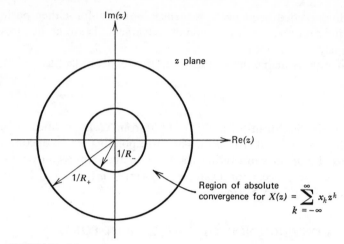

FIGURE 4.2

z. The region of convergence is shown in Figure 4.2. The inner circle bounds the terms in negative powers of z away from the origin. The outer circle bounds the terms in positive powers of z away from large $|z|$ values.

Case 1. Finite Sequences: Consider the sequence $\{x_k\}_{k=a}^{b}$. In this case, $X(z) = \sum_{k=a}^{b} x_k z^k$ is a polynomial in z, and so $X(z)$ converges absolutely for all z except 0 if $a < 0$, and infinity if $b > 0$. For example, suppose $\{x_k\} = \{1, -1, 3, 5, -2\}$ for $k = 0, \ldots, 4$, respectively. Then $X(z) = 1 - z + 3z^2 + 5z^3 - 2z^4$, which exists for all $|z| < \infty$.

Case 2. Semi-infinite Sequences: Consider the sequence $\{x_k\}_{k=a}^{\infty}$. The Z-transform is $X(z) = \sum_{k=a}^{\infty} x_k z^k$. Thus $X(z)$ converges absolutely for all z such that $|z| < 1/R_+$ except 0 if $a < 0$. For example, suppose $\{x_k\} = \{2^k\}_{k=0}^{\infty}$. In this case,

$$X(z) = 1 + 2z + (2z)^2 + (2z)^3 + \cdots = \frac{1}{1 - 2z} \qquad (4.9)$$

which converges absolutely for all $|z| < \frac{1}{2}$. Here $R_+ = 2$.

EXAMPLE 4.4.   Suppose $\{x_k\} = \{2^k\}$, $k = 1, 3, 5, \ldots$
$$= \{5^k\},\ k = 0, 2, 4, \ldots \qquad (4.10)$$

Then the Z-transform of $\{x_k\}$ is

$$X(z) = \sum_{k=0}^{\infty} x_k z^k = \sum_{k=0}^{\infty} x_{2k} z^{2k} + \sum_{j=0}^{\infty} x_{2j+1} z^{2j+1}$$

$$= \sum_{k=0}^{\infty} (5z)^{2k} + \sum_{j=0}^{\infty} (2z)^{2j+1}$$

$$= \frac{1}{1 - 25z^2} + \frac{2z}{1 - 4z^2} \qquad (4.11)$$

Because the sequence $\{x_k\}$ is bounded above by the sequence R_1^k for any $R_1 > 5$, the lower limit of all such R_1 is $R_+ = 5$, and hence $X(z)$ converges absolutely for all $|z| < \frac{1}{5}$. In other words, the first sum in (4.11) converges absolutely for $|z| < \frac{1}{5}$ and the second sum for $|z| < \frac{1}{2}$. Thus $X(z)$ converges absolutely for $|z| < \frac{1}{5}$.

EXAMPLE 4.5. Consider the sequence $\{x_k\}_{k=-\infty}^{b}$. The Z-transform of $\{x_k\}$ is $X(z) = \sum_{k=-\infty}^{b} x_k z^k$. Thus $X(z)$ converges absolutely for all z such that $|z| > 1/R_-$ except for $z = \infty$ in the case $b > 0$. Suppose, for example, that $\{x_k\} = \{-3^k\}_{k=-\infty}^{-1}$. Then

$$X(z) = \cdots + \frac{-1}{(3z)^3} + \frac{-1}{(3z)^2} + \frac{-1}{3z} = \frac{1}{1 - 3z} \qquad (4.12)$$

Thus $R_- = 3$ and $X(z)$ converges absolutely for all $|z| > \frac{1}{3}$.

Case 3. Infinite Sequences: Consider the sequence $\{x_k\}$ nonzero for all k. In this case, the Z-transform of $\{x_k\}$ contains both negative and positive powers of z. This means that $X(z)$ converges in some annulus in the z-plane. For example, suppose $\{x_k\}$ is the sequence

$$x_k = \begin{cases} -3^k, & k < 0 \\ 2^k, & k \geq 0 \end{cases} \qquad (4.13)$$

The Z-transform of (4.13) is

$$X(z) = \frac{1}{1 - 3z} + \frac{1}{1 - 2z} \qquad (4.14)$$

Thus $X(z)$ converges absolutely for all z in the annulus $\frac{1}{3} < |z| < \frac{1}{2}$.

Suppose in (4.13) that x_k has the same functional form for both positive and negative indices. That is,

$$x_k = 2^k, \qquad -\infty < k < \infty$$

In this case $X(z)$ is given by

$$X(z) = \frac{1}{2z - 1} + \frac{1}{1 - 2z}$$

The region of absolute convergence is $\frac{1}{2} < |z| < \frac{1}{2}$: i.e., $X(z)$ does not converge absolutely in any region of the z plane. For two-sided sequences, x_k must have different functional forms for positive and negative indices in order that $X(z)$ converge absolutely in the z plane.

4.4 PROPERTIES OF THE Z-TRANSFORM

There are many methods of finding closed-form expressions for the Z-transform of a given sequence. In this section, we develop certain useful properties of the Z-transform that can be used both in determining $X(z)$ from $\{x_k\}$ and in the inverse problem of finding $\{x_k\}$ given $X(z)$.

The simplest method of finding $X(z)$ is the obvious one, by direct summation. The use of the formula for the sum of a geometric series is often the essential step in direct summation. Recall that for a geometric series we have

$$a + ar + ar^2 + \cdots + ar^k + \cdots = \frac{a}{1 - r} \tag{4.15}$$

The sum in (4.15) converges absolutely for $|r| < 1$. Thus, for example, if $\{x_k\} = \{2^k\}$, $k = 0, 1, 2, \ldots$, the transform $X(z)$ is

$$X(z) = \sum_{k=0}^{\infty} (2z)^k = 1 + 2z + (2z)^2 + \cdots = \frac{1}{1 - 2z} \tag{4.16}$$

which converges absolutely for $|z| < \frac{1}{2}$.

We can use (4.15) to generate a whole family of Z-transforms by differentiating both sides of (4.15). Thus, we see that

$$a + 2ar + 3ar^2 + 4ar^3 + \cdots = \frac{a}{(1 - r)^2} \tag{4.17}$$

which implies that the sequence $\{a, 2a, 3a, 4a, \ldots\}$ has Z-transform

$$\{a, 2a, 3a, 4a, \ldots\} \leftrightarrow \frac{a}{(1 - z)^2} \tag{4.18}$$

The notation "$\leftrightarrow$" is used as a shorthand for "transformation of."

Linearity

The Z-transform is a *linear* operation: i.e.,

$$\begin{aligned}
Z\{a\{x_k\} + b\{w_k\}\} &= \sum_{k=-\infty}^{\infty} (ax_k + bw_k)z^k \\
&= a \sum_{k=-\infty}^{\infty} x_k z^k + b \sum_{k=-\infty}^{\infty} w_k z^k \\
&= aZ\{\{x_k\}\} + bZ\{\{w_k\}\} \\
&= aX(z) + bW(z)
\end{aligned} \tag{4.19}$$

Shifting

a. Let $\{x_k\} \leftrightarrow X(z)$. Consider the Z-transform of $\{x_{k+1}\}$. By definition,

$$Z\{\{x_{k+1}\}\} = \sum_{k=-\infty}^{\infty} x_{k+1} z^k = z^{-1} \sum_{k=-\infty}^{\infty} x_{k+1} z^{k+1} = z^{-1}X(z) \tag{4.20}$$

If we are considering only one-sided sequences, with nonzero values for $k \geq 0$, we must write the shifted sequence as $\{x_{k+1}\} \cdot \{u_k\}$, where $\{u_k\}$ is the unit step sequence

$$u_k = \begin{cases} 1, & k \geq 0 \\ 0, & k < 0 \end{cases} \tag{4.21,}$$

The Z-transform of the one-sided shifted sequence is thus

$$Z\{\{x_{k+1}\}\{u_k\}\} = \sum_{k=0}^{\infty} x_{k+1} z^k = z^{-1} \sum_{k=0}^{\infty} x_{k+1} z^{k+1}$$

$$= z^{-1} \sum_{m=1}^{\infty} x_m z^m = z^{-1}[X(z) - x_0] \tag{4.22}$$

b. Consider the Z-transform of $\{x_{k-1}\}$. By definition

$$Z\{\{x_{k-1}\}\} = \sum_{k=0}^{\infty} x_{k-1} z^k = z \sum_{k=0}^{\infty} x_{k-1} z^{k-1} = z \sum_{m=-1}^{\infty} x_m z^m$$

$$= z\left(\sum_{m=0}^{\infty} x_m z^m + x_{-1} z^{-1} \right) = z(X(z) + X_{-1} z^{-1}) \tag{4.23}$$

c. The results for arbitrary integer delay and advance are easily obtained as indicated above. The results are:

$$\{x_{k \pm a}\} \leftrightarrow z^{\mp a} X(z), \text{ two-sided sequences} \tag{4.24}$$

$$\{x_{k+a}\} \leftrightarrow z^{-a}[X(z) - x_0 - z x_1 - \cdots - z^{a-1} x_{a-1}] \tag{4.25}$$

$$\{x_{k-a}\} \leftrightarrow z^a[X(z) + x_{-1} z^{-1} + \cdots + x_{-a} z^{-a}] \tag{4.26}$$

where a is a positive integer.

The shifting property is very useful in solving linear difference equations. Compare the following approach with the methods used in Chapter 1.

EXAMPLE 4.6. Consider a system described by the difference equation

$$f_{n+1} + 2f_n = u_n, \qquad f_0 = 1, \qquad n = 0, 1, 2, \ldots \tag{4.27}$$

where, again, $\{u_n\}$ is the unit step sequence. Let $F(z) \leftrightarrow \{f_n\}$. Taking Z-transforms of both sides of (4.27), we have

$$z^{-1}[F(z) - f_0] + 2F(z) = \frac{1}{1 - z}$$

Because $f_0 = 1$, we can solve for $F(z)$ as

$$F(z) = \frac{z}{(1 - z)(1 + 2z)} + \frac{1}{1 + 2z} \tag{4.28}$$

Expanding the first term on the right-hand side of (4.28) in partial fractions, we have

$$\frac{z}{(1 - z)(1 + 2z)} = \frac{A}{1 - z} + \frac{B}{1 + 2z} \qquad (4.29)$$

To evaluate the constant A, multiply both sides of (4.29) by $(1 - z)$ and let $z = 1$. Thus

$$A = \frac{(1 - z)z}{(1 - z)(1 + 2z)}\Bigg|_{z=1} = \tfrac{1}{3}$$

In the same way, B is calculated as

$$B = \frac{(1 + 2z)z}{(1 - z)(1 + 2z)}\Bigg|_{z=-\frac{1}{2}} = -\tfrac{1}{3}$$

Thus

$$F(z) = \frac{\tfrac{1}{3}}{1 - z} + \frac{-\tfrac{1}{3}}{1 + 2z} + \frac{1}{1 + 2z} = \frac{\tfrac{1}{3}}{1 - z} + \frac{\tfrac{2}{3}}{1 + 2z}$$

Recall that $1/(1 - az) \leftrightarrow \{a^n\}_{n=0}^{\infty}$. Thus, the inverse transform of $F(z)$ is

$$f_n = \tfrac{1}{3} + \tfrac{2}{3}(-2)^n, \qquad n \geq 0 \qquad (4.30)$$

In this example, we did not have to concern ourselves with the convergence region of $F(z)$ because the sequence f_n is nonzero for $n \geq 0$ only. If the forcing function in (4.27) is nonzero for both positive and negative n, then the convergence region of $F(z)$ is needed to find f_n.

EXAMPLE 4.7. The Z-transform method in many problems leads to an easy method of determining a system's impulse response. Recall that the impulse response is the output of a relaxed system resulting from an impulse input. Thus, for the system of (4.27), we have

$$h_{n+1} + 2h_n = \delta_n, \qquad h_0 = 0$$

Taking Z-transforms, we obtain

$$z^{-1}H(z) + 2H(z) = 1$$

Solving for $H(z)$ yields

$$H(z) = \frac{z}{1 + 2z}$$

Recall that $zF(z) \leftrightarrow \{f_{n-1}\}$ and $1/(1 + az) \leftrightarrow \{-a^n\}\{u_n\}$. Thus the impulse-response function is

$$h_n = (-2)^{n-1}u_{n-1} \qquad (4.31)$$

EXAMPLE 4.8.   Consider a system described by the second-order difference equation

$$2f_{n+2} - f_{n+1} - f_n = 0, \qquad n \geq 0 \qquad (4.32)$$

with initial conditions $f_0 = 2$ and $f_1 = 1$. Taking Z-transforms on both sides of (4.32), we have

$$2z^{-2}[F(z) - f_0 - zf_1] - z^{-1}[F(z) - f_0] - F(z) = 0$$

Substituting for the initial conditions and solving for $F(z)$ yields

$$F(z) = \frac{-4}{z^2 + z - 2}$$

We can break $F(z)$ into partial fractions as in Example (4.6) to obtain

$$F(z) = \frac{-\frac{4}{3}}{z - 1} + \frac{\frac{4}{3}}{2 + z} = \frac{\frac{4}{3}}{1 - z} + \frac{\frac{2}{3}}{1 + z/2}$$

Using the transform pair $1/(1 - az) \leftrightarrow \{a^n\}\{u_n\}$, we obtain

$$f_n = \tfrac{4}{3} + \tfrac{2}{3}(-\tfrac{1}{2})^n, \qquad n \geq 0 \qquad (4.33)$$

Multiplication by k

Let $\{x_k\} \leftrightarrow X(z)$. Consider the transform of $\{kx_k\}$.

$$Z\{\{kx_k\}\} = \sum_{k=-\infty}^{\infty} kx_k z^k = z \sum_{k=-\infty}^{\infty} x_k(kz^{k-1})$$

$$= z \sum_{k=-\infty}^{\infty} x_k \frac{d}{dz} z^k = z \frac{d}{dz} \sum_{k=-\infty}^{\infty} x_k z^k$$

$$= z \frac{d}{dz} X(z) \qquad (4.34)$$

The interchange of summation and differentiation in (4.34) is valid because one can always differentiate and integrate a power series term-by-term to obtain the derivative and integral of the series within its radius of convergence. The derivative and integral of a power series are other power series with the same convergence region.

One can generalize the above result to multiplication by any positive power of k. Thus, by a similar argument, we obtain the transform pair[2]

$$\{k^n x_k\} \leftrightarrow \left[z \frac{d}{dz}\right]^n X(z) \qquad (4.35)$$

The above transform pair is valid for both one-sided and two-sided sequences.

[2] The notation $\left[z\dfrac{d}{dz}\right]^n$ means $z\dfrac{d}{dz}\left(\cdots\left(z\dfrac{d}{dz}\left(z\dfrac{d}{dz}\left(\right)\right)\right)\cdots\right)$.

EXAMPLE 4.9.   The sequence $\{1, a, a^2, \ldots\}$ has the Z-transform

$$X_1(z) = \sum_{k=0}^{\infty} a^k z^k = \frac{1}{1 - az}, \qquad |z| < \frac{1}{|a|} \qquad (4.36)$$

Thus the sequence $\{0, a, 2a^2, 3a^3, \ldots\}$ has the Z-transform

$$X_2(z) = z \frac{d}{dz} [(1 - az)^{-1}] = \frac{az}{(1 - az)^2}, \qquad |z| < \frac{1}{|a|} \qquad (4.37)$$

Division by $k + a$

We shall find it useful also to have a transform pair for $\{x_k/(k + a)\}$, where a is an integer. Let $X(z)$ be the Z-transform of $\{x_k\}$. Then, by definition,

$$Z\left\{\left\{\frac{x_k}{k + a}\right\}\right\} = \sum_{k=-\infty}^{\infty} \frac{x_k}{k + a} z^k = \sum_{k=-\infty}^{\infty} x_k z^{-a} \int z^{k+a-1} \, dz$$

$$= z^{-a} \int \sum_{k=-\infty}^{\infty} x_k z^k z^{a-1} \, dz = z^{-a} \int z^{a-1} X(z) \, dz \qquad (4.38)$$

Thus, we obtain the transform pair

$$\left\{\frac{x_k}{k + a}\right\} \leftrightarrow z^{-a} \int z^{a-1} X(z) \, dz \qquad (4.39)$$

Scale change

Let $\{x_k\} \leftrightarrow X(z)$. Consider the inverse Z-transform of $X(z/a)$. We have

$$Z^{-1}\left\{X\left(\frac{z}{a}\right)\right\} = Z^{-1}\left\{\sum_{k=-\infty}^{\infty} x_k \left(\frac{z}{a}\right)^k\right\}$$

$$= Z^{-1}\left\{\sum_{k=-\infty}^{\infty} a^{-k} x_k z^k\right\} = \{a^{-k} x_k\} \qquad (4.40)$$

Thus, we have the transform pair

$$\{a^{-k} x_k\} \leftrightarrow X(z/a) \qquad (4.41)$$

This pair is valid for both one-sided and two-sided sequences.

Initial and final values (sequences nonzero for $n \geq 0$)

a. We wish to develop a formula for finding the initial value of the sequence $\{f_n\}_{n=0}^{\infty}$ from $F(z)$. By definition,

$$F(z) = \sum_{n=0}^{\infty} f_n z^n = f_0 + z f_1 + z^2 f_2 + \cdots \qquad (4.42)$$

Taking the limit as z approaches zero, we obtain

$$f_0 = \lim_{z \to 0} F(z) \tag{4.43}$$

b. Perhaps a more useful formula involves the determination of the final value of $\{f_n\}$ from $F(z)$. Consider the Z-transform of $\{f_{n+1} - f_n\}$. Using the definition, we have

$$Z\{\{f_{n+1} - f_n\}\} = \frac{F(z) - f_0}{z} - F(z) = \sum_{n=0}^{\infty} (f_{n+1} - f_n) z^n$$

Thus

$$F(z)\left(\frac{1-z}{z}\right) = \frac{f_0}{z} + \lim_{N \to \infty} \sum_{n=0}^{N} (f_{n+1} - f_n) z^n \tag{4.44}$$

Taking the limit of (4.44) as z approaches unity, we obtain

$$\lim_{z \to 1} F(z)\left(\frac{1-z}{z}\right) = f_0 + \lim_{N \to \infty} f_N - f_0$$

Thus

$$\lim_{N \to \infty} f_N = \lim_{z \to 1} \left(\frac{1-z}{z}\right) F(z) \tag{4.45}$$

Summation

The summation of sequences corresponds to integration in the continuous time domain. Consider the Z-transform of $\sum_{k=0}^{n} f_k$. Define g_n as the partial sum

$$g_n = \sum_{k=0}^{n} f_k \tag{4.46}$$

Then we can write g_n as

$$g_n = g_{n-1} u_{n-1} + f_n \tag{4.47}$$

where $\{u_n\}$ is the unit step sequence. Taking Z-transforms on both sides of (4.47) with respect to the index n,

$$G(z) = z G(z) + F(z)$$

Solving for $G(z)$, we have

$$G(z) = \frac{F(z)}{1 - z}$$

Thus, we obtain the transform pair

$$\left\{\sum_{k=0}^{n} f_k\right\} \leftrightarrow \frac{F(z)}{1-z} \tag{4.48}$$

We can use the summation and final value formulas to find closed-form expressions for infinite series. In this case, we are interested in

$$\lim_{n \to \infty} g_n = \lim_{z \to 1} \frac{1-z}{z} G(z)$$

$$= \lim_{z \to 1} \frac{1-z}{z} \cdot \frac{F(z)}{1-z} = F(1) \tag{4.49}$$

that is,

$$\sum_{k=0}^{\infty} f_k = F(1) \tag{4.50}$$

EXAMPLE 4.10. Suppose we wish to determine a closed-form expression for the infinite series $\sum_{n=0}^{\infty} (-1)^n x^{n+1}/(n+1)$. The solution process is to find the transform of the summand, in this case $(-1)^n x^{m+1}/(n+1)$, and then use (4.50) to find an expression for the sum.

The Z-transform of $\{x^n\}$ is, by direct summation,

$$Z\{\{x^n\}\} = \sum_{n=0}^{\infty} x^n = \frac{1}{1-xz} \tag{4.51}$$

Thus

$$Z\{\{x^{n+1}\}\} = xZ\{\{x^n\}\} = \frac{x}{1-xz} \tag{4.52}$$

(because x is merely some constant). From (4.39), the transform of $\{f_n/(n+1)\}$ is $z^{-1}\int F(z)\, dz$. Thus,

$$Z\left\{\left\{\frac{x^{n+1}}{n+1}\right\}\right\} = Z\left\{x\left\{\frac{x^n}{n+1}\right\}\right\} = z^{-1} \int \frac{x}{1-xz}\, dz$$

$$= -z^{-1} \log_e (1 - xz)$$

Using this result, we see that

$$\left\{\frac{(-1)^n x^{n+1}}{n+1}\right\} \leftrightarrow \frac{1}{z} \log_e (1 + xz) \tag{4.53}$$

because $\{a^n f_n\} \leftrightarrow F(az)$. Using (4.50) in (4.53), we obtain

$$\sum_{n=0}^{\infty} \frac{(-1)^n x^{n+1}}{n+1} = \frac{1}{z} \log_e (1 + xz)\bigg|_{z=1} = \log_e (1 + x) \tag{4.54}$$

Convolution

One of the most important properties of Z-transforms in applications involves the convolution property. Let $\{x_k\}_{k=0}^{\infty} \leftrightarrow X(z)$ and $\{h_k\}_{k=0}^{\infty} \leftrightarrow H(z)$. Then the convolution of $\{x_k\}$ and $\{h_k\}$ is $\{y_k\}_{k=0}^{\infty}$, where

$$y_k = \sum_{m=0}^{k} x_m h_{k-m} \tag{4.55}$$

The Z-transform of y_k is $Y(z)$ and is equal to

$$Y(z) = X(z)H(z) \tag{4.56}$$

A simple way to show (4.56) is by induction. Consider the product $X(z)H(z)$.

$$X(z)H(z) = [x_0 + x_1 z + x_2 z^2 + \cdots][h_0 + h_1 z + h_2 z^2 + \cdots]$$

$$= x_0 h_0 + (x_0 h_1 + x_1 h_0)z + (x_0 h_2 + x_1 h_1 + x_2 h_0)z^2 + \cdots$$

$$= Y(z) = y_0 + y_1 z + y_2 z^2 + \cdots \tag{4.57}$$

By equating like coefficients of z, we obtain

$$y_0 = x_0 h_0$$
$$y_1 = x_0 h_1 + x_1 h_0$$
$$y_2 = x_0 h_2 + x_1 h_1 + x_2 h_0$$
$$\cdot$$
$$\cdot$$
$$\cdot$$
$$y_k = x_0 h_k + x_1 h_{k-1} + \cdots + x_{k-1} h_1 + x_k h_0$$
$$= \sum_{m=0}^{k} x_m h_{k-m} \tag{4.58}$$

In the case of general sequences which are nonzero for both positive and negative k, we have

$$y_k = \sum_{m=-\infty}^{\infty} x_m h_{k-m} \tag{4.59}$$

Taking the Z-transform of (4.59), we obtain

$$Y(z) = \sum_{k=-\infty}^{\infty} \sum_{m=-\infty}^{\infty} x_m h_{k-m} z^k \tag{4.60}$$

Any power series which converges absolutely also converges uniformly within its radius of convergence. Thus, we can interchange the summation

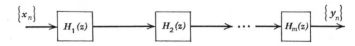

FIGURE 4.3

signs in (4.60) to give

$$Y(z) = \sum_{m=-\infty}^{\infty} x_m z^m \sum_{k=-\infty}^{\infty} h_{k-m} z^{k-m}$$

$$= X(z)H(z) \tag{4.61}$$

$Y(z)$ converges in the common region of convergence of $X(z)$ and $H(z)$, provided that such a region exists. $H(z)$ is called the *system transfer function*.

If a system can be broken down into a cascade of noninteracting systems, as shown in Figure 4.3, then the *Z*-transform of the output takes on a particularly simple form. If we have m systems in cascade with transfer functions $H_1(z)$, $H_2(z)$, ... , $H_m(z)$, then the output in the transform domain is

$$Y(z) = X(z)H_1(z)H_2(z) \cdots H_m(z) \tag{4.62}$$

The overall system transfer function is

$$H(z) = \frac{Y(z)}{X(z)} = H_1(z)H_2(z) \cdots H_m(z) \tag{4.63}$$

Other Techniques for Obtaining Transform Pairs

One can obtain transforms for the sequences $\{\sin n\omega\}$ and $\{\cos n\omega\}$ by using the *Z*-transform of $\{e^{j\omega n}\}$. Thus

$$Z\{\{e^{j\omega n}\}\{u_n\}\} = \frac{1}{1 - e^{j\omega} z}$$

$$= \frac{1}{1 - z\cos\omega - jz\sin\omega}$$

$$= \frac{1 - z\cos\omega + jz\sin\omega}{1 - 2z\cos\omega + z^2} \tag{4.64}$$

Equating real and imaginary parts in (4.64), we obtain

$$\{\cos n\omega\}\{u_n\} \leftrightarrow \frac{1 - z\cos\omega}{1 - 2z\cos\omega + z^2} \tag{4.65}$$

$$\{\sin n\omega\}\{u_n\} \leftrightarrow \frac{z\sin\omega}{1 - 2z\cos\omega + z^2} \tag{4.66}$$

The transforms of hyperbolic functions can be obtained by writing these functions in terms of their exponential representations. For example, we can write the $\sinh n\omega$ as $\frac{1}{2}[\exp(n\omega) - \exp(-n\omega)]$. Thus the sequence $\{\sinh n\omega\} \cdot \{u_n\}$ has transform

$$Z\{\{\sinh n\omega\}\{u_n\}\} = \frac{\frac{1}{2}}{1 - ze^\omega} + \frac{-\frac{1}{2}}{1 - ze^{-\omega}}$$

$$= \frac{2\sinh\omega}{1 - 2z\cosh\omega + z^2} \qquad (4.67)$$

The function $\cosh n\omega$ can also be written in terms of exponentials as $\frac{1}{2}[\exp(n\omega) + \exp(-n\omega)]$. Thus we have

$$Z\{\{\cosh n\omega\}\{u_n\}\} = \frac{\frac{1}{2}}{1 - ze^\omega} + \frac{\frac{1}{2}}{1 - ze^{-\omega}}$$

$$= \frac{1 - z\cosh\omega}{1 - 2z\cosh\omega + z^2} \qquad (4.68)$$

Tables 4.1 and 4.2 summarize a number of transform pairs.

TABLE 4.1

Z-Transform Pairs; f_k nonzero for $-\infty < k < \infty$

	$\{f_k\}$	Z-Transform
1	$\{f_k\}$	$F(z)$
2	$\{f_k\} + \{g_k\}$	$F(z) + G(z)$
3	$\{f_{k-n}\}$	$z^n F(z)$
4	$\{f_{k+n}\}$	$z^{-n} F(z)$
5	$\{k^n f_k\}$	$\left(z\dfrac{d}{dz}\right)^n F(z)$
6	$\{a^k f_k\}$	$F(az)$
7	$\{f_k\} * \{g_k\}$	$F(z)G(z)$
8	$\left\{\dfrac{f_k}{k + a}\right\}$	$z^{-a} \int z^{a-1} F(z)\,dz$
9	$\{af_k\}$	$aF(z)$

TABLE 4.2

Z-Transforms for $\{f_k\}$, $k \geq 0$

	$\{f_k\}$	Z-Transform
1	$\{f_k\}$	$F(z)$
2	$a\{f_k\} + b\{g_k\}$	$aF(z) + bG(z)$
3	$\{f_k\} * \{g_k\}$	$F(z)G(z)$
4	$\{f_{k+n}\}$	$z^{-n}\left(F(z) - \sum_{i=0}^{n-1} z^i f_i\right)$
5	$\{f_{k-n}\}$	$z^n F(z)$
6	$\{k^n f_k\}$	$\left(z\dfrac{d}{dz}\right)^n F(z)$
7	$\{f_k/(k+a)\}$	$z^{-a} \displaystyle\int z^{a-1} F(z)\, dz$
8	$\{a^k f_k\}$	$F(az)$
9	$\{\delta_k\}$	1
10	$\{u_k\}$	$(1-z)^{-1}$
11	$\{k\}$	$z(1-z)^{-2}$
12	$\{k^2\}$	$z(1+z)(1-z)^{-3}$
13	$\{k^n\}$	$\left(z\dfrac{d}{dz}\right)^n (1-z)^{-1}$
14	$\left\{\binom{k}{n}\right\}$	$z^n(1-z)^{n+1}$
15	$\left\{\binom{n}{k}\right\}$	$(1+z)^n$
16	$\{\alpha^k\}$	$(1-\alpha z)^{-1}$
17	$\{\alpha^k/k!\}$	$e^{\alpha z}$
18	$\{k^n \alpha^k\}$	$\left(z\dfrac{d}{dz}\right)^n (1-\alpha z)^{-1}$
19	$\{\cos(\alpha k)\}$	$(1 - z\cos\alpha)(1 - 2z\cos\alpha + z^2)^{-1}$
20	$\{\sin(\alpha k)\}$	$(z\sin\alpha)(1 - 2z\cos\alpha + z^2)^{-1}$
21	$\{\cosh(\alpha k)\}$	$(1 - z\cosh\alpha)(1 - 2z\cosh\alpha + z^2)^{-1}$
22	$\{\sinh(\alpha k)\}$	$(z\sinh\alpha)(1 - 2z\cosh\alpha + z^2)^{-1}$

4.5 INVERSION OF THE Z-TRANSFORM

We turn now to the important problem of recovering the sequence $\{x_k\}$ from its Z-transform $X(z)$. As we have discussed previously, because we have not restricted the domain of the index set to be either positive or negative integers, we must have knowledge of the region of convergence of $X(z)$ to determine $\{x_k\}$ uniquely. We consider three methods of inversion: inversion by series expansion, by partial fraction expansion, and by use of an inversion integral. We assume that $X(z)$ is a rational function, as given in (4.69).

$$X(z) = \frac{a_0 + a_1 z + a_2 z^2 + \cdots + a_m z^m}{b_0 + b_1 z + b_2 z^2 + \cdots + b_n z^n} \qquad (4.69)$$

Inversion of X(z) by Direct Division

We can generate a series in z by dividing the numerator of (4.69) by the denominator. For example, if $X(z)$ converges for $|z| < 1/R_+$, we can obtain the series

$$X(z) = \frac{a_0}{b_0} + \left(\frac{a_1}{b_0} + \frac{a_0 b_1}{b_0^2}\right) z + \cdots \qquad (4.70)$$

which converges for the same range of z. Equation (4.70) is obtained by beginning the division with the lowest order of z: i.e., b_0. We then identify the coefficient of z^k as x_k. If, however, $X(z)$ were known to converge for $1/R_- < |z|$, then the series in (4.70) would *diverge* for some z in the region of convergence of $X(z)$, and hence cannot be the correct representation. In this case, we should begin our division with the highest order of z to obtain

$$X(z) = \frac{a_m}{b_n} z^{m-n} + \left(\frac{a_{m-1}}{b_n} + \frac{a_m b_{n-1}}{b_n^2}\right) z^{m-n-1} + \cdots \qquad (4.71)$$

From (4.71), we can identify the nonzero terms of the sequence.

EXAMPLE 4.11.   Let $X(z) = a/(a - z)$ for $|z| < a$

Here we expand in positive powers of z because $X(z)$ converges for $|z|$ less than some radius of convergence. Divide $a - z$ into a to obtain

$$
\begin{array}{r}
1 + \dfrac{z}{a} + \dfrac{z^2}{a^2} + \cdots \\[4pt]
a - z \,\big|\, \overline{a } \\
\underline{a - z} \\
z \\
\underline{z - \dfrac{z^2}{a}} \\
\dfrac{z^2}{a} \cdots
\end{array}
$$

and so

$$X(z) = 1 + \frac{z}{a} + \left(\frac{z^2}{a}\right) + \cdots$$

from which

$$\{x_k\} = \left\{1, \frac{1}{a}, \frac{1}{a^2}, \ldots, \frac{1}{a^k}, \ldots\right\}$$

The series $X(z)$ converges absolutely for $|z| < a$.

EXAMPLE 4.12. Let

$$X(z) = \frac{a}{a - z}, \qquad |z| > a$$

In this case, we expand $X(z)$ in negative powers of z. The division is

$$
\begin{array}{r}
-\dfrac{a}{z} - \dfrac{a^2}{z^2} - \dfrac{a^3}{z^3} - \cdots \\[4pt]
-z + a \overline{\smash{\big)}\, a } \\[4pt]
a - \dfrac{a^2}{z} \\[4pt]
\hline
\dfrac{a^2}{z} \\[4pt]
\dfrac{a^2}{z} - \dfrac{a^3}{z^2} \\[4pt]
\hline
\dfrac{a^3}{z^2} \cdots
\end{array}
$$

Thus

$$X(z) = -\frac{a}{z} - \left(\frac{a}{z}\right)^2 - \left(\frac{a}{z}\right)^3 - \cdots$$

from which

$$\{x_k\} = \{\ldots, -a^{-k}, \ldots, -a^3, -a^2, -a\}$$

EXAMPLE 4.13. Consider inversion of the *Z*-transform

$$X(z) = \frac{5z}{1 + z - 6z^2}, \qquad \tfrac{1}{3} < |z| < \tfrac{1}{2}$$

Because $X(z)$ converges in an annulus, it is implied that the series expansion for $X(z)$ consists of both negative and positive powers of z. Thus, the sequence $\{x_k\}$ is nonzero for both positive and negative k. If we expand $X(z)$ in a power series in positive powers of z, we obtain

$$X(z) = 5z - 5z^2 + 35z^3 - 65z^4 + - \cdots$$

This series converges absolutely for $|z| < \frac{1}{3}$. Because these values are not in the stated region of convergence, this expansion cannot be used to represent $X(z)$. If we expand $X(z)$ in negative powers of z, we obtain

$$X(z) = -\tfrac{5}{6}z^{-1} - \tfrac{5}{36}z^{-2} - \tfrac{35}{216}z^{-3} - \cdots$$

This series converges absolutely for $|z| > \frac{1}{2}$. Again, this series cannot be used to represent $X(z)$ because the series does not converge in the stated region of convergence.

To obtain the correct expansion, we write $X(z)$ as a sum of two functions $X_1(z)$ and $X_2(z)$ so that $X_1(z)$ converges for $|z| < \frac{1}{2}$ and $X_2(z)$ converges for $|z| > \frac{1}{3}$.

$$X(z) = \frac{5z}{1 + z - 6z^2} = X_1(z) + X_2(z) = \frac{-1}{1 + 3z} + \frac{1}{1 - 2z}$$

Expanding $X_1(z)$ in negative powers of z, we obtain

$$X_1(z) = -\tfrac{1}{3}z^{-1} + \tfrac{1}{9}z^{-2} - \tfrac{1}{27}z^{-3} + - \cdots + \frac{(-1)^k}{(3z)^k} + \cdots, \qquad |z| > \tfrac{1}{3}$$

Expanding $X_2(z)$ in positive powers of z, we obtain

$$X_2(z) = 1 + 2z + 4z^2 + \cdots + (2z)^k + \cdots, \qquad |z| < \tfrac{1}{2}$$

Thus the sum $X_1(z) + X_2(z) = X(z)$ converges for $\frac{1}{3} < |z| < \frac{1}{2}$, and the sequence $\{x_k\}$ is

$$\{x_k\} = \begin{cases} \{2^k\}, & k \geq 0 \\ \{(-3)^k\}, & k < 0 \end{cases}$$

It is usually difficult to determine the correct decomposition of $X(z)$ by inspection. The search for a more general procedure for finding $\{x_k\}$ from $X(z)$ leads us to the method of partial fraction expansions.

Inversion of X(z) by Partial Fractions

Let us assume that the degree of the numerator in (4.69) is less than the degree of the denominator: i.e., $m < n$. If this is not the case, then we can write $X(z)$ as the sum of a polynomial $Q(z)$ of degree $m - n$ plus a ratio of polynomials with a degree of the numerator one less than the degree of the denominator. That is

$$X(z) = q_0 + q_1 z + \cdots + q_{m-n} z^{m-n} + \frac{\hat{a}_0 + \hat{a}_1 z + \cdots + \hat{a}_{n-1} z^{n-1}}{b_0 + b_1 z + \cdots + b_n z^n}$$

$$(4.72)$$

EXAMPLE 4.14.   Let

$$X(z) = \frac{33.5 + 44.5z + 29.5z^2 + 4.5z^3 - 3.5z^4 - z^5}{15 + 8.5z - 1.5z^2 - z^3}$$

By long division, beginning with the highest orders of z, we obtain

$$X(z) = 1 + 2z + z^2 + \frac{18.5 + 6z - z^2}{15 + 8.5z - 1.5z^2 - z^3}$$

We now consider the problem of determining the sequence $\{x_k\}$, whose Z-transform is this ratio of polynomials. Let us determine the roots of the denominator polynomial, $p_1, p_2, p_3, \ldots, p_n$. These roots are called the *poles* of $X(z)$. If no root is repeated, then we can write $X(z)$ as

$$\begin{aligned}
X(z) &= \frac{a_0 + a_1 z + a_2 z^2 + \cdots + a_m z^m}{b_0 + b_1 z + b_2 z^2 + \cdots + b_n z^n} \\
&= \frac{a_0 + a_1 z + a_2 z^2 + \cdots + a_m z^m}{b_n (z - p_1)(z - p_2) \cdots (z - p_n)} \\
&= \frac{c_1}{(z - p_1)} + \frac{c_2}{(z - p_2)} + \cdots + \frac{c_n}{(z - p_n)}
\end{aligned} \qquad (4.73)$$

where the coefficient c_i is given by

$$c_i = (z - p_i)X(z)\big|_{z=p_i} \qquad (4.74)$$

In the case of a multiple root, say p_i repeated r times, the expansion (4.73) must include the terms

$$\frac{c_{i1}}{(z - p_i)} + \frac{c_{i2}}{(z - p_i)^2} + \cdots + \frac{c_{ir}}{(z - p_i)^r} \qquad (4.75)$$

where

$$\begin{aligned}
c_{ir} &= (z - p_i)^r X(z)\big|_{z=p_i} \\
c_{ir-1} &= \frac{d}{dz}\{(z - p_i)^r X(z)\big|_{z=p_i}\} \\
&\quad \cdots \\
c_{ir-k} &= \frac{1}{k!}\frac{d^k}{dz^k}\{(z - p_i)^r X(z)\}\big|_{z=p_i} \\
&\quad \cdots \\
c_{i1} &= \frac{1}{(r-1)!}\frac{d^{r-1}}{dz^{r-1}}\{(z - p_i)^r X(z)\}\big|_{z=p_i}
\end{aligned} \qquad (4.76)$$

This procedure for obtaining the partial fraction expansion of a (rational) Z-transform is identical to that commonly used with Laplace transforms. The general form, assuming that the pole p_i is repeated r_i times ($r_i = 1$ for an isolated pole) for $i = 1, \cdots, k$ (with $r_1 + r_2 + \cdots + r_k = n$) is

$$X(z) = \frac{c_1}{z - p_1} + \cdots + \frac{c_{i_1}}{z - p_i} + \frac{c_{i_2}}{(z - p_i)^2} + \cdots$$

$$+ \frac{c_{i_{r_i}}}{(z - p_i)^{r_i}} + \cdots + \frac{c_k}{(z - p_k)^{r_k}} \quad (4.77)$$

Now, knowing the region of convergence of $X(z)$, we can easily find the sequence corresponding to each term in (4.77). The sum of these sequences is the sequence $\{x_k\}$ with transform $X(z)$. Table 4.3 is helpful in finding the individual sequences.

EXAMPLE 4.15.   Find the sequence $\{x_k\}$ for $X(z)$ as given in (4.78) by using a partial fraction expansion of $X(z)$.

$$X(z) = \frac{18.5 + 6z - z^2}{15 + 85z - 1.5z^2 - z^3}, \quad |z| < 2 \quad (4.78)$$

$X(z)$ is given in correct form for a partial fraction expansion. To find the poles of $X(z)$, we find the roots of the denominator by solving

$$15 + 8.5z - 1.5z^2 - z^3 = 0 \quad (4.79)$$

The roots of (4.79) are $p_1 = -2$, $p_2 = -2.5$, and $p_3 = 3$. Thus, we have

$$X(z) = \frac{18.5 + 6z - z^2}{15 + 8.5z - 1.5z^2 - z^3} = \frac{A}{z + 2} + \frac{B}{z + 2.5} + \frac{C}{z - 3}$$

from which we find

$$X(z) = \frac{-1}{z + 2} + \frac{1}{z + 2.5} - \frac{1}{z - 3}, \quad |z| < 2$$

Hence, the sequence $\{x_k\}$, corresponding to $X(z)$, has kth term

$$X_k = \left(\frac{-1}{2}\right)^{k+1} - \left(\frac{-1}{2.5}\right)^{k+1} + \left(\frac{1}{3}\right)^{k+1}, \quad k = 0, 1, 2, \ldots$$

EXAMPLE 4.16.   Let $X(z)$ be the Z-transform

$$X(z) = \frac{5.45 - 11.2z + 7z^2}{.605 + .11z - 1.7z^2 + z^3}, \quad \tfrac{1}{2} < |z| < 1.1$$

TABLE 4.3
Inverse Transforms of the Partial Fractions of $X(z)$

Partial Fraction	If $X(z)$ converges absolutely for $\|z\| < \|\alpha\|$
$\dfrac{1}{z - \alpha}$	$x_k = -\left(\dfrac{1}{\alpha}\right)^{k+1}, \; k = 0, 1, 2, \cdots$
$\dfrac{1}{(z - \alpha)^2}$	$x_k = (k + 1)\left(\dfrac{1}{\alpha}\right)^{k+2}, \; k = 0, 1, 2, \cdots$
$\dfrac{1}{(z - \alpha)^3}$	$x_k = \dfrac{-(k + 1)(k + 2)}{2!}\left(\dfrac{1}{\alpha}\right)^{k+3}, \; k = 0, 1, 2, \cdots$
$\cdots$	$\cdots$
$\dfrac{1}{(z - \alpha)^n}$	$x_k = \dfrac{(-1)^n(k + 1)(k + 2) \cdots (k + n - 1)}{(n - 1)!}\left(\dfrac{1}{\alpha}\right)^{k+n}$ $k = 0, 1, 2, \cdots$ and $n \geq 2$

	If $X(z)$ converges absolutely for $\|z\| > \|\alpha\|$
$\dfrac{1}{z - \alpha}$	$x_k = \left(\dfrac{1}{\alpha}\right)^{k+1}, \; k = -1, -2, -3, \cdots$
$\dfrac{1}{(z - \alpha)^2}$	$x_k = -(k + 1)\left(\dfrac{1}{\alpha}\right)^{k+2}, \; k = -1, -2, -3, \cdots$
$\dfrac{1}{(z - \alpha)^3}$	$x_k = \dfrac{(k + 1)(k + 2)}{2!}\left(\dfrac{1}{\alpha}\right)^{k+3}, \; k = -1, -2, -3, \cdots$
$\cdots$	$\cdots$
$\dfrac{1}{(z - \alpha)^n}$	$x_k = \dfrac{(-1)^{n+1}(k + 1)(k + 2) \cdots (k + n - 1)}{(n - 1)!}\left(\dfrac{1}{\alpha}\right)^{k+n}$ $k = -1, -2, -3, \cdots$ and $n \geq 2$

Factoring the denominator polynomial, we find a double pole a $z = 1.1$ and a single pole at $z = -.5$. Thus,

$$X(z) = \frac{c_1}{z + .5} + \frac{c_2}{(z - 1.1)^2} + \frac{c_3}{z - 1.1}$$

with

$$c_1 = \frac{5.45 - 11.2z + 7z^2}{(z - 1.1)^2}\bigg|_{z=-.5} = 5.0$$

$$c_2 = \frac{5.45 - 11.2z + 7z^2}{(z + .5)}\bigg|_{z=1.1} = 1.0$$

$$c_3 = \frac{d}{dz}\left(\frac{5.45 - 11.2z + 7z^2}{z + .5}\right)\bigg|_{z=1.1} = 2.0$$

Because we know that $X(z)$ converges for some $|z| > .5$ and some $|z| < 1.1$, we find from Table 4.3 that the corresponding $\{x_k\}$ has kth term

$$x_k = \begin{cases} 5(-2)^{k+1}, & k < 0 \\ (k + 1)\left(\dfrac{1}{1.1}\right)^{k+2} - 2\left(\dfrac{1}{1.1}\right)^{k+1}, & k \geq 0 \end{cases}$$

This expression can be simplified to

$$x_k = \begin{cases} -10(-2)^k, & k < 0 \\ \left(\dfrac{1}{1.1}\right)^{k+2}(k - 1.2), & k \geq 0 \end{cases} \tag{4.80}$$

Inversion of $X(z)$ by Use of the Inversion Integral

We mention only briefly inversion by use of the inversion integral. Consider the Z-transform

$$X(z) = \sum_{k=-\infty}^{\infty} x_k z^k \tag{4.81}$$

for z within the radius of convergence. Choose a contour Γ which is within the radius of convergence of $X(z)$ and which encircles the origin. Multiply both sides of (4.81) by $z^{-(n+1)}$ and integrate along Γ with respect to z. This procedure gives

$$\oint_\Gamma z^{-(n+1)}X(z)\,dz = \sum_{k=-\infty}^{\infty} x_k \oint_\Gamma z^{k-n-1}\,dz \tag{4.82}$$

Because the series converges uniformly, it is allowable to interchange the order of integration and summation.

From complex variable theory,[3] the integral on the right-hand side of

[3] See, for example, Churchill, R. V., *Complex Variables and Applications*, McGraw-Hill, New York, 1960 or Freeman, H., *Discrete-Time Systems*, Wiley, New York, 1965.

(4.82) can be evaluated by

$$
\oint z^m \, dz = \begin{cases} 2\pi j, & m = -1 \\ 0, & m \neq -1 \end{cases} \tag{4.83}
$$

Thus, using (4.83) in (4.82), we can solve for x_k

$$
x_k = \frac{1}{2\pi j} \oint z^{-(k+1)} X(z) \, dz \tag{4.84}
$$

Equation (4.84) is evaluated by calculating the residues of the integrand at the poles of $z^{-(k+1)} X(z)$ located within Γ, including for $k > 0$ a multiple pole at $z = 0$. The interested reader is referred to Freeman or Churchill for a further discussion of the evaluation of (4.84). We shall most often use the method of partial fractions for inversions that follow.

4.6 APPLICATIONS OF THE Z-TRANSFORM

The Z-transform is a useful tool in analyzing stationary linear discrete-variable systems. The applications in which this transform can be used are as varied as the range of phenomena one can model using linear difference equations. We begin with a classical example involving the determination of currents and voltages in an electrical circuit.

EXAMPLE 4.17.   Consider the resistor ladder network shown in Figure 4.4. The resistors all have value R ohms. The ladder is infinitely long. The problem is to determine the input resistance and the currents in all the branches. Sum the voltage drops around loop $k + 1$ to obtain the difference equation

$$
-Ri_k + 3Ri_{k+1} - Ri_{k+2} = 0 \tag{4.85}
$$

This equation is valid for $k = 1, 2, \ldots$. The boundary conditions are

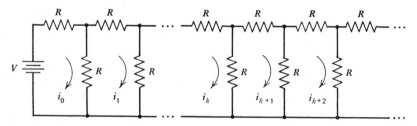

FIGURE 4.4

$i_0 = V/R_{in}$, $i_\infty = 0$ (otherwise we must have infinite power input), and $i_1 = 2i_0 - V/R$. Taking the Z-transform of (4.85), we obtain

$$-I(z) + 3\left(\frac{I(z) - i_0}{z}\right) - \left(\frac{I(z) - i_0 - zi_1}{z^2}\right) = 0 \qquad (4.86)$$

Solving for $I(z)$ and substituting the initial conditions, (4.86) becomes

$$I(z) = \frac{V}{-z^2 + 3z - 1}\left[\frac{z-1}{R_{in}} + \frac{z}{R}\right]$$

$$= \frac{V}{z^2 - 3z + 1}\left[-zR' + \frac{1}{R_{in}}\right]$$

$$= \frac{V}{(z - z_1)(z - z_2)}\left[-zR' + \frac{1}{R_{in}}\right] \qquad (4.87)$$

where $R' = (R + R_{in})/RR_{in}$, and z_1 and z_2, the roots of the denominator polynomial, are given by

$$z_1 = \frac{3 + \sqrt{5}}{2} \qquad z_2 = \frac{3 - \sqrt{5}}{2}$$

We can expand (4.87) in partial fractions as

$$I(z) = \frac{z_1}{z - z_1}\left[\frac{-VR'}{z_1 - z_2}\right] + \frac{V/R_{in}}{(z - z_1)(z_1 - z_2)}$$

$$+ \frac{z_2}{z - z_2}\left[\frac{-VR'}{z_2 - z_1}\right] + \frac{V/R_{in}}{(z - z_2)(z_2 - z_1)} \qquad (4.88)$$

Now we can use the Z-transform pairs

$$\frac{\alpha z}{z - a} \longleftrightarrow -\alpha\{a^{-n}\}\{u_{n-1}\}$$

$$\frac{\alpha}{z - a} \longleftrightarrow -\alpha\{a^{-(n+1)}\}\{u_n\} \qquad (4.89)$$

In (4.88), consider the terms containing the factor $1/(z - z_2)$. The inverse Z-transform of this factor is of the form $-(1/z_2)^{n+1}$. Because $z_2 < 1$, this term $\to \infty$ as $n \to \infty$. Thus, in order for $\lim_{n\to\infty} i_n = 0$, all terms containing the factor $1/(z - z_2)$ must sum to zero. Thus, from (4.88) we see that

$$\frac{z_2 VR'}{z_2 - z_1} = \frac{V}{R_{in}(z_2 - z_1)}$$

Substituting for z_1, z_2, and R' and solving for R_{in} yields

$$R_{in} = \left(\frac{1 + \sqrt{5}}{2}\right)R \qquad (4.90)$$

The inverse z-transform of (4.88), omitting the terms involving $1/(z - z_2)$ yields the branch currents as

$$i_n = \frac{V}{R_{in}}\left(\frac{3 - \sqrt{5}}{2}\right)^n \qquad (4.91)$$

A problem that we have previously discussed in Chapter 3 involves the determination of the state vector of a system. Recall that in the state variable formulation, the essential calculation concerned the evolution of the state vector. We can use Z-transforms to perform this calculation, as the next example demonstrates.

EXAMPLE 4.18. Consider the state variable representation for a linear stationary system. The state of the system is governed by the equation

$$\mathbf{x}_{k+1} = \mathbf{A}\mathbf{x}_k + \mathbf{B}\mathbf{u}_k \qquad (4.92)$$

where $\mathbf{u}_k$ is the input, $\mathbf{x}_k$ is the state, and $\mathbf{A}$ and $\mathbf{B}$ characterize the system as we discussed in Chapter 3. Now (4.92) is a difference equation with constant coefficients. We can, therefore, transform (4.92) to obtain

$$z^{-1}[\mathbf{X}(z) - \mathbf{x}_0] = \mathbf{A}\mathbf{X}(z) + \mathbf{B}\mathbf{U}(z) \qquad (4.93)$$

Multiplying both sides of (4.93) by z and collecting terms gives

$$[\mathbf{I} - z\mathbf{A}]\mathbf{X}(z) = \mathbf{x}_0 + z\mathbf{B}\mathbf{U}(z)$$

Thus, solving for $\mathbf{X}(z)$, we have

$$\mathbf{X}(z) = [\mathbf{I} - z\mathbf{A}]^{-1}\mathbf{x}_0 + [\mathbf{I} - z\mathbf{A}]^{-1}z\mathbf{B}\mathbf{U}(z) \qquad (4.94)$$

By taking inverse Z-transforms on both sides of (4.94), we can obtain $\mathbf{x}_k$. Equation (4.94) is thus a sometimes convenient method of finding $\mathbf{x}_k$ given the initial state $\mathbf{x}_0$ and the sequence of inputs $\{\mathbf{u}_k\}$. Suppose in (4.94) that $\mathbf{U}(z)$ is zero, so that we have an unforced system. From previous discussions, the state $\mathbf{x}_k$ in this case is related to $\mathbf{x}_0$ by

$$\mathbf{x}_k = \mathbf{A}^k\mathbf{x}_0 \qquad (4.95)$$

In (4.94), the second term is zero, and so (4.94) reduces to

$$\mathbf{X}(z) = [\mathbf{I} - z\mathbf{A}]^{-1}\mathbf{x}_0 \qquad (4.96)$$

If we take inverse Z-transforms on both sides of (4.96), we obtain (4.95). Therefore, we have an important result. The inverse Z-transform of

$[\mathbf{I} - z\mathbf{A}]^{-1}$ is identically $\mathbf{A}^k$! This result furnishes us with another and powerful way of calculating the function of the matrix $\mathbf{A}^k$. The next example demonstrates this calculation.

EXAMPLE 4.19. As an application of the previous concept, let us consider the generation of Fibonacci numbers. Fibonacci numbers have been studied since at least 1202[4] and were perhaps known by the early Greeks before the time of Christ. The Fibonacci numbers occur in such unsuspected places as the number of ancestors in succeeding generations of the male bee, the input impedance of a resistor ladder network, and the spacing of buds on the branch of a tree.[5] A generalized Fibonacci sequence is a sequence of real numbers $\{a_n\}$ satisfying the difference equation

$$a_{n+2} = a_n + a_{n+1}, \quad n \geq 0 \qquad (4.97)$$

The classical Fibonacci sequence $\{f_n\}$ satisfies (4.97) with $f_0 = 0$ and $f_1 = 1$. Alternatively, if one uses as initial conditions in (4.97) $l_1 = 1$ and $l_2 = 3$, one generates the so called Lucas sequence $\{l_n\}$.

We can easily solve for nth term of the sequence $\{f_n\}$ by the use of Z-transforms. The Z-transform of (4.97) results in

$$z^{-2}[F(z) - f_0 - zf_1] = F(z) + z^{-1}[F(z) - f_0] \qquad (4.98)$$

Substituting the initial conditions $f_0 = 0$, $f_1 = 1$ and solving for $F(z)$ yields

$$F(z) = \frac{-z}{z^2 + z - 1} = \frac{\frac{z}{\sqrt{5}}}{z + \frac{1}{2} + \sqrt{\frac{5}{2}}} + \frac{\frac{-z}{\sqrt{5}}}{z + \frac{1}{2} - \sqrt{\frac{5}{2}}} \qquad (4.99)$$

Using the transform pair $z/(z - a) \leftrightarrow \{-(1/a)^n\}$, we obtain the nth term of the sequence as

$$f_n = \frac{1}{\sqrt{5}}\left[\left(\frac{1 + \sqrt{5}}{2}\right)^n - \left(\frac{1 - \sqrt{5}}{2}\right)^n\right], \quad n \geq 0 \qquad (4.100)$$

Because (4.97) is a second-order difference equation, we can construct a set of two first-order difference equations which generate the Fibonacci sequence. Thus let $\mathbf{x}_n$ be a two-dimensional vector. Then the sequence

[4] "Fibonacci Numbers: Their History Through 1900," Maxey Brooke, *Fibonacci Quarterly* **2**, 149–152.

[5] "The Fibonacci Sequence as it Appears in Nature," S. L. Basin *Fibonacci Quarterly* **1**, 53–56.

$\{f_n\}$ can be generated by

$$\mathbf{x}_{n+1} = \mathbf{A}\mathbf{x}_n \tag{4.101}$$

where the components of $\mathbf{x}_n$ are defined as

$$\mathbf{x}_n = \begin{bmatrix} f_n \\ f_{n+1} \end{bmatrix} = \begin{bmatrix} x_n^1 \\ x_n^2 \end{bmatrix}$$

To identify the matrix $\mathbf{A}$, we express the relationships between $\mathbf{x}_{n+1}$ and $\mathbf{x}_n$. Thus,

$$x_{n+1}^1 = f_{n+1} = x_n^2$$

and

$$x_{n+2}^2 = f_{n+2} = f_{n+1} + f_n = x_n^1 + x_n^2$$

Thus

$$\mathbf{x}_{n+1} = \mathbf{A}\mathbf{x}_n \tag{4.102}$$

where

$$\mathbf{A} = \begin{bmatrix} 0 & 1 \\ 1 & 1 \end{bmatrix}$$

and

$$\mathbf{x}_n = \mathbf{A}^n \mathbf{x}_0 \tag{4.103}$$

Thus, to generate the nth term of the Fibonacci sequence, we need only to find $\mathbf{A}^n$ and use (4.103) with the correct initial conditions. To find $\mathbf{A}^n$, we proceed in the same manner as we did for the vector linear system problem. If we take Z-transforms of (4.102), we obtain

$$\mathbf{X}(z) = [\mathbf{I} - z\mathbf{A}]^{-1}\mathbf{x}_0 \tag{4.104}$$

Comparing (4.104) and (4.103), we argue that the inverse Z-transform of $[\mathbf{I} - z\mathbf{A}]^{-1}$ is identically $\mathbf{A}^n$. Thus,

$$Z\{\mathbf{A}^n\} = [\mathbf{I} - z\mathbf{A}]^{-1} = \frac{-1}{z^2 + z - 1}\begin{bmatrix} 1 - z & z \\ z & 1 \end{bmatrix} \tag{4.105}$$

We can break (4.105) into partial fractions as we have discussed previously. In this case, the undetermined coefficients are matrices rather than real numbers.

$$\frac{1}{z^2 + z - 1}\begin{bmatrix} z - 1 & -z \\ -z & -1 \end{bmatrix} = \frac{\mathbf{A}}{z - z_1} + \frac{\mathbf{B}}{z - z_2}$$

where

$$\mathbf{A} = (z - z_1) \frac{1}{z^2 + z - 1} \begin{bmatrix} z - 1 & -z \\ -z & -1 \end{bmatrix}\Bigg|_{z=z_1}, \qquad z_1 = \frac{1 + \sqrt{5}}{2}$$

$$\mathbf{B} = (z - z_2) \frac{1}{z^2 + z - 1} \begin{bmatrix} z - 1 & -z \\ -z & -1 \end{bmatrix}\Bigg|_{z=z_2}, \qquad z_2 = \frac{1 - \sqrt{5}}{2}$$

Evaluating the matrices $\mathbf{A}$ and $\mathbf{B}$ gives

$$[\mathbf{I} - z\mathbf{A}]^{-1} = \frac{\begin{bmatrix} 1 - z_1 & z_1 \\ z_1 & 1 \end{bmatrix}}{(z - z_1)(z_2 - z_1)} + \frac{\begin{bmatrix} 1 - z_2 & z_2 \\ z_2 & 1 \end{bmatrix}}{(z - z_2)(z_1 - z_2)} \qquad (4.106)$$

Using the transform pair $k/(z - \alpha) \leftrightarrow \{-k(1/\alpha)^{n+1}\}$, we can write the inverse transform of (4.106) as

$$\mathbf{A}^n = -\frac{\begin{bmatrix} 1 - z_1 & z_1 \\ z_1 & 1 \end{bmatrix}}{z_2 - z_1} \left(\frac{1}{z_1}\right)^{n+1} - \frac{\begin{bmatrix} 1 - z_2 & z_2 \\ z_2 & 1 \end{bmatrix}}{z_1 - z_2} \left(\frac{1}{z_2}\right)^{n+1} \qquad (4.107)$$

Substituting the values of z_1 and z_2 into (4.107) results in

$$\mathbf{A}^n = \frac{\sqrt{5}}{10} \begin{bmatrix} 3 - \sqrt{5} & -1 + \sqrt{5} \\ -1 + \sqrt{5} & 2 \end{bmatrix} \left(\frac{1 + \sqrt{5}}{2}\right)^{n+1}$$
$$- \frac{\sqrt{5}}{10} \begin{bmatrix} 3 + \sqrt{5} & -1 - \sqrt{5} \\ -1 - \sqrt{5} & 2 \end{bmatrix} \left(\frac{1 - \sqrt{5}}{2}\right)^{n+1}$$

Now, with $\mathbf{x}_0 = \begin{bmatrix} 0 \\ 1 \end{bmatrix}$, by definition

$$\mathbf{x}_n = \begin{bmatrix} f_n \\ f_{n+1} \end{bmatrix} = \mathbf{A}^n \begin{bmatrix} 0 \\ 1 \end{bmatrix} = \begin{bmatrix} \dfrac{\sqrt{5}}{5}\left[\left(\dfrac{1 + \sqrt{5}}{2}\right)^n - \left(\dfrac{1 - \sqrt{5}}{2}\right)^n\right] \\ \dfrac{5 + \sqrt{5}}{10}\left(\dfrac{1 + \sqrt{5}}{2}\right)^n + \dfrac{5 - \sqrt{5}}{10}\left(\dfrac{1 - \sqrt{5}}{2}\right)^n \end{bmatrix}$$

So the nth term of the Fibonacci sequence is

$$f_n = \frac{\sqrt{5}}{5}\left[\left(\frac{1 + \sqrt{5}}{2}\right)^n - \left(\frac{1 - \sqrt{5}}{2}\right)^n\right]$$

as before. One can also show that

$$\mathbf{A}^n = \begin{bmatrix} f_{n-1} & f_n \\ f_n & f_{n+1} \end{bmatrix}$$

The formulation that expresses the state at stage $n + 1$ in terms of a matrix times the state at stage n occurs in many problems. The next example uses this formulation in a Markov chain application.

*EXAMPLE 4.20. Suppose we are dating two girls, say Jane and Mary. If we take Jane out, we shall call the state of our dating system (1), and if we take Mary out, we shall label the system state as (2). If we date Jane one weekend, system in state (1), there is a 50% chance of dating Jane the next weekend and a 50% chance of dating Mary: i.e., of going to state (2). Dating Mary, however, tends to be more enjoyable, and so there is a probability of only 2/5 of returning to state (1). A graphical representation of this dating system is shown in Figure 4.5. The numbers on the branches in Figure 4.5 are the state transition probabilities p_{ij}. These state transition probabilities are the probability that the system now in state i will be in state j after one transition. Thus, e.g., $p_{21} = 2/5$ is the probability of going to state (1) in one transition, given that the system is presently in state (2). It is convenient to summarize the state transition probabilities in the matrix $\mathbf{P}$ as

$$\mathbf{P} = \|p_{ij}\| = \begin{bmatrix} \frac{1}{2} & \frac{1}{2} \\ \frac{2}{5} & \frac{3}{5} \end{bmatrix}$$

The matrix $\mathbf{P}$ is a complete description of the Markov process. This matrix plays the same role as the matrix $\mathbf{A}$ in the previous two examples. For example, if we define $\mathbf{x}_n$ as

$$\mathbf{x}_n = \begin{bmatrix} x_n^1 \\ x_n^2 \end{bmatrix}$$

where x_n^i = probability that system is in state i at stage n, $i = 1, 2$, then the dating probabilities at time $n + 1$ are

$$\mathbf{x}_{n+1} = \mathbf{P}^T \mathbf{x}_n \tag{4.108}$$

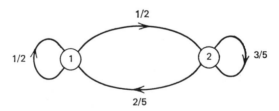

FIGURE 4.5

Thus if $\mathbf{x}_n = \begin{bmatrix} 0 \\ 1 \end{bmatrix}$, which implies with probability one we have a date with Mary, then the next week the dating probabilities are

$$\mathbf{x}_{n+1} = \begin{bmatrix} \frac{1}{2} & \frac{2}{5} \\ \frac{1}{2} & \frac{3}{5} \end{bmatrix} \begin{bmatrix} 0 \\ 1 \end{bmatrix} = \begin{bmatrix} \frac{2}{5} \\ \frac{3}{5} \end{bmatrix}$$

One might ask what is the dating policy after a large number of weeks, given some initial dating probabilities $\mathbf{x}_0$? The answer to this question involves essentially the same calculation that we used in the previous two examples.

If we apply the transition matrix to the initial state $\mathbf{x}_0$ several times, we can, of course, calculate the final state probabilities. We can also make use of (4.108). Thus applying $\mathbf{P}^T$ to $\mathbf{x}_0$ over and over, we obtain

$$\mathbf{x}_n = (\mathbf{P}^T)^n \mathbf{x}_0$$

Now consider the Z-transform of (4.108). We see that

$$z^{-1}[X(z) - \mathbf{x}_0] = \mathbf{P}^T X(z)$$

If we solve for $X(z)$, we obtain

$$X(z) = [\mathbf{I} - z\mathbf{P}^T]^{-1}\mathbf{x}_0$$

(4.109)

Hence, the inverse Z-transform of $[\mathbf{I} - z\mathbf{P}^T]^{-1}$ must be identically $(\mathbf{P}^T)^n$. Because the calculations are essentially of the same form as those of the previous example, we leave the details for the reader. The result of these calculations is that

$$(\mathbf{P}^T)^n = Z^{-1}\{[\mathbf{I} - z\mathbf{P}^T]^{-1}\} = \begin{bmatrix} \frac{4}{9} & \frac{4}{9} \\ \frac{5}{9} & \frac{5}{9} \end{bmatrix} + \left(\frac{1}{10}\right)^n \begin{bmatrix} \frac{5}{9} & -\frac{4}{9} \\ -\frac{5}{9} & \frac{4}{9} \end{bmatrix}$$

Thus the state probability vector $\mathbf{x}_n$ can be found for any n by multiplying $(\mathbf{P}^T)^n$ by the initial state probability vector $\mathbf{x}_0$. The (i, j)th element of $(\mathbf{P}^T)^n$ represents the probability that the system will be in state i at time n, given that the initial state is j at time $n = 0$. Notice that $(\mathbf{P}^T)^n$ consists of a steady-state matrix independent of n and a transient matrix that dies away as n increases. The columns of the steady-state matrix will be identical and represent the limiting-state probabilities, in our case the steady-state dating probabilities. Thus, the steady-state dating probabilities for this example are that $\frac{4}{9}$ of the time we date Jane and $\frac{5}{9}$ of the time we date Mary. The effect of the initial dating policy dies away with increasing n.

The following example demonstrates the use of Z-transforms in solving discrete probability problems.

*EXAMPLE 4.21. Two players are matching pennies. The first player begins with m pennies and the second player with n pennies. The play ends when either player wins all the pennies. What is the probability that the first player wins all the pennies? Let $p_k = Pr\{\text{first player wins all pennies} \mid \text{first player has } k \text{ coins}\}$. Now, the probability the 1st player wins on any throw is $1/2$. If he wins, then k is increased to $k + 1$. If he loses, k is decreased to $k - 1$. Thus, he can arrive at k pennies in two ways:

(1) He can win with $k - 1$ pennies;
(2) He can lose with $k + 1$ pennies.

Thus we have the following recurrence relation (difference equation)

$$p_k = \tfrac{1}{2} \cdot p_{k+1} + \tfrac{1}{2} \cdot p_{k-1} \tag{4.110}$$

where the $1/2$'s are the probability of occurrence of each of the events. The boundary conditions for (4.110) are $p_0 = 0$ and $p_{n+m} = 1$. We can rewrite (4.110) as

$$2p_{k+1} = p_{k+2} + p_k \tag{4.111}$$

Transforming (4.111) yields

$$2z^{-1}[P(z) - p_0] = z^{-2}[P(z) - p_0 - zp_1] + P(z)$$

Solving for $P(z)$ and using the boundary condition for p_0 gives

$$P(z) = \frac{zp_1}{z^2 - 2z + 1}$$

$$= \frac{A}{z - 1} + \frac{B}{(z - 1)^2}$$

A and B can be evaluated as

$$A = \frac{d}{dz}\{(z - 1)^2 P(z)\}\big|_{z=1} = p_1$$

$$B = (z - 1)^2 P(z)\big|_{z=1} = p_1$$

and so the transform $P(z)$ is

$$P(z) = \frac{p_1}{z - 1} + \frac{p_1}{(z - 1)^2}$$

Taking inverse transforms yields

$$p_k = -p_1(1)^{k+1} + p_1(k+1)(1)^{k+2}$$
$$= p_1(-1 + k + 1)$$
$$= kp_1 \tag{4.112}$$

The boundary condition $p_{n+m} = 1$ can now be used in (4.112) to evaluate p_1

$$p_{n+m} = (n + m)p_1 = 1$$

Solving for p_1, we have

$$p_1 = \frac{1}{n + m}$$

and

$$p_k = \frac{k}{n + m}$$

which is the same result we obtained for this problem in Examples 1.18 and 1.23 by use of the direct method.

Another technique that is useful in two dimensional problems involves taking two discrete transforms. This same kind of calculation can be used in the numerical solution of partial differential equations, because partial differential equations always have two or more independent variables.

EXAMPLE 4.22. Consider the following difference equation on the integer variables n and m

$$c(n + 1, m + 1) = c(n, m + 1) + c(n, m) \tag{4.113}$$

with boundary data

$$c(n, 0) = 1, \qquad n \geq 0$$
$$c(0, m) = 0, \qquad m > 0$$

Let

$$\Gamma(n, z) = Z\{\{c(n, m)\}|_{n \text{ fixed}}\}$$

Taking Z-transforms on the discrete variable m in (4.113) yields

$$z^{-1}[\Gamma(n + 1, z) - c(n + 1, 0)]$$
$$= z^{-1}[\Gamma(n, z) - c(n, 0)] + \Gamma(n, z) \tag{4.114}$$

Now let

$$\Omega(y, z) = Z\{\{\Gamma(n, z)\}|_{z \text{ fixed}}\} \tag{4.115}$$

Substitute the boundary data in (4.114) and take Z-transforms in (4.114) on the variable n.

$$z^{-1}\{y^{-1}[\Omega(y, z) - \Gamma(0, z)] - \Omega(y, z)\} = \Omega(y, z)$$

However, the boundary data yield

$$\Gamma(0, z) = Z\{\{c(0, m)\}\} = 1$$

Thus we have for $\Omega(y, z)$

$$\Omega(y, z) = \frac{1}{1 - y(1 + z)}$$

This implies that

$$\Gamma(n, z) = (1 + z)^n$$

$$= 1 + nz + \frac{n(n - 1)}{2!} z^2 + \frac{n(n - 1)(n - 2)}{3!} z^3 + \cdots$$

$$= \sum_{k=0}^{\infty} \binom{n}{k} z^k$$

Thus

$$c(n, m) = \binom{n}{m} \tag{4.116}$$

Notice that in this two-dimensional problem, the boundary consists of two lines rather than points. Thus, the boundary data are specified along two lines rather than two points.

EXAMPLE 4.23. As another example, consider the following application to social structures.[6] This example is concerned with the marriage and descent rules for the Natchez Indians. The Natchez Indians consist of four classes, the Suns, the Nobles, the Honoreds, and the Stinkards. Table 4.4 summarizes the marriage and descent rules. The dash indicates the marriage is not allowed. The table indicates that, for example, Suns, Nobles, and Honoreds must marry Stinkards. Also, children born of a

TABLE 4.4
Marriage and Descent Rules of the Natchez Indians

		Father			
		1 Sun	2 Noble	3 Honored	4 Stinkard
Mother	1 Sun	—	—	—	Sun
	2 Noble	—	—	—	Noble
	3 Honored	—	—	—	Honored
	4 Stinkard	Noble	Honored	Stinkard	Stinkard

[6] The anthropological background can be found in C. W. Hart, "A Reconsideration of the Natchez Social Structure," *American Anthropologist*, New Series, **45**, 1943.

Sun mother and Stinkard father are Suns but children of a Sun father and a Stinkard mother are Nobles. The question we wish to answer is whether this system permits a stable distribution in which each of the four classes maintains a constant number of people. To answer this question, we must make some assumptions on the initial distribution and on the birth rate. We assume that:

(1) each class has an equal number of men and women in each generation n.

(2) each individual marries only once.

(3) each married couple has a single son and daughter.

Let $x(n)$ be the state of this system, where the components $x_i(n)$ are the number of men in class i in generation n. The classes are numbered as in Table 4.4. Notice that by the second assumption $x_i(n)$ is also the number of women in class i in generation n.

From Table 4.4 each Sun mother has one son and mothers of other classes have no Sun children. Thus

$$x_1(n + 1) = x_1(n), \qquad n = 0, 1, 2, \cdots$$

A Noble son is produced from marriages in which the father is a Sun or the mother is a Noble. Therefore

$$x_2(n + 1) = x_1(n) + x_2(n), \qquad n = 0, 1, 2, \ldots$$

Similarly,

$$x_3(n + 1) = x_2(n) + x_3(n), \qquad n = 0, 1, 2, \ldots$$

The recurrence equation for the number of Stinkards can be obtained by noting that the total number of men in each generation is the same. Thus

$$x_1(n + 1) + x_2(n + 1) + x_3(n + 1) + x_4(n + 1)$$
$$= x_1(n) + x_2(n) + x_3(n) + x_4(n) \quad (4.117)$$

Using the relationships in the previous three equations in (4.117) gives

$$x_4(n + 1) = -x_1(n) - x_2(n) + x_4(n), \qquad n = 0, 1, 2, \ldots$$

We can combine these results into a single vector equation as

$$\mathbf{x}(n + 1) = \mathbf{A}\mathbf{x}(n)$$

where

$$\mathbf{A} = \begin{bmatrix} 1 & 0 & 0 & 0 \\ 1 & 1 & 0 & 0 \\ 0 & 1 & 1 & 0 \\ -1 & -1 & 0 & 1 \end{bmatrix}$$

Thus the state of our system at generation n in terms of some initial state $\mathbf{x}(0)$ is

$$\mathbf{x}(n) = \mathbf{A}^n \mathbf{x}(0)$$

To answer the original question, we must therefore calculate $\mathbf{A}^n$. We leave as an exercise the question as to whether this social system can persist as described.

EXAMPLE 4.24.   As a final example, we return to a difference equation we have previously considered in Examples 1.32, 2.6, and 3.14.

$$y(k + 2) - 5y(k + 1) + 6y(k) = u(k) \qquad (4.118)$$

where $\{u(k)\}$ is the unit step sequence and the initial conditions are $y(0) = y(1) = 0$. Taking transforms of (4.118) gives

$$z^{-2}[Y(z) - y(0) - zy(1)] - 5z^{-1}[Y(z) - y(0)] + 6Y(z) = \frac{1}{1 - z}$$

$$(4.119)$$

Substituting the initial conditions into (4.119) and solving for $Y(z)$ yields

$$Y(z) = \frac{z^2}{(1 - z)(6z^2 - 5z + 1)} = \frac{A}{1 - z} + \frac{B}{3z - 1} + \frac{C}{2z - 1}$$

The constants in the partial fraction expansion are easily evaluated to give

$$Y(z) = \frac{\frac{1}{2}}{1 - z} + \frac{\frac{1}{2}}{1 - 3z} - \frac{1}{1 - 2z}$$

which means

$$y(k) = \begin{cases} 0, & k < 0 \\ \frac{1}{2} + \frac{1}{2}(3)^k - (2)^k, & k \geq 0 \end{cases}$$

For this particular example, the Z-transform method of solution is probably the easiest. The reader should compare the three solution processes used in solving the system represented by (4.118) as presented previously in Examples 1.32, 2.6, and 3.14.

4.7   SUMMARY

This chapter has presented yet another formulation for the solution of linear, stationary, discrete-time problems. This formulation is basically different from those presented in the first three chapters in that the formulation is given in a transform domain rather than in the time domain. The properties of the Z-transform that we derived are remarkably similar to properties that occur in other transform domains (like the Fourier and

Laplace domains). One of the advantages of the transform method is that the convolution process in the time domain is transformed into a multiplication process. Thus, the output of a stationary linear system can be calculated by taking the inverse transform of the product of the transform of the input and the system impulse response. This method of calculating the output is often simpler than a direct convolution of the input and the impulse-response function. By transforming the impulse-response sequence to obtain the system transfer function, we obtain another characterization of the input-output relation that is useful in analyzing cascaded systems.

PROBLEMS

1. Solve the first-order difference equations

   a. $f_{n+1} + 3f_n = n$, $f_0 = 1$, $n \geq 0$

   b. $f_{n+1} - 5f_n = \sin n$, $f_0 = 0$, $n \geq 0$

2. Sum the series

   a. $\sum\limits_{n=0}^{\infty} e^{-x(2n+1)}$

   b. $\sum\limits_{k=0}^{\infty} \alpha^k \sinh(kx)$

3. Find the general solution to the second-order difference equation

   $$f_{n+1} - 2f_n + f_{n-1} = \phi_n$$

   where

   a. $\phi_n = a^n$, $a \neq 1$, $n \geq 0$

   b. $\phi_n = n$, $n \geq 0$

   c. $\phi_n = u_n$

4. Find the general solution of each of the following equations

   a. $f_{n+4} + f_n = 0$, $n \geq 0$

   b. $f_{n+2} + 2f_n + f_{n-2} = 0$, $n \geq 0$

5. Show that the ratio of f_n/f_{n+1}, where $\{f_n\}$ is the Fibonacci sequence, has a limit as $n \to \infty$ of $2/(1 + \sqrt{5})$. This number is often known as the "golden mean" and is said to be the ratio of the sides of that rectangle of most pleasing proportions.

*6. Let $\{x_n\}$ be a sequence of independent identically distributed integer-valued random variables with

   Prob. $\{x_n = k\} \triangleq p_k = (1 - b)b^k$,

   $$k = 0, 1, 2, \ldots \quad \text{and} \quad 0 < b < 1$$

Let $y_n = \sum_{m=0}^{n} x_m$ and Prob. $\{y_n = k\} \triangleq q_k(n)$. Let $\Pi(z) = \sum_{k=0}^{\infty} p_k z^k$ be the generating function for the sequence $\{p_k\}$. What is the generating function $\Phi(z)$ for $\{q_k(n)\}$?

7. Let $\{f_n\}$ be the Fibonacci sequence. Verify the following:

a. $f_n^2 + f_{n+1}^2 = f_{2n+1}^2$

b. $f_{n+2}^2 - f_{n+1}^2 = f_n f_{n+3}$

c. Verify the following parlor trick. Turn your back and tell someone to write two positive integers in a column. Increase the column to ten numbers by adding each two successive numbers to obtain the next one. You turn around and write the sum by inspection. (The sum is 11 times the 7th number.)

8.

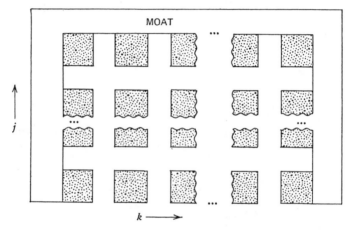

A blindfolded prisoner is placed in a square maze consisting of N equally spaced interior passageways extending in the x-direction and crossed at right angles by N passageways extending in the y-direction. Along 3 sides of the maze there is a deep moat. The 4th side is open to freedom ($y = 0$). Let the passageways in the x direction be denoted by $k = 1, 2, \ldots, N$ and those in the y direction be denoted by $j = 1, 2, \ldots, N$.

a. If the probability of eventual escape for the prisoner at (k, j) is p_{kj}, then show that

$$p_{kj} = \tfrac{1}{4}\{p_{k+1,j} + p_{k-1,j} + p_{k,j+1} + p_{k,j-1}\}, \quad k, j = 1, 2, \ldots, N$$
$$p_{0,j} = 0, \quad p_{N+1,j} = 0, \quad p_{k,N+1} = 0, \quad p_{k,0} = 1$$

b. Determine the probability of eventual escape for all k, j for $N = 3$.

9. Consider the series

$$\frac{1}{1 - x^m} = 1 + x^m + x^{2m} + \cdots$$

Examine the coefficient of x^k in the series for $[1/(1 - x^m)][1/(1 - x^n)]$. Show that the coefficient of x^k is the number of ways one can solve the equation

$$k = l_1 m + l_2 n$$

for l_1 and l_2 nonnegative integers. Thus, for example, the number of ways a dollar can be changed is the 100th coefficient of the series

$$\frac{1}{1 - x} \cdot \frac{1}{1 - x^5} \cdot \frac{1}{1 - x^{10}} \cdot \frac{1}{1 - x^{25}} \cdot \frac{1}{1 - x^{50}}$$

Let x_n be the number of solutions to the equation

$$n = l_1 + 2l_2$$

a. What is the Z-transform of $\{x_n\}$?
b. What difference equation does x_n satisfy?
c. What is the solution to this equation?

10. A conducting rod of cross-sectional area A, thermal conductivity k, and length $L = (n + 1)h$ connects n mass points of equal mass M and specific heat c. The masses are located at a constant separation h, and the rod extends a distance h beyond each of the extreme masses (see below):

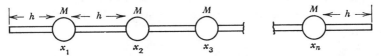

a. If q_k is the rate of flow of heat into the kth mass and T_k is the temperature of the kth mass, show that q_k satisfies

$$q_k = cM \frac{dT_k(t)}{dt}$$

$$q_k = \frac{kA}{h} \{(T_{k+1} - T_k) - (T_k - T_{k-1})\}$$

Thus deduce that T_k satisfies

$$\frac{dT_k(t)}{dt} = \frac{kA}{hcM} (T_{k+1}(t) - 2T_k(t) + T_{k-1}(t))$$

with the appropriate boundary conditions.
b. Find the general solution to the temperature equation.

11. Two identical systems, H_1 and H_2, with impulse response $\{h_k\} = \{2^{-k}\}\{u_k\}$ are cascaded. If the input to the first system is a step, $\{x_k\} = \{u_k\}$,
    a. What is the input to the second?
    b. What is the output from the second? (You may wish to check your answers using convolution.)

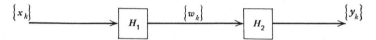

*12. A coin is thrown n times. The probability that it will turn up heads on the first throw is p'. On any subsequent throw, the probability the coin shows the same face as on the previous toss is p. Derive a difference equation that can be solved to obtain the probability the coin shows heads at the nth toss. Solve this difference equation using Z-transforms.

*13. N integers are selected at random (it is equally likely to select any integer). These N integers are multiplied together. Show that the probability of obtaining a two in the *units* digit is $\frac{1}{4}\{(\frac{4}{5})^N - (\frac{2}{5})^N\}$.

*14. A random variable x has a geometric probability distribution if it can assume any positive integer value with probabilities given by

$$Pr(x = k) = p^{k-1}q, \qquad k = 1, 2, \ldots$$

where $p = 1 - q$ and p and q are positive constants. If $P(z)$ is the Z-transform of $Pr(x = k)$, then show that $P'(1)$ is $E(x)$, and thus evaluate $E(x)$. Assume $Pr(x = 0) = 0$. (Prime denotes differentiation.)

*15. The variance of a random variable x, σ^2, is given by $\sigma^2 = E(x^2) - E^2(x)$ where

$$E(x^2) = \sum_{k=0}^{\infty} k^2 Pr(x = k)$$

If $P(z)$ is the Z-transform of the sequence $\{Pr(x = k)\}$, then show that

$$E(x^2) = P''(1) + P'(1)$$

and thus obtain the formula

$$\sigma^2 = P''(1) + P'(1) - (P'(1))^2$$

*16. Generalize Example 4.17 by the following. The first player's probability of winning at any throw is p (instead of $1/2$) and the second player's probability of winning is, therefore, $1 - p$. The game terminates when either of the two has l pennies. What is the probability that the first player wins?

17. Answer the question posed in Example 4.23.

18. Consider the following problem taken from economics.[7] Let $r(n)$ be the amount of money demanded for a commodity and $p(n)$ the price of the commodity in period $(n - 1, n)$. Assume a market equation of the form $p(n) = kr(n)$, where k is a constant. Suppose further that "people extrapolate the trend of prices so that rising prices lead to the expectation of further rise, and thus to an increase in demand, while falling prices lead to an expectation of further fall, and thus to a decrease in demand." To model this last statement, assume $r(n)$ is equal to some "rest" level $r(0)$ plus a factor proportional to the increase of the price $p(n)$ over $p(n - 1)$: i.e., $r(n) = r(0) + \alpha[p(n) - p(n - 1)]$. Derive a difference equation for the price at $n + 1$. Show that if $k\alpha > 1$, then $\{p(n)\}$ diverges to $+\infty$ for $p(0) > kr(0)$ or diverges to $-\infty$ for $p(0) < kr(0)$. For $0 < k\alpha < 1$, find the behavior of $\{p(n)\}$. What conditions are needed to obtain a limiting (equilibrium) price of $kr(0)$?

*19. John Good-Fisher is an avid fisherman, and he has found through many days of fishing that his fishing is either good or bad. If fishing is good, he has found that 60% of the time the next day is also good and 40% of the time the next day is bad. If fishing is bad, his data indicate that the next day is also bad 30% of the time and is good 70% of the time. John is beginning a 20-day fishing trip, and his first day's fishing is bad. What is the probability John will have good fishing on his last day?

20. Find the output of the following discrete-time system for an input sequence $\{u_n\} = \{2^{-n}\}$, $n \geq 0$.

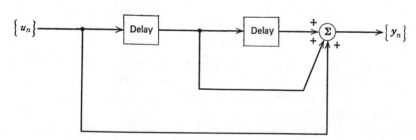

21. What is the output sequence $\{y_n\}$ if two systems as shown in Problem 20 are cascaded? Assume the same input sequence, $\{u_n\} = \{2^{-n}\}$, $n \geq 0$.

22. Throw n infinitely long straight sticks down on a plane. How many possible separate regions can be defined by these n sticks if we assume that no more than two sticks cross at any one point?

23. How many possible regions can be formed by n planes in three-dimensional space? (Use the results of the previous problem.)

[7] K. E. Boulding, *Economic Analysis*, Harper, New York, 1955.

24. Assume gerbils reproduce at a rate in which one pair each month is born to each pair of adults, provided the adults are two or more months old. Assume one pair is present initially and none of the gerbils die. How many pairs of gerbils are there at the end of the first year? How many pairs of gerbils are there at the end of the nth month?

25. If the ratio (f_k/f_{k+1}) of successive Fibonacci numbers (see Example 4.19) is denoted by r_k, then show that r_k satisfies the nonlinear difference equation

$$r_k(r_{k-1} + 1) = 1, \qquad r_0 = 0$$

26. Ivan Evileye is a fallen-away mathematician who has turned to gambling in order to earn his livelihood. Ivan writes down the first part of a sequence of numbers $x_1, x_2, \ldots, x_k, \ldots$ such that each term of the sequence (after x_2) exceeds the average of the previous two terms by a constant m. He then offers the following proposition: to divide into two equal shares a pot formed by your addition of m dollars and Ivan's addition of $(x_{k+1} - x_k)$ dollars, where k is an index $\rightarrow \infty$. Would you take Ivan's proposition?

27. a. What is the transfer function of the following system?

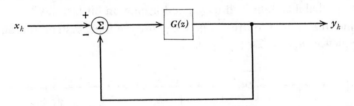

[Hint: Let the input be $\{x_k\}$ and the output be $\{y_k\}$. Find an equation relating $\{x_k\}$, $\{y_k\}$, and the impulse-response sequence $\{g_k\}$. Now take Z-transforms on both sides of this equation.]
b. Repeat for the system

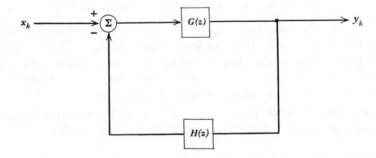

28. Find the impulse response of the system below by using Z-transforms. Compare with your solution to Problem 28 of Chapter 3.

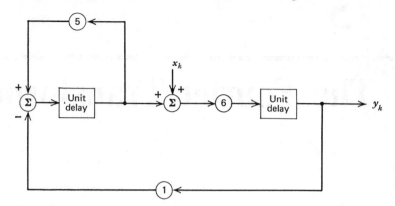

29. For what values of the parameter K is this system stable?

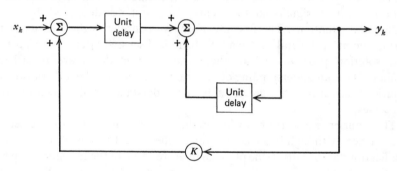

SUGGESTED READINGS

Freeman, H., *Discrete-Time Systems*, Wiley, New York, 1965.

Chapter 3 of this book contains a good discussion of Z-transforms at a level somewhat above this text.

Jury, E. I., *Theory and Application of the Z-Transform Method*, Wiley, New York, 1964.

This book is a complete and detailed discussion of Z-transforms at a senior-first-year-graduate level.

Schwarz, R. J., and B. Friedland, *Linear Systems*, McGraw-Hill, New York, 1965.

Chapter 8 of this book contains a discussion of Z-transforms at a level comparable to that of Freeman's text.

5

The Fourier Transform

In the last chapter, we have considered a transform method for discrete-time signals and systems. We wish to extend this method of transform analysis to continuous-time signals and systems: i.e., systems described by linear differential equations. The first transform we consider is the Fourier transform. The more general Laplace transform is discussed later. The properties and solution processes of all these transform methods are remarkably similar. Thus, an understanding of one transform method, be it Z-transforms, Fourier transforms, etc., greatly aids our understanding of other transform processes.

The Fourier transform has wide application in many areas of engineering and science. Although it is true that the mathematical details of these various applications are identical, the physical interpretation of the transform depends on the application (as we have already noted in Chapter 4). We call the transform of a function (usually a time function) the *spectrum* of the function. The spectrum of a waveform is equally as physical as the waveform itself. We can, with a spectrum analyzer, for example, see and measure the spectrum of an electrical or optical signal waveform. Acoustic waveforms have spectra which are equally as real as the time waveform—in fact, the hearing process of the ear is more easily understood in terms of spectra (rather than time waveforms).

To introduce the Fourier integral transform, we first consider generalized Fourier series. The generalized Fourier series is, in some sense, an extreme case of the Fourier integral transform. However, for those familiar with harmonic analysis, such as occurs in circuit analysis, the Fourier series seems to be a good starting point. The interpretation of the generalized Fourier coefficients as the discrete spectrum of the function is analogous to the continuous spectrum that arises in the Fourier Integral.

214

5.1   GENERALIZED FOURIER SERIES: ORTHOGONAL FUNCTIONS

As we have mentioned previously, the key to the analysis of a linear system is often the correct representation of the input signal. The Fourier series can be used as a method of representing a certain class of functions by a set of so-called orthogonal functions. This representation of functions in terms of an orthogonal set of functions is analogous to the representation of a vector in terms of an orthogonal coordinate system.

For vectors, we know that n vectors $\mathbf{v}_1, \mathbf{v}_2, \ldots, \mathbf{v}_n$, are mutually orthogonal if the dot or scalar product of $\mathbf{v}_i$ and $\mathbf{v}_j$ is zero for all i and j. That is,

$$\mathbf{v}_i \cdot \mathbf{v}_j = [v_i^1 \quad v_i^2 \quad \cdots \quad v_i^n]\begin{bmatrix} v_j^1 \\ v_j^2 \\ \cdot \\ \cdot \\ \cdot \\ v_j^n \end{bmatrix} = \sum_{k=1}^{n} v_i^k v_j^k = \begin{cases} 0, & i \neq j \\ k_i, & i = j \end{cases} \tag{5.1}$$

To represent a vector $\mathbf{f}$ in the n-dimensional space spanned by the set $\{\mathbf{v}_i\}_{i=1}^n$, we express $\mathbf{f}$ as a linear combination of the basis set $\{\mathbf{v}_i\}_{i=1}^n$ as in (5.2)

$$\mathbf{f} = c_1\mathbf{v}_1 + c_2\mathbf{v}_2 + \cdots + c_n\mathbf{v}_n \tag{5.2}$$

The constants c_i, $i = 1, 2, \ldots, n$ can be found by taking the dot product of (5.2) with $\mathbf{v}_1, \mathbf{v}_2, \ldots, \mathbf{v}_n$. For example,

$$\mathbf{f} \cdot \mathbf{v}_i = c_1\mathbf{v}_1 \cdot \mathbf{v}_i + \cdots + c_i\mathbf{v}_i \cdot \mathbf{v}_i + \cdots + c_n\mathbf{v}_n \cdot \mathbf{v}_i = c_i\mathbf{v}_i \cdot \mathbf{v}_i$$

The orthogonality of the basis vectors $\{\mathbf{v}_i\}$ leaves only one nonzero term in $\mathbf{f} \cdot \mathbf{v}_i$. Solving for c_i we obtain

$$c_i = \frac{\mathbf{f} \cdot \mathbf{v}_i}{\mathbf{v}_i \cdot \mathbf{v}_i} = \frac{\mathbf{f} \cdot \mathbf{v}_i}{\|\mathbf{v}_i\|} \tag{5.3}$$

where $\|\mathbf{v}_i\|$ = norm or length squared of $\mathbf{v}_i$. Figure 5.1 depicts a typical representation for $\mathbf{f}$ in terms of the basis set $\{\mathbf{v}_i\}$.

This representation in (5.2) is exact as long as $\mathbf{f}$ is contained in the space spanned by the basis set $\{\mathbf{v}_i\}$, $i = 1, 2, \ldots, n$. If the basis set $\{\mathbf{v}_i\}$ does not span the space containing $\mathbf{f}$, then the representation of (5.2) is only an approximation to $\mathbf{f}$ and contains some error. In this latter case, the question arises

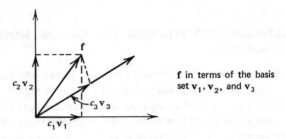

FIGURE 5.1

as to how one might reduce this approximation error by the correct choice of the c_i's.

We can construct a structure for representing functions that is similar to that in the preceding discussion on vector representation. Suppose we have a function $f(t)$ which we wish to represent on a finite interval $[t_1, t_2]$ by a set of n functions. Assume these n functions $\phi_1(t), \phi_2(t), \ldots, \phi_n(t)$ are orthogonal on $[t_1, t_2]$. The set of functions $\{\phi_i(t)\}$ is orthogonal on $[t_1, t_2]$ if

$$\int_{t_1}^{t_2} \phi_i(t)\phi_j(t)\, dt = \begin{cases} 0, & i \neq j \\ k_i, & i = j \end{cases} \tag{5.4}$$

Now we ask, how should we represent $f(t)$ on $[t_1, t_2]$ in terms of the set of functions $\{\phi_i(t)\}_{i=1}^{n}$? Let us assume a representation or approximation of $f(t)$ by a linear combination of the functions $\phi_i(t)$, $i = 1, 2, \ldots, n$. That is, the representation for $f(t)$ on $[t_1, t_2]$ is of the form

$$f(t) \sim c_1\phi_1(t) + c_2\phi_2(t) + \cdots + c_n\phi_n(t) \tag{5.5}$$

We do not use an equal sign in (5.5) because, in general, the representation $\sum_{i=1}^{n} c_i\phi_i(t)$ is in error. We want the representation or approximation to be "close" to $f(t)$ in some sense. One criterion that is often used is to require that the approximation be chosen so as to minimize the mean square error between the true value of $f(t)$ and the approximation $\sum_{i=1}^{n} c_i\phi_i(t)$. In symbols, the c_i, $i = 1, 2, \ldots, n$ are chosen to minimize

$$\text{MSE} = \frac{1}{t_2 - t_1} \int_{t_1}^{t_2} \left(f(t) - \sum_{i=1}^{n} c_i\phi_i(t) \right)^2 dt \tag{5.6}$$

The integrand of (5.6) is, of course, the error squared. The integral and constant $1/(t_2 - t_1)$ average this square error over the interval $[t_1, t_2]$. We can rewrite (5.6) as

$$\text{MSE} = \frac{1}{t_2 - t_1} \int_{t_1}^{t_2} (f(t) - c_1\phi_1(t) - c_2\phi_2(t) - \cdots - c_n\phi_n(t))^2 \, dt \tag{5.7}$$

In (5.7), we now square the integrand to obtain the following expression for the MSE.

$$
\begin{aligned}
\text{MSE} &= \frac{1}{t_2 - t_1} \int_{t_1}^{t_2} [f^2(t) + c_1^2 \phi_1^2(t) + c_2^2 \phi_2^2(t) + \cdots + c_n^2 \phi_n^2(t) - 2c_1 f(t)\phi_1(t) \\
&\qquad\qquad - 2c_2 f(t)\phi_2(t) - \cdots - 2c_n f(t)\phi_n(t)]\, dt \\
&= \frac{1}{t_2 - t_1} \left\{ \int_{t_1}^{t_2} f^2(t)\, dt + c_1^2 k_1 + c_2^2 k_2 + \cdots + c_n^2 k_n \right.\\
&\qquad\qquad\qquad \left. - 2c_1\gamma_1 - 2c_2\gamma_2 - \cdots - 2c_n\gamma_n \right\} \quad (5.8)
\end{aligned}
$$

where we have defined γ_i, $i = 1, 2, \ldots, n$ to be

$$
\gamma_i = \int_{t_1}^{t_2} f(t)\phi_i(t)\, dt \qquad (5.9)
$$

In the expression given in (5.8), we complete the square in each of the terms $(c_i^2 k_i - 2c_i\gamma_i)$ by adding and subtracting γ_i^2/k_i. That is, we write

$$
c_i^2 k_i - 2c_i\gamma_i = (c_i\sqrt{k_i} - \gamma_i/\sqrt{k_i})^2 - \frac{\gamma_i^2}{k_i} \qquad (5.10)
$$

The expression for the MSE in (5.8) can thus be written as

$$
\text{MSE} = \frac{1}{t_2 - t_1} \left\{ \int_{t_1}^{t_2} f^2(t)\, dt + \sum_{i=1}^{n} (c_i\sqrt{k_i} - \gamma_i/\sqrt{k_i})^2 - \sum_{i=1}^{n} \frac{\gamma_i^2}{k_i} \right\} \qquad (5.11)
$$

It is clear from (5.6) that the MSE is always greater than or equal to zero: i.e., $\text{MSE} \geq 0$. From (5.11), it follows that the MSE has its least value when

$$
c_i\sqrt{k_i} = \frac{\gamma_i}{\sqrt{k_i}}, \qquad i = 1, 2, \ldots, n \qquad (5.12)
$$

That is, the c_i's should be chosen as

$$
c_i = \frac{\gamma_i}{k_i} = \frac{\displaystyle\int_{t_1}^{t_2} f(t)\phi_i(t)\, dt}{\displaystyle\int_{t_1}^{t_2} \phi_i^2(t)\, dt} \qquad (5.13)
$$

Notice the similarity between (5.13) and the corresponding vector equation of (5.2).

To summarize: given n mutually orthogonal functions $\phi_1(t), \phi_2(t), \ldots, \phi_n(t)$ on an interval $[t_1, t_2]$, the best approximation of an arbitrary function $f(t)$ on $[t_1, t_2]$ of the form $\sum_{i=1}^{n} c_i\phi_i(t)$ is given by choosing the c_i's according to

(5.13). The criterion used in choosing this approximation is to minimize the mean square error between $f(t)$ and $\sum_{i=1}^{n} c_i \phi_i(t)$.

In (5.13), the coefficient c_i can be interpreted, loosely speaking, as the projection of $f(t)$ in the "direction" of the function $\phi_i(t)$. The analogous vector interpretation is that of a vector, say $\mathbf{f}$, projected onto some basis set of orthogonal vectors $\mathbf{v}_i$, $i = 1, 2, \ldots, n$. If, for example, the coefficient of some c_i is zero, this fact implies that $f(t)$ and $\phi_i(t)$ are orthogonal on $[t_1, t_2]$. Thus the representation or approximation we have discussed is a decomposition of a function $f(t)$ in terms of an orthogonal basis set of functions $\{\phi_i(t)\}_{i=1}^{n}$.

The Mean Square Error

The coefficients c_i chosen according to (5.13) guarantee a minimum mean square error between $f(t)$ and its approximation $\sum_{i=1}^{n} c_i \phi_i(t)$. How small is the error? To answer this question, we need only to consider (5.11) with the optimal values of the c_i, $i = 1, 2, \ldots, n$. The minimum MSE is

$$\text{MSE} = \frac{1}{t_2 - t_1} \left\{ \int_{t_1}^{t_2} f^2(t)\, dt - \sum_{i=1}^{n} \frac{\gamma_i^2}{k_i} \right\}$$

From (5.12), $\gamma_i^2/k_i = c_i^2 k_i$. Therefore

$$\text{MSE} = \frac{1}{t_2 - t_1} \left\{ \int_{t_1}^{t_2} f^2(t)\, dt - [c_1^2 k_1 + c_2^2 k_2 + \cdots + c_n^2 k_n] \right\} \quad (5.14)$$

Equation (5.14) suggests that as we increase n, the number of orthogonal functions, the minimum value of the mean square error decreases. This idea certainly seems reasonable, because as we increase n, we "fill out more directions" in the space containing $f(t)$. Equation (5.6) implies that MSE is always non-negative. Thus, as n increases without limit, the sum $\sum_{i=1}^{n} c_i \phi_i(t)$ may converge to the function $f(t)$, in which case the MSE is zero and

$$\int_{t_1}^{t_2} f^2(t)\, dt = \sum_{i=1}^{\infty} c_i^2 k_i \quad (5.15)$$

Equation (5.15) is known as Parseval's Theorem, and if (5.15) is true, then the sum $\sum_{i=1}^{\infty} c_i \phi_i(t)$ is said to *converge in the mean* to $f(t)$ and the set $\{\phi_i(t)\}$ is said to be *complete*. A complete set of functions $\{\phi_i(t)\}$ is one in which there does not exist a function outside the set which is orthogonal to each member of the set. For every complete set, Parseval's theorem (5.15) holds. Notice that complete is defined here with respect to convergence in the mean, and this type of convergence does not guarantee ordinary convergence at any point. That is, $\lim_{n \to \infty} \int_{t_1}^{t_2} [f(t) - \sum_{i=1}^{n} c_i \phi_i(t)]^2\, dt = 0$ is quite different from ordinary convergence, which implies that

$$\lim_{n \to \infty} \sum_{i=1}^{n} c_i \phi_i(t) = f(t) \tag{5.16}$$

This is a technical point. In most engineering applications, the functions we encounter which converge in the mean also converge pointwise.

Other Types of Orthogonality

Our discussion thus far has been restricted to the case where the basis functions $\{\phi_i(t)\}$ are real. If the basis functions are instead complex functions of the real variable t, orthogonality is defined as

$$\int_{t_1}^{t_2} \phi_i(t)\phi_j^*(t)\, dt = \begin{cases} 0, & i \neq j \\ k_i, & i = j \end{cases} \tag{5.17}$$

where $\phi_j^*(t)$ is the complex conjugate of $\phi_j(t)$. The generalized Fourier coefficients in this case are

$$c_i = \frac{\displaystyle\int_{t_1}^{t_2} f(t)\phi_i^*(t)\, dt}{\displaystyle\int_{t_1}^{t_2} \phi_i(t)\phi_i^*(t)\, dt} \tag{5.18}$$

One can also define orthogonality with respect to a weight function $w(t)$. The set of functions $\{\phi_i(t)\}$ is said to be orthogonal with respect to $w(t)$ on $[t_1, t_2]$ if

$$\int_{t_1}^{t_2} \phi_i(t)\phi_j(t)w(t)\, dt = \begin{cases} 0, & i \neq j \\ k_i, & i = j \end{cases} \tag{5.19}$$

This type of orthogonality can be reduced to the ordinary kind with weight function 1 by defining the orthogonal set to be the set of functions $\{\phi_i(t)\sqrt{w(t)}\}$.

A set of orthogonal functions $\{\phi_i(t)\}$ is said to be *orthonormal* if the constant k_i is 1 for all i. An orthonormal set can be obtained from an orthogonal set of functions by normalization of each function $\phi_i(t)$ in the set so that its norm (or length squared) is unity. This is accomplished by dividing each function $\phi_i(t)$ by its length, $(\|\phi_i(t)\|)^{1/2}$, where

$$\|\phi_i(t)\| = \int_{t_1}^{t_2} \phi_i^2(t)\, dt$$

EXAMPLE 5.1. Consider a representation of the following periodic waveform, shown in Figure 5.2, by the set of mutually orthogonal

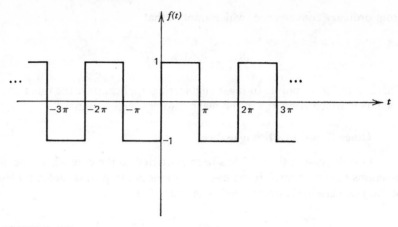

FIGURE 5.2

functions $\{\sin n\omega_0 t\}$, $n = 1, 2, \ldots$. This set of functions is orthogonal on $(t_0, t_0 + 2\pi/\omega_0)$, where t_0 is any number. This fact can be seen by the following calculation.

$$\int_{t_0}^{t_0+2\pi/\omega_0} \sin n\omega_0 t \sin m\omega_0 t \, dt$$

$$= \int_{t_0}^{t_0+2\pi/\omega_0} [\tfrac{1}{2} \cos (n - m)\omega_0 t - \tfrac{1}{2} \cos (n + m)\omega_0 t] \, dt$$

$$= \frac{1}{2\omega_0}\left\{\frac{1}{n - m} \sin (n - m)\omega_0 t - \frac{1}{n + m} \sin (n + m)\omega_0 t\right\}\Bigg|_{t_0}^{t_0+2\pi/\omega_0}$$

$$= \begin{cases} 0, & n \neq m \\ \pi, & n = m \end{cases} \tag{5.20}$$

The set of functions $\{\sin nt\}$ $n = 1, 2, \ldots$, is thus orthogonal on $[0, 2\pi]$. Because $f(t)$ is periodic of period 2π, this set is one possible basis set to use in representing $f(t)$. In order to see how the approximation error varies with the number of basis functions, consider the approximation of $f(t)$ based on one term, then two terms, three terms, etc. The approximation $\hat{f}(t)$ is of the form

$$\hat{f}(t) = c_1 \sin t + c_2 \sin 2t + \cdots + c_n \sin nt \tag{5.21}$$

The values of the coefficients are

$$c_n = \frac{\displaystyle\int_0^{2\pi} f(t) \sin nt\, dt}{\displaystyle\int_0^{2\pi} \sin^2 nt\, dt} = \frac{1}{\pi}\left\{\int_0^{\pi} \sin nt\, dt - \int_{\pi}^{2\pi} \sin nt\, dt\right\}$$

$$= \begin{cases} \dfrac{4}{\pi n}, & n \text{ odd} \\[2mm] 0, & n \text{ even} \end{cases} \tag{5.22}$$

The mean square error is obtained from (5.14) as

$$\text{MSE} = \frac{1}{2\pi}\left\{\int_0^{2\pi} f^2(t)\, dt - c_1^2 k_1 - c_2^2 k_2 - \cdots\right\} \tag{5.23}$$

where $k_1 = k_2 = \cdots = \pi$ in this case. Thus, the mean square error is

$$\text{MSE} = \frac{1}{2\pi}\{2\pi - \pi(c_1^2 + c_2^2 + \cdots)\}$$

$$= 1 - \tfrac{1}{2}(c_1^2 + c_2^2 + \cdots) \tag{5.24}$$

where the c_i's are given by (5.22). Thus, if $n = 1$ (one term), the mean square error is

$$\text{MSE} = 1 - \frac{1}{2}\left(\frac{4}{\pi}\right)^2 = .19$$

For two terms, the mean square error is

$$\text{MSE} = 1 - \frac{1}{2}\left(\frac{4}{\pi}\right)^2 - \frac{1}{2}\left(\frac{4}{3\pi}\right)^2 = .10$$

For three terms, the mean square error is

$$\text{MSE} = 1 - \frac{1}{2}\left(\frac{4}{\pi}\right)^2 - \frac{1}{2}\left(\frac{4}{3\pi}\right)^2 - \frac{1}{2}\left(\frac{4}{5\pi}\right)^2 = .0675$$

Graphically, the approximations appear as shown in Figure 5.3.

5.2 EXAMPLES OF ORTHOGONAL FUNCTIONS

There are many sets of orthogonal functions which can be used to expand or represent a given function on an interval $[t_1, t_2]$. One of the most important classes is the set of exponential functions $\{e^{jn\omega_0 t}\}$ $n = 0, \pm 1, \pm 2, \ldots$ which we shall examine in detail later. Another useful basis set is the class of sinc functions, which are used in the reconstruction of time-sampled

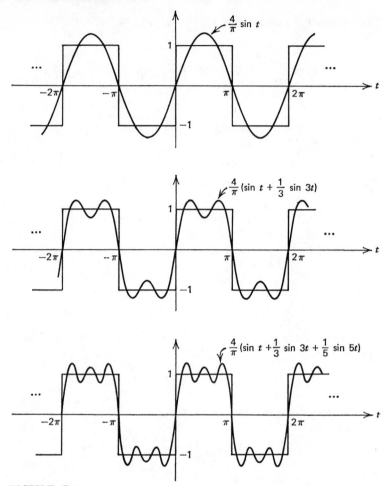

FIGURE 5.3

waveforms. The choice of a basis set depends on the application and the function to be represented. If, for example, one wishes to represent a given function $f(t)$ by a finite number of basis functions so that the mean square error is minimized, then the choice of the basis set is indeed critical.

Suppose, for example, we wish to represent the rectangular waveform of Example 5.1 by three orthogonal functions which are better suited to representing rectangular waveforms than are sinusoids. There is a basis set which is ideally suited for such waveforms. These orthogonal functions are called Walsh functions.[1]

[1] Walsh, J. L., "A Closed Set of Orthogonal Functions," *American Journal of Mathematics*, **55,** 1923, 5–24.

EXAMPLE 5.2. The function we wish to approximate is shown in Figure 5.2. Because $f(t)$ is periodic, we shall use a periodic basis set of the same period and concern ourselves with an approximation over a single period. We can construct the basis set we want in the following way. Let the first basis function be

$$\phi_1(t) = \frac{1}{\sqrt{2\pi}}, \qquad 0 \le t \le 2\pi \tag{5.25}$$

We choose $\phi_2(t)$ so that it is orthogonal to $\phi_1(t)$ and has length or norm 1. Thus, $\phi_2(t)$ must satisfy the equations

$$\int_0^{2\pi} \phi_1(t)\phi_2(t)\,dt = 0 \tag{5.26a}$$

$$\int_0^{2\pi} \phi_2^2(t)\,dt = 1 \tag{5.26b}$$

There is an infinite number of functions that satisfy equations (5.26). However, if we ask for the "simplest" such function (in some sense), we might choose $\phi_2(t)$ as

$$\phi_2(t) = \begin{cases} \dfrac{1}{\sqrt{2\pi}}, & 0 \le t \le \pi \\[2ex] \dfrac{-1}{\sqrt{2\pi}}, & \pi \le t \le 2\pi \end{cases} \tag{5.27}$$

Likewise we choose $\phi_3(t)$ to satisfy the equations

$$\int_0^{2\pi} \phi_1(t)\phi_3(t)\,dt = 0 \tag{5.28a}$$

$$\int_0^{2\pi} \phi_2(t)\phi_3(t)\,dt = 0 \tag{5.28b}$$

$$\int_0^{2\pi} \phi_3^2(t)\,dt = 1 \tag{5.28c}$$

Again choosing the next simplest function to satisfy equations (5.28), we obtain $\phi_3(t)$ as

$$\phi_3(t) = \begin{cases} \dfrac{1}{\sqrt{2\pi}}, & t \in \left[0, \dfrac{\pi}{2}\right], \quad \left[\pi, \dfrac{3\pi}{2}\right] \\[2ex] \dfrac{-1}{\sqrt{2\pi}}, & t \in \left[\dfrac{\pi}{2}, \pi\right], \quad \left[\dfrac{3\pi}{2}, 2\pi\right] \end{cases} \tag{5.29}$$

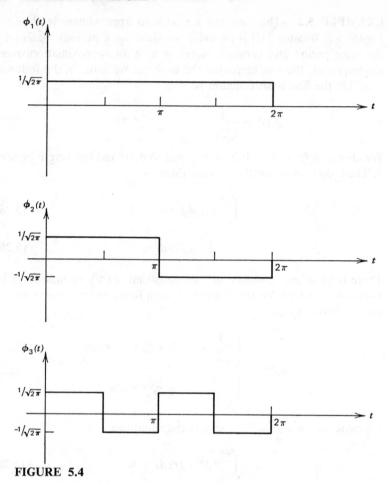

FIGURE 5.4

These three functions are orthonormal on $[0, 2\pi]$. They are Walsh functions and are shown in Figure 5.4.

Using these functions as a basis set for representing $f(t)$, we calculate the generalized Fourier coefficients to be

$$c_1 = \int_0^{2\pi} f(t)\phi_1(t)\, dt = 0 \tag{5.30a}$$

$$c_2 = \int_0^{2\pi} f(t)\phi_2(t)\, dt = \int_0^{\pi} \frac{1}{\sqrt{2\pi}}\, dt + \int_\pi^{2\pi} \frac{1}{\sqrt{2\pi}}\, dt = \sqrt{2\pi} \tag{5.30b}$$

$$c_3 = \int_0^{2\pi} f(t)\phi_3(t)\, dt = 0 \tag{5.30c}$$

The mean square approximation error for a representation based on the functions of Figure 5.4 is given by (5.23).

$$\text{MSE} = \frac{1}{2\pi}\left\{\int_0^{2\pi} f^2(t)\,dt - c_1^2 k_1 - c_2^2 k_2 - c_3^2 k_3\right\}$$

$$= \frac{1}{2\pi}\left\{\int^{2\pi} 1^2\,dt - (\sqrt{2\pi})^2\right\} = 0 \tag{5.31}$$

The mean square error for this set of three basis functions is zero! This occurrence is unusual in that the function $f(t)$ just happens to lie completely in a "direction" of one of the basis set functions. Thus the representation of $f(t)$ is exact even though only three basis functions are used. In general, of course, there is some error no matter which finite basis set is used to expand $f(t)$. This example does point out the importance of choosing an appropriate finite basis set in order to obtain a small mean square error. Roughly speaking, one wants the finite basis set to span the most important dimensions or directions in $f(t)$.

The class of orthogonal functions known as Walsh functions consists of piecewise constant functions. These functions are defined by

$$\phi_0(t) = 1, \quad 0 \leq t \leq 1$$

$$\phi_1(t) = \begin{cases} 1, & 0 \leq t < \tfrac{1}{2} \\ -1, & \tfrac{1}{2} < t \leq 1 \end{cases}$$

$$\phi_2^{(1)}(t) = \begin{cases} 1, & 0 \leq t < \tfrac{1}{4}, \quad \tfrac{3}{4} < t \leq 1 \\ -1, & \tfrac{1}{4} < t < \tfrac{3}{4} \end{cases}$$

$$\phi_2^{(2)}(t) = \begin{cases} 1, & 0 \leq t < \tfrac{1}{4}, \quad \tfrac{1}{2} < t < \tfrac{3}{4} \\ -1, & \tfrac{1}{4} < t < \tfrac{1}{2}, \quad \tfrac{3}{4} < t \leq 1 \end{cases} \tag{5.32}$$

$$\phi_{m+1}^{2k-1}(t) = \begin{cases} \phi_m^{(k)}(2t), & 0 \leq t < \tfrac{1}{2} \\ (-1)^{k+1}\phi_m^{(k)}(2t-1), & \tfrac{1}{2} < t \leq 1 \end{cases} \quad m = 1, 2, 3, \ldots$$

$$\phi_{m+1}^{(2k)}(t) = \begin{cases} \phi_m^{(k)}(2t), & 0 \leq t < \tfrac{1}{2} \\ (-1)^k\phi_m^{(k)}(2t-1), & \tfrac{1}{2} < t \leq 1 \end{cases} \quad k = 1, 2, \ldots, 2^{m-1}$$

These functions are of considerable practical importance because they can easily be generated by digital logic circuitry. Multiplication using these functions is simple. All that is needed is a polarity-reversing switch. Graphs for the first six Walsh functions are shown in Figure 5.5.

The following examples are some of the better known sets of complete orthogonal functions.[2]

[2] R. V. Churchill, *Fourier Series and Boundary Value Problems*, McGraw-Hill, New York, 1941.

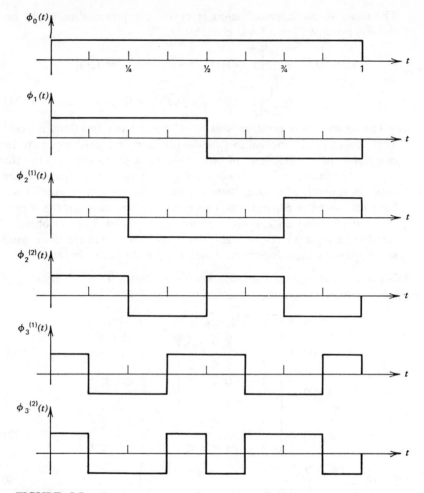

FIGURE 5.5

Legendre Polynomials

The Legendre polynomials form an orthogonal set on the interval $[-1, 1]$ with a weight function $w(t) = 1$. The corresponding normalized basis functions are

$$\phi_0(t) = \frac{1}{\sqrt{2}}, \qquad \phi_1(t) = \sqrt{\tfrac{3}{2}}\,t, \qquad \phi_2(t) = \sqrt{\tfrac{5}{2}}(\tfrac{3}{2}t^2 - \tfrac{1}{2}),$$

$$\phi_3(t) = \sqrt{\tfrac{7}{2}}(\tfrac{5}{2}t^3 - \tfrac{3}{2}t), \dots, \phi_n(t) = \sqrt{\frac{2n+1}{2}}\,P_n(t) \tag{5.33}$$

The $P_n(t)$ are the Legendre polynomials. These polynomials can be generated by the formula

$$P_n(t) = \frac{1}{2^n n!} \frac{d^n}{dt^n} (t^2 - 1)^n \qquad (5.34)$$

or by using the difference equation

$$nP_n(t) = (2n - 1)tP_{n-1}(t) - (n - 1)P_{n-2}(t) \qquad (5.35)$$

Laguerre Functions

For the interval $[0, \infty)$ and with a weight function $w(t) = 1$, the Laguerre functions

$$\phi_n(t) = \frac{e^{-t^2/2}}{n!} L_n(t), \qquad n = 0, 1, 2, \ldots \qquad (5.36)$$

form an orthonormal set. In (5.36), $L_n(t)$ are the Laguerre polynomials given by

$$L_n(t) = e^t \frac{d^n}{dt^n} (t^n e^{-t}) \qquad (5.37)$$

The Laguerre functions can also be generated by the difference equation

$$L_n(t) = (2n - 1 - t)L_{n-1}(t) - (n - 1)^2 L_{n-2}(t) \qquad (5.38)$$

Laguerre functions are interesting in that they can also be generated as the impulse responses of relatively simple networks.[3]

EXAMPLE 5.3. Consider again a representation of the function $f(t)$ of Figure 5.2. In this example, we represent $f(t)$ over one period by a set of Legendre functions. Because the Legendre functions defined in (5.33) are orthonormal on $[-1, 1]$, we need to redefine $f(t)$ as shown in Figure 5.6 by letting $t' = t/\pi$. The generalized Fourier

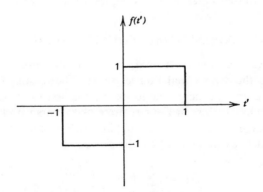

FIGURE 5.6

[3] Y. W. Lee, "Statistical Theory of Communication," Wiley, New York, 1960.

coefficients are given by

$$c_n = \int_{-1}^{1} f(t)\phi_n(t)\,dt, \qquad n = 0, 1, 2, \ldots \qquad (5.39)$$

where the functions $\phi_n(t)$ are given by (5.33). Thus we obtain the generalized Fourier coefficients as

$$
\begin{aligned}
c_0 &= \int_{-1}^{1} \frac{1}{\sqrt{2}} f(t)\,dt = 0 \\[4pt]
c_1 &= \int_{-1}^{1} \sqrt{\tfrac{3}{2}}\, t f(t)\,dt = -\sqrt{\tfrac{3}{2}} \\[4pt]
c_2 &= \int_{-1}^{1} \sqrt{\tfrac{5}{2}}(\tfrac{3}{2}t^2 - \tfrac{1}{2})f(t)\,dt = 0 \\[4pt]
c_3 &= \int_{-1}^{1} \sqrt{\tfrac{7}{2}}(\tfrac{5}{2}t^3 - \tfrac{3}{2}t)f(t)\,dt = \sqrt{\tfrac{7}{16}}
\end{aligned}
\qquad (5.40)
$$

In general, the even-numbered coefficients c_n, $n = 0, 2, 4, \ldots$ are zero. Thus we have representation for $f(t)$ in terms of Legendre polynomials as

$$
\begin{aligned}
f(t) &= \sum_{i=0}^{\infty} c_i \phi_i(t) \\[4pt]
&= -\tfrac{3}{2}t + \left(\frac{7}{4\sqrt{2}}(t^3 - \tfrac{3}{2}t) \right) + \cdots + c_n \phi_n(t) + \cdots \qquad (5.41)
\end{aligned}
$$

The reader should calculate the MSE for a finite expansion and compare this error to that obtained in Examples 5.1 and 5.2.

5.3   THE EXPONENTIAL FOURIER SERIES

In applications, the most used orthogonal functions are the complex sinusoids: i.e., the exponential Fourier series. The reasons for this fact are partly historical in that these functions were the first to be used to represent an arbitrary function. Joseph Fourier presented a paper on heat conduction to the Paris Academy of Science in 1807[4] in which he stated that any bounded function $f(t)$ defined on $(-a, a)$ can be expressed as

$$f(t) = \sum_{n=-\infty}^{\infty} F_n \exp\left(\frac{jn\pi t}{a} \right) \qquad (5.42)$$

[4] Thomas Hawkins, *Lebesgue's Theory of Integration*, University of Wisconsin Press, Madison, 1970.

where the coefficients F_n are given by

$$F_n = \frac{1}{2a} \int_{-a}^{a} f(t) \exp\left(\frac{-jn\pi t}{a}\right) dt \qquad (5.43)$$

However, historical reasons are not the primary grounds for using the exponential functions $\{e^{jn\pi t/a}\}$ as a basis set. The exponential functions possess the very useful property of being closed under multiplication. In other words, multiplying any two functions in the set results in another function of the set. This property simplifies the mathematical manipulation of these functions.

Consider the set of exponential functions $\{e^{jn\omega_0 t}\}$, $n = 0, \pm1, \pm2, \ldots$. These functions are orthogonal on the interval $[t_0, t_0 + 2\pi/\omega_0]$ for any value of t_0. We can easily demonstrate the orthogonality by calculating the integral I below.

$$I = \int_{t_0}^{t_0 + 2\pi/\omega_0} e^{jn\omega_0 t}(e^{jm\omega_0 t})^* \, dt = \int_{t_0}^{t_0 + 2\pi/\omega_0} e^{jn\omega_0 t} e^{-jm\omega_0 t} \, dt \qquad (5.44)$$

If $n = m$, the integral is

$$I = \int_{t_0}^{t_0 + 2\pi/\omega_0} dt = \frac{2\pi}{\omega_0}$$

If $n \neq m$, then the integral I is

$$I = \frac{1}{j(n-m)\omega_0} e^{j(n-m)\omega_0 t} \Big|_{t_0}^{t_0 + 2\pi/\omega_0}$$

$$= \frac{1}{j(n-m)\omega_0} e^{j(n-m)\omega_0 t_0}[e^{j2\pi(n-m)} - 1] = 0$$

The integral I is equal to zero because $e^{j2\pi k}$ is always unity for any integer k. Thus we have

$$\int_{t_0}^{t_0 + 2\pi/\omega_0} e^{jn\omega_0 t} e^{-jm\omega_0 t} \, dt = \begin{cases} 0, & n \neq m \\ \dfrac{2\pi}{\omega_0}, & n = m \end{cases} \qquad (5.45)$$

The set of functions $\{e^{jn\omega_0 t}\}$, $n = 0, \pm1, \pm2, \ldots$ forms a complete orthogonal set on $(t_0, t_0 + T)$ where $T = 2\pi/\omega_0$. We can represent a function $f(t)$ on this interval $(t_0, t_0 + T)$ as

$$f(t) = \sum_{n=-\infty}^{\infty} F_n e^{jn\omega_0 t} \qquad (5.46)$$

where the coefficients are given by

$$F_n = \frac{\displaystyle\int_{t_0}^{t_0+T} f(t)e^{-jn\omega_0 t} \, dt}{\displaystyle\int_{t_0}^{t_0+T} e^{jn\omega_0 t} e^{-jn\omega_0 t} \, dt} = \frac{1}{T}\int_{t_0}^{t_0+T} f(t)e^{-jn\omega_0 t} \, dt \qquad (5.47)$$

We can obtain the coefficients F_n directly from (5.46) by multiplying both sides of (5.46) by $e^{-jn\omega_0 t}$ and integrating with respect to t over the interval $(t_0, t_0 + T)$. The orthogonality condition applied to the sum on the right-hand side eliminates all terms in the sum except the mth term. This expression is then solved for F_m.

To summarize: any function $f(t)$ may be represented on the interval $(t_0, t_0 + T)$ as an infinite sum of exponentials.

$$f(t) = \sum_{n=-\infty}^{\infty} F_n e^{jn\omega_0 t}, \qquad t_0 < t < t_0 + T \qquad (5.48)$$

The coefficients F_n, $n = 0, \pm 1, \pm 2, \ldots$ are found by projecting the function $f(t)$ onto the basis set $\{e^{jn\omega_0 t}\}$, $n = 0, \pm 1, \pm 2, \ldots$.

$$F_n = \frac{1}{T} \int_{t_0}^{t_0+T} f(t) e^{-jn\omega_0 t} \, dt \qquad (5.49)$$

EXAMPLE 5.4. Consider an expansion of the function $f(t) = e^{-t}$ in the interval $(-1, 1)$ by an exponential Fourier Series. Because the period $T = 2$, the Fourier coefficients are

$$F_n = \frac{1}{2} \int_{-1}^{1} f(t) e^{-jn\omega_0 t} \, dt, \qquad \omega_0 = \frac{2\pi}{T} = \pi$$

Thus

$$F_n = \frac{1}{2} \int_{-1}^{1} e^{-t} e^{-jn\pi t} \, dt = \frac{1}{2} \left[\frac{e^{-(1+jn\pi)t}}{-(1 + jn\pi)} \right]_{-1}^{1}$$

$$= \frac{e^{-(1+jn\pi)} - e^{(1+jn\pi)}}{-2(1 + jn\pi)}$$

$$= \frac{e e^{jn\pi} - e^{-1} e^{-jn\pi}}{2(1 + jn\pi)} \qquad (5.50)$$

We can simplify the form of F_n by using the fact that

$$e^{j\pi} = \cos \pi + j \sin \pi = -1$$

which implies that

$$e^{jn\pi} = (-1)^n \qquad (5.51)$$

Therefore

$$F_n = \frac{(-1)^n}{1 + jn\pi} \left(\frac{e - e^{-1}}{2} \right) = \frac{(-1)^n \sinh 1}{(1 + jn\pi)}$$

$$= \frac{(-1)^n (1 - jn\pi) \sinh 1}{1 + n^2 \pi^2} \qquad (5.52)$$

The expansion for $f(t)$ is therefore

$$f(t) = \sum_{n=-\infty}^{\infty} F_n e^{jn\pi t} = \sum_{n=-\infty}^{\infty} \frac{(-1)^n(1 - jn\pi)\sinh 1 \, e^{jn\pi t}}{1 + n^2\pi^2} \tag{5.53}$$

EXAMPLE 5.5. In Example 5.1, we expanded the rectangular waveform of Figure 5.2 in terms of the mutually orthogonal set of functions $\{\sin n\omega_0 t\}$. Suppose that instead of using the set $\{\sin n\omega_0 t\}$ we were to use the set $\{\cos n\omega_0 t\}$. It is easy to show (see Example 5.1) that this set is also mutually orthogonal on $(t_0, t_0 + (2\pi/\omega_0))$. If we calculate the Fourier coefficients for the function $f(t)$ as given in Figure 5.2, we obtain

$$F_n = \frac{\displaystyle\int_0^{2\pi} f(t)\cos nt \, dt}{\displaystyle\int_0^{2\pi} \cos^2 nt \, dt} = \frac{1}{\pi}\left\{\int_0^{\pi}\cos nt \, dt - \int_{\pi}^{2\pi}\cos nt \, dt\right\}$$

$$= \frac{1}{\pi}\left\{\frac{-\sin nt}{n}\Big|_0^{\pi} + \frac{\sin nt}{n}\Big|_{\pi}^{2\pi}\right\} = 0, \qquad \text{for all } n \tag{5.54}$$

Equation (5.54) implies that $f(t)$ is orthogonal to all functions in the set $\cos nt$, $n = 0, 1, 2, \ldots$. This example points out that in order to obtain an accurate representation of a function $f(t)$, one must use a *complete* set of orthogonal functions. The exponential set of functions $\{e^{jn\omega_0 t}\}$, $n = 0, \pm 1$, $\pm 2, \ldots$ is complete, whereas the sets $\{\sin n\omega_0 t\}$ and $\{\cos n\omega_0 t\}$, $n = 0, 1$, $2, \ldots$, are not complete. The exponential set is a convenient way of combining the sines and cosines into a complete set. One can, of course, use the trigonometric form of the exponential set $\{e^{jn\omega_0 t}\}$. In this case, the representation for $f(t)$ is given by

$$f(t) = \frac{a_0}{2} + \sum_{n=1}^{\infty}(a_n\cos n\omega_0 t + b_n\sin n\omega_0 t) \tag{5.55}$$

where

$$a_n = \frac{\omega_0}{\pi}\int_{t_0}^{t_0+2\pi/\omega_0} f(t)\cos n\omega_0 t \, dt \tag{5.56a}$$

$$b_n = \frac{\omega_0}{\pi}\int_{t_0}^{t_0+2\pi/\omega_0} f(t)\sin n\omega_0 t \, dt \tag{5.56b}$$

One can easily convert from the exponential to the trigonometric form (and vice versa) by using the conversion formulas given in Table 5.1.

Convergence of the Fourier Expansion

We must be careful at this point not to delude ourselves into thinking we have proved that a function $f(t)$ has a Fourier expansion which is a valid

TABLE 5.1
Fourier Series Representation of a Periodic Function $f(t)$, $T = 2\pi/\omega_0$

Form	Series Representation	Fourier Coefficients	Conversion Formulas
Exponential	$f(t) = \sum\limits_{n=\infty}^{\infty} F_n e^{jn\omega_0 t}$	$F_n = \dfrac{1}{T} \int_{-T/2}^{T/2} f(t) e^{-jn\omega_0 t}\, dt$	$F_0 = a_0$ $\quad F_n = \frac{1}{2}(a_n - jb_n)$
Trigono-metric	$f(t) = \dfrac{a_0}{2}$ $+ \sum\limits_{n=1}^{\infty} (a_n \cos n\omega_0 t + b_n \sin n\omega_0 t)$	$a_n = \dfrac{2}{T} \int_{-T/2}^{T/2} f(t) \cos n\omega_0 t\, dt,$ $b_n = \dfrac{2}{T} \int_{-T/2}^{T/2} f(t) \sin n\omega_0 t\, dt,$	$a_n = F_n + F_-$ $b_n = j(F_n - F_-$
	$f(t) = \dfrac{a_0}{2} + \sum\limits_{n=1}^{\infty} A_n \cos (n\omega_0 t + \phi_n)$		$A_n = (a_n^2 + b_n^2)^{1}$ $\quad = 2|F_n|$ $\phi_n = -\tan^{-1}\left(\dfrac{b_n}{a_n}\right.$

representation of it. What we have shown thus far is that *if* a function $f(t)$ can be represented by a Fourier expansion, *then* the coefficients are calculated according to (5.13). The questions as to whether the series converges to the function itself and the necessary and sufficient conditions under which the expansions will converge have just recently been answered.[5] These conditions are somewhat technical. For engineering applications, the following sufficient conditions proposed by Dirichlet are usually all we need. The Dirichlet conditions are as follows: if $f(t)$ is bounded and of period T and if $f(t)$ has at most a finite number of maxima and minima in one period and a finite number of discontinuities, then the Fourier series for $f(t)$ converges to $f(t)$ at all points where $f(t)$ is continuous, and converges to the average of the right-hand and left-hand limits of $f(t)$ at each point where $f(t)$ is discontinuous.

This theorem of Dirichlet indicates that the function need not be continuous in order to possess a valid Fourier expansion. If, as in Figure 5.7, $f(t)$ is piecewise continuous, then at points of discontinuity, the value the Fourier series expansion takes on is the value halfway between the right and left hand limits of $f(t)$.

[5] L. Carleson, "On Convergence and Growth of Partial Sums by Fourier Series," *Acta Mathematica*, June 1966, 135–157. Carleson shows that if $\int_{-T/2}^{T/2} |f(t)|^p\, dt < \infty$, p a real number, $p > 1$, then the Fourier series for $f(t)$ converges pointwise to $f(t)$ almost everywhere.

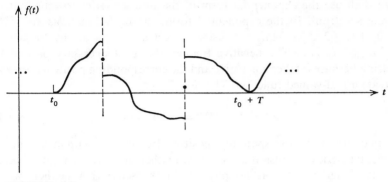

FIGURE 5.7

5.4   THE COMPLEX FOURIER SPECTRUM

The Fourier series expansion of a function $f(t)$ can be used for two classes of functions. One can represent an aperiodic function on a finite interval, say $[0, T]$. In this case, the Fourier series converges to the periodic extension of $f(t)$ outside of $[0, T]$: i.e., to $f(t + nT)$, $n = \pm1, \pm2, \ldots$ One also uses the Fourier series expansion to represent a periodic function $f(t)$ over any interval of interest. Let us, for the moment, consider this latter use of the Fourier series expansion.

One interpretation of the Fourier series expansion for a periodic function is that we are decomposing the function $f(t)$ in terms of its harmonics: i.e., its various frequency components. If $f(t)$ is periodic of period T, then it has frequency components at the radian frequencies $n\omega_0$, $n = 1, 2, \ldots$ where $\omega_0 = 2\pi/T$. The set or collection of these frequency components that make up $f(t)$ is called the *frequency spectrum* or *spectrum* of $f(t)$. In the case of a periodic function, this spectrum is discrete: i.e., the spectrum has nonzero value only for the frequencies $n\omega_0$, $n = 1, 2, \ldots$ The spectrum is an alternate representation for $f(t)$ and given the spectrum we can specify $f(t)$. Thus we have two methods of specifying a periodic function $f(t)$. We can define $f(t)$ in the time domain by some kind of time waveform description or we can specify $f(t)$ in the frequency domain using its frequency spectrum. The frequency spectrum is often displayed graphically by a so-called *line spectrum*. The amplitude of each harmonic is represented by a vertical line proportional to the amplitude of the harmonic and located along the abscissa of the graph at $\omega = \omega_0, 2\omega_0, \ldots$ The discrete frequency spectrum is thus a graph of equally spaced lines with heights proportional to the amplitudes of the component frequencies contained in $f(t)$. Note that the phase of each harmonic must also be specified.

We shall use the exponential form of the Fourier series to represent the discrete spectrum. In the exponential form, the basis functions are $e^{jn\omega_0 t}$, $n = 0, \pm 1, \pm 2, \ldots$ Thus the discrete spectrum exists at the frequencies $0, \pm\omega_0, \pm 2\omega_0, \ldots$ The negative frequencies are a necessary part of the spectrum because it takes a negative and the corresponding positive frequency to create a real-valued function. That is,

$$e^{jn\omega_0 t} + e^{-jn\omega_0 t} = 2\cos n\omega_0 t \tag{5.57}$$

The negative and corresponding positive frequency components in the exponential series expansion constitute a mathematical description of a real sinusoid. Each by itself is nonphysical in the sense that neither can be physically measured—only the combination $(e^{jn\omega_0 t} + e^{-jn\omega_0 t})$ is physical.

For a periodic function of period T, the exponential Fourier series is

$$f(t) = \sum_{n=-\infty}^{\infty} F_n e^{jn\omega_0 t} \tag{5.58}$$

where

$$F_n = \frac{1}{T} \int_{-T/2}^{T/2} f(t) e^{-jn\omega_0 t}\, dt \tag{5.59}$$

In general, the F_n, $n = 0, \pm 1, \pm 2, \ldots$ are complex. Thus, we write $F_n = |F_n| e^{j\theta_n}$ where $|F_n|$ is the magnitude of F_n and θ_n is the angle of F_n. Therefore, in general, we need two discrete or line spectra to represent $f(t)$ in the frequency domain. The *amplitude spectrum* of $f(t)$ is a specification of $|F_n|$ as a function of $n\omega_0$. Similarly, the *phase spectrum* is a specification of θ_n as a function of $n\omega_0$. In many cases, F_n is not complex, and so we can use a single spectrum to represent $f(t)$.

EXAMPLE 5.6.   Find the frequency spectrum of the periodic function shown. This periodic function often is called the periodic gate function.

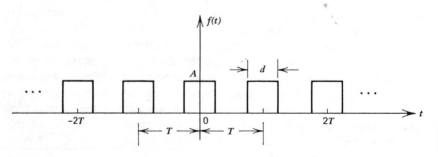

FIGURE 5.8

To find the spectrum, we calculate F_n from

$$F_n = \frac{1}{T} \int_{-T/2}^{T/2} f(t)e^{-jn\omega_0 t}\, dt, \qquad \omega_0 = \frac{2\pi}{T}$$

$$= \frac{1}{T} \int_{-d/2}^{d/2} Ae^{-jn\omega_0 t}\, dt$$

$$= \frac{-A}{jn\omega_0 T} [e^{-jnw_0 t}]_{-d/2}^{d/2}$$

$$= \frac{2A}{n\omega_0 T}\left[\frac{e^{jn\omega_0 d/2} - e^{-jn\omega_0 d/2}}{2j}\right] = \frac{2A}{n\omega_0 T}\sin\left(\frac{n\omega_0 d}{2}\right)$$

$$= \frac{Ad}{T}\left[\frac{\sin\left(\dfrac{n\omega_0 d}{2}\right)}{\dfrac{n\omega_0 d}{2}}\right] = \frac{Ad}{T}\operatorname{sinc}\left(\frac{n\omega_0 d}{2}\right) \tag{5.60}$$

where we define sinc $x \stackrel{\Delta}{=} (\sin x)/x$. This function plays an important role in signal representation. A plot of sinc x is shown in Figure 5.9. Appendix A tabulates values of the sinc function. The sinc function oscillates with period 2π and decays with increasing x. It has zeros at $n\pi$, $n = \pm 1, \pm 2, \ldots$ and is an even function of x. The spectrum of the gate function is thus

$$F_n = \frac{Ad}{T}\operatorname{sinc}\left(\frac{n\omega_0 d}{2}\right)$$

$$= \frac{Ad}{T}\operatorname{sinc}\left(\frac{n\left(\dfrac{2\pi}{T}\right)d}{2}\right)$$

$$= \frac{Ad}{T}\operatorname{sinc}\left(\frac{n\pi d}{T}\right) \tag{5.61}$$

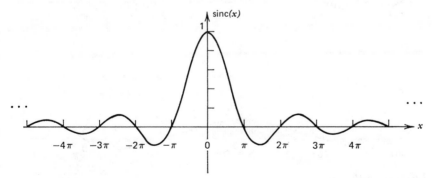

FIGURE 5.9

Thus we can represent $f(t)$ as

$$f(t) = \sum_{n=-\infty}^{\infty} \frac{Ad}{T} \operatorname{sinc}\left(\frac{n\pi d}{T}\right) e^{(jn2\pi/T)t} \tag{5.62}$$

From (5.61), we can see that the spectrum F_n is real. Thus, we need only a single spectrum. To plot the spectrum, we consider three cases for d and T.

Case 1: Suppose $d = \frac{1}{10}$ and $T = \frac{1}{2}$. In this case, the spectrum is $F_n = (A/5) \operatorname{sinc}(n\pi/5)$ and a plot is shown in Figure 5.10a. The fundamental frequency is $\omega_0 = 2\pi/T = 4\pi$, and so the harmonic spacing $\Delta\omega_0$ is also 4π.

Case 2: Assume $d = \frac{1}{10}$ as before, and that we increase the period to 1. Now the spectrum is $F_n = (A/10) \operatorname{sinc}(n\pi/10)$ which is depicted in

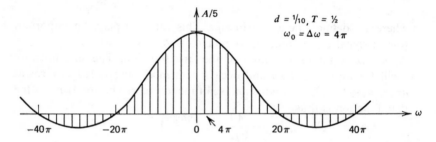

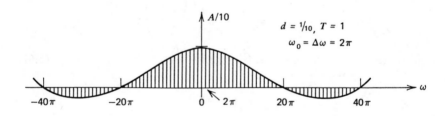

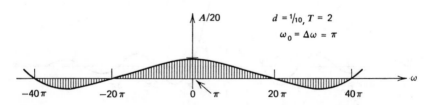

FIGURE 5.10

Figure 5.10b. The fundamental frequency is decreased to $\omega_0 = 2\pi$ and so also the harmonic spacing to 2π.

Case 3: Assume $d = \frac{1}{10}$ and we further increase the period to 2. The spectrum is $F_n = (A/20)$ sinc $(n\pi/20)$ as shown in Figure 5.10c. The spectrum now exists at $\omega = 0, \pm\pi, \pm2\pi, \ldots$.

In this example, we observe that as the period T increases, two qualitative features of the spectrum are changed. The spacing or separation of the component frequencies of $f(t)$ decreases and the amplitudes of the frequency harmonics decrease. The spectrum thus becomes denser and the amplitudes smaller as T increases. Notice, however, that the *shape* of the frequency spectrum does not change with the period T. The shape of the envelope depends only on the periodic pulse shape which repeats every T seconds.

Properties of the Discrete Frequency Spectrum

The discrete frequency spectrum possesses certain properties that are useful in understanding the spectral representation. For example, the magnitude spectrum of every real periodic function is an even function of $n\omega_0$ and the phase spectrum is an odd function of $n\omega_0$. This relationship can be easily shown as follows. The coefficient F_n is given by

$$F_n = \frac{1}{T} \int_{-T/2}^{T/2} f(t)e^{-jn\omega_0 t}\, dt \tag{5.63}$$

and

$$F_{-n} = \frac{1}{T} \int_{-T/2}^{T/2} f(t)e^{+jn\omega_0 t}\, dt \tag{5.64}$$

Thus, from (5.63) and (5.64) we see that F_n and F_{-n} are complex conjugates. That is, $F_n = F_{-n}^*$. This fact implies that $|F_n| = |F_{-n}|$, which in turn implies that $|F_n|$ is an even function of n (or $n\omega_0$).

The phase of F_n is θ_n and the phase of F_{-n} is θ_{-n}, which equals $-\theta_n$. Thus, the phase spectrum is an odd function of n (or $n\omega_0$). These two symmetry properties suggest that we need plot the spectrum only for positive n.

Suppose we have a periodic function $f(t)$ representable by the exponential expansion

$$f(t) = \sum_{n=-\infty}^{\infty} F_n e^{jn\omega_0 t} \tag{5.65}$$

What happens to the spectrum of $f(t)$ if we shift the time origin of $f(t)$ by an amount τ? That is, what is the spectrum of $f(t \pm \tau)$? Intuitively, we might argue that the amplitude spectrum should not change because the amplitude spectrum depends on the waveform or shape of $f(t)$. However, the phase spectrum of the shifted version should change in some manner, because shifting the time origin changes the phase of all the harmonics that make up $f(t)$. From (5.65), we have

$$f(t \pm \tau) = \sum_{n=-\infty}^{\infty} F_n e^{jnw_0(t\pm\tau)} = \sum_{n=-\infty}^{\infty} F_n e^{\pm jn\omega_0\tau} e^{jn\omega_0 t} = \sum_{n=-\infty}^{\infty} \hat{F}_n e^{jn\omega_0 t} \quad (5.66)$$

where $\hat{F}_n = F_n e^{\pm jn\omega_0\tau}$. Represent the original spectrum in the form $F_n = |F_n| e^{j\theta_n}$. The spectrum of the shifted version is therefore

$$\hat{F}_n = |F_n| e^{j(\theta_n \pm n\omega_0\tau)} \quad (5.67)$$

Equation (5.67) indicates that the amplitude spectrum of $f(t \pm \tau)$ is identical to the amplitude spectrum of $f(t)$: i.e., $|F_n|$. The harmonic frequencies are identical, as can be seen from (5.66). The phase spectrum, however, is changed. The time shift of $\pm\tau$ causes a phase shift of $\pm n\omega_0\tau$ radians in the nth harmonic: i.e., the frequency component at $n\omega_0$. Thus, a time shift of $\pm\tau$ seconds does not affect the amplitude spectrum, but changes the phase spectrum by an amount of $\pm n\omega_0\tau$ radians in the nth harmonic of $f(t)$.

The Power Spectrum of a Periodic Function

The power of a time signal waveform is often an important characterization of the signal. We define the power associated with the periodic signal $f(t)$ as the integral

$$P = \frac{1}{T} \int_{-T/2}^{T/2} f^2(t)\, dt \quad (5.68)$$

Equation (5.68) expresses the power of $f(t)$ in the time domain. We can also express the power in $f(t)$ in the frequency domain by calculating the power associated with each frequency component. This leads us to the idea of a *power spectrum* for $f(t)$ in which we calculate the power associated with each harmonic in $f(t)$.

Let the Fourier series for $f(t)$ be

$$f(t) = \sum_{n=-\infty}^{\infty} F_n e^{jn\omega_0 t} \quad (5.69)$$

Because the exponential set $\{e^{jn\omega_0 t}\}$, $n = 0, \pm1, \pm2, \ldots$ is a complete orthogonal set, Parseval's theorem, (5.15), holds. That is,

$$\int_{-T/2}^{T/2} f^2(t)\, dt = \sum_{n=-\infty}^{\infty} |F_n|^2 \cdot T \quad (5.70)$$

or

$$\frac{1}{T} \int_{-T/2}^{T/2} f^2(t)\, dt = \sum_{n=-\infty}^{\infty} |F_n|^2 \tag{5.71}$$

The power in $f(t)$ can thus be calculated from

$$P = \cdots + |F_{-n}|^2 + \cdots + |F_{-1}|^2 + |F_0|^2 + |F_1|^2 + \cdots + |F_n|^2 + \cdots \tag{5.72}$$

Equation (5.72) indicates that the power in $f(t)$ can be calculated by adding together the power associated with each frequency component in $f(t)$. The power associated with the frequency component at $n\omega_0$ radians is $|F_n|^2$ and that of $-n\omega_0$ is $|F_{-n}|^2$. Recall in this representation that it takes both frequency components at $\pm n\omega_0$ to form a single real harmonic. We know that

$$F_n = F^*_{-n}$$

which implies that

$$|F_n|^2 = |F_{-n}|^2$$

Thus the power in $f_n(t)$, the nth (real) harmonic of $f(t)$, is

$$p_n = |F_n|^2 + |F_{-n}|^2 = 2\,|F_n|^2 \tag{5.73}$$

Equation (5.73) is the power in the time function

$$\begin{aligned}
f_n(t) &= F_n e^{+jn\omega_0 t} + F_{-n} e^{-jn w_0 t} \\
&= |F_n|\, e^{j(n\omega_0 t + \theta_n)} + |F_n|\, e^{-j(n\omega_0 t + \theta_n)} \\
&= 2\,|F_n| \cos(n\omega_0 t + \theta_n)
\end{aligned} \tag{5.74}$$

Because the power in $A \cos(\omega t + \phi)$ is $A^2/2$, it follows that the power in $2\,|F_n| \cos(n\omega_0 t + \theta_n)$ is $2\,|F_n|^2$, as given by (5.73). To summarize: we can calculate the power in $f(t)$ in the frequency domain by calculating the power in each frequency component of $f(t)$. We then add these component powers to find the total power P. The collection of component powers, $|F_n|^2$, as a function of $n\omega_0$ is called the power spectrum of $f(t)$.

EXAMPLE 5.7. What percentage of the total power is contained within the first zero crossing of the spectrum envelope for $f(t)$ as given in Figure 5.11? In this example, we have

$$f(t) = \sum_{n=-\infty}^{\infty} F_n e^{jn\omega_0 t}$$

where

$$\begin{aligned}
F_n &= \frac{1}{T} \int_{-T/2}^{T/2} f(t) e^{-jn\omega_0 t}\, dt \\
&= \frac{Ad}{T} \operatorname{sinc}\left(\frac{n\pi d}{T}\right) \\
&= \frac{1}{5} \operatorname{sinc}\left(\frac{n\pi}{5}\right)
\end{aligned} \tag{5.75}$$

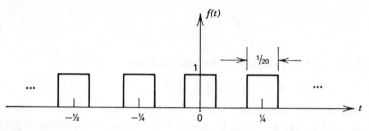

FIGURE 5.11

The spectrum of $f(t)$ is shown in Figure 5.12. The first zero crossing occurs at 40π rad/sec, and there are four harmonics plus the d.c. value within the first zero crossing. The total power in $f(t)$ is

$$P = \frac{1}{T} \int_{-T/2}^{T/2} f^2(t)\, dt$$

$$= 4 \int_{-1/40}^{1/40} (1)^2\, dt = 4[\tfrac{1}{40} + \tfrac{1}{40}] = .20 \qquad (5.76)$$

The power contained within the first zero crossing of the spectrum envelope is

$$P_{fzc} = 2\{|F_1|^2 + |F_2|^2 + |F_3|^2 + |F_4|^2\} + |F_0|^2$$

$$= \left(\frac{1}{5}\right)^2 + \frac{2}{5^2}\Big\{ \operatorname{sinc}^2\left(\frac{\pi}{5}\right) + \operatorname{sinc}^2\left(\frac{2\pi}{5}\right) $$
$$ + \operatorname{sinc}^2\left(\frac{3\pi}{5}\right) + \operatorname{sinc}^2\left(\frac{4\pi}{5}\right) \Big\}$$

$$= \left(\frac{1}{5}\right)^2 + \frac{2}{5^2}\{.875 + .756 + .255 + .055\}$$

$$\cong .04 + .155 = .195 \qquad (5.77)$$

Comparing (5.76) and (5.77), we see that 97.5% of the total power in $f(t)$ is contained within the first zero crossing of the spectrum for $f(t)$.

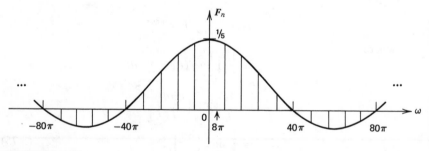

FIGURE 5.12

For applications involving the power of $f(t)$, one could then reasonably represent $f(t)$ in terms of its first four harmonics and its average value.

The complex Fourier series furnishes us with a method of decomposing a signal in terms of a sum of elementary signals of the form $\{e^{jn\omega_0 t}\}$. This decomposition is valuable in analyzing complicated linear systems excited by arbitrary signals because the response of linear systems to exponentials or sinusoids is easy to calculate and measure. The total response of a system to an arbitrary signal is easily calculated by summing the responses that result from each elementary signal. This is essentially the idea involved in convolution. In convolution, we break up an arbitrary signal in terms of a continuum of impulses. The total response is then calculated by summing the responses resulting from each elementary signal: i.e., the response from each impulse. There are an infinite number of decompositions of a function $f(t)$ in terms of elementary functions. The choice of a particular decomposition depends primarily on two factors:

(1) the properties of the function $f(t)$ and the interval over which $f(t)$ is to be represented

(2) the properties of the system to be analyzed and the model used to represent the system

For example, if the system is linear and time invariant and the exciting function is periodic, then a natural choice for representing $f(t)$ is a Fourier series. However, if $f(t)$ is not periodic and we wish to use a representation valid on $(-\infty, \infty)$, a decomposition of $f(t)$ in terms of a Fourier series is no longer valid. Recall that a Fourier series can be used for signals $f(t)$ that are

(1) periodic, $f(t) = f(t + T)$, in which case the representation is valid on $(-\infty, \infty)$.

(2) aperiodic, in which case the representation is valid on a finite interval (a, b). The periodic extension of $f(t)$ is obtained outside of (a, b).

If we wish to extend the class of functions to include aperiodic functions represented on $(-\infty, \infty)$, we must use another type of decomposition for $f(t)$. One useful decomposition is to represent $f(t)$ by a continuum of complex sinusoids of the form $e^{j\omega t}$. The representation is then in terms of the *Fourier transform* of $f(t)$. This decomposition is useful because, as mentioned previously, linear systems excited by sinusoids are easy to analyze. In the remainder of this chapter, we consider signal representations in terms of a continuum of complex sinusoids of the form $e^{j\omega t}$.

5.5 THE FOURIER TRANSFORM

To obtain a representation for a function $f(t)$ over the interval $(-\infty, \infty)$ in terms of a continuum of complex exponentials $e^{j\omega t}$, we consider the function $f(t)$ defined over a finite interval $(-T/2, T/2)$ and then allow $T \to \infty$.

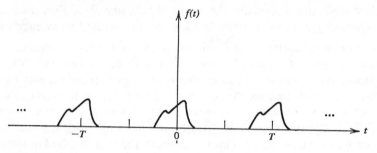

FIGURE 5.13

In the limit, we obtain the desired representation. Thus, suppose $f(t)$ is as shown in Figure 5.13. Consider now the representation $f(t)$ in terms of a complex Fourier series over the interval $(-T/2, T/2)$.

$$f(t) = \sum_{n=-\infty}^{\infty} F_n e^{jn\omega_0 t}, \qquad \omega_0 = \frac{2\pi}{T} \tag{5.78}$$

where

$$F_n = \frac{1}{T} \int_{-T/2}^{T/2} f(t) e^{-jn\omega_0 t} \, dt \tag{5.79}$$

The spectrum of $f(t)$ has a fundamental harmonic frequency of $\omega_0 = 2\pi/T$ and the separation of the harmonics is also ω_0. If we substitute (5.79) into (5.78), we see that $f(t)$ is given by

$$f(t) = \sum_{n=-\infty}^{\infty} \left(\frac{1}{T} \int_{-T/2}^{T/2} f(t) e^{-jn\omega_0 t} \, dt \right) e^{jn\omega_0 t} \tag{5.80}$$

Suppose now we allow $T \to \infty$. The density of the harmonics increases and the spacing $\omega_0 = 2\pi/T$ approaches a differential $d\omega$. The number of harmonics n grows without limit and $(n\omega_0)$ becomes a continuous variable ω. The sum in (5.80) becomes an integral and we see that

$$f(t) = \int_{-\infty}^{\infty} \frac{d\omega}{2\pi} \left(\int_{-\infty}^{\infty} f(t) e^{-j\omega t} \, dt \right) e^{j\omega t}$$

$$= \frac{1}{2\pi} \int_{-\infty}^{\infty} \left(\int_{-\infty}^{+\infty} f(t) e^{-j\omega t} \, dt \right) e^{j\omega t} \, d\omega \tag{5.81}$$

In (5.81), the inner integral is a function only of ω because the time variable is the integration variable. This integral is known as the Fourier transform of $f(t)$ and is denoted by $F(\omega)$:

$$F(\omega) = \int_{-\infty}^{+\infty} f(t) e^{-j\omega t} \, dt \tag{5.82}$$

Thus we can write (5.81) as

$$f(t) = \frac{1}{2\pi} \int_{-\infty}^{\infty} F(\omega)e^{j\omega t} \, d\omega \tag{5.83}$$

We interpret (5.83) as a decomposition of $f(t)$ in terms of the continuum of elementary basis functions $e^{j\omega t}$. The function $F(\omega)$ plays the same role as the F_n's in the Fourier series representation. $(F(\omega)/2\pi) \, d\omega$ is the "coefficient" associated with the elementary basis function $e^{j\omega t}$. $F(\omega)$ is called the continuous *frequency spectrum* of $f(t)$. $F(\omega)$ is, in general, a complex function of the real variable ω and so can be written in the form

$$F(\omega) = |F(\omega)| \, e^{j\theta(\omega)} \tag{5.84}$$

where $|F(\omega)|$ is the continuous *amplitude spectrum* of $f(t)$ and $\theta(\omega)$ is the continuous *phase spectrum* of $f(t)$. If $F(\omega)$ is finite (as it is unless $f(t)$ contains a periodic component), then these coefficients are infinitesimal. $F(\omega)$ can be interpreted as the distribution of the signal in the frequency space ω.

Not all functions can be expanded in a continuum of complex exponentials $e^{j\omega t}$. However, if a function $f(t)$ has a Fourier transform, then this transform and its inverse are unique. Given a time function, there is only one Fourier transform for this time function, and conversely, given a Fourier transform, there is only one corresponding time function. One set of sufficient conditions for representing a function $f(t)$ in terms of a Fourier integral are the so-called Dirichlet conditions:

(1) $f(t)$ must be absolutely integrable: i.e., $\int_{-\infty}^{+\infty} |f(t)| \, dt < \infty$

(2) $f(t)$ must have a finite number of maxima and minima and finite discontinuities in any finite interval.

These conditions include all useful finite energy signals: i.e., signals $f(t)$ for which $\int_{-\infty}^{+\infty} f^2(t) \, dt < \infty$. However, there are a number of important signals like the unit step function which are not absolutely integrable. We can obtain Fourier transforms for these functions by using the theory of distributions[6] and allowing delta functions in the Fourier transforms. We shall classify signals $f(t)$ as being either

(1) "energy signals": i.e., $f(t)$ has finite energy, or
(2) "power signals": i.e., $f(t)$ has infinite energy but finite power. This means $\int_{-\infty}^{+\infty} f^2(t) \, dt$ does not exist, but $\lim_{T \to \infty} (1/T) \int_{-T/2}^{T/2} f^2(t) \, dt$ does exist.

[6] Lighthill, M. J., "Fourier Analysis and Generalized Functions," Cambridge University Press, Cambridge, England, 1964.

5.6 TRANSFORMS OF SOME SIMPLE ENERGY SIGNALS

To develop some insight into the continuous frequency spectrum for a function $f(t)$, we consider the Fourier transform for several simple energy functions.

Rectangular Pulse

Consider the rectangular pulse shown in Figure 5.14. This pulse is given by (5.85). We often call it a gate function.

$$g_T(t) = \begin{cases} 1, & |t| < \dfrac{T}{2} \\ 0, & \text{otherwise} \end{cases} \tag{5.85}$$

The transform of $g_T(t)$ is $G_T(\omega)$ given by

$$G_T(\omega) = \mathscr{F}\{g_T(t)\} = \int_{-\infty}^{\infty} g_T(t)e^{-j\omega t}\, dt$$

$$= \int_{-T/2}^{T/2} 1 \cdot e^{-j\omega t}\, dt$$

$$= \frac{e^{-j\omega t}}{-j\omega}\bigg|_{-T/2}^{T/2}$$

$$= \frac{1}{-j\omega}(e^{-j\omega T/2} - e^{+j\omega T/2})$$

$$= T \cdot \frac{\sin\left(\dfrac{\omega T}{2}\right)}{\left(\dfrac{\omega T}{2}\right)} = T \operatorname{sinc}\left(\frac{\omega T}{2}\right) \tag{5.86}$$

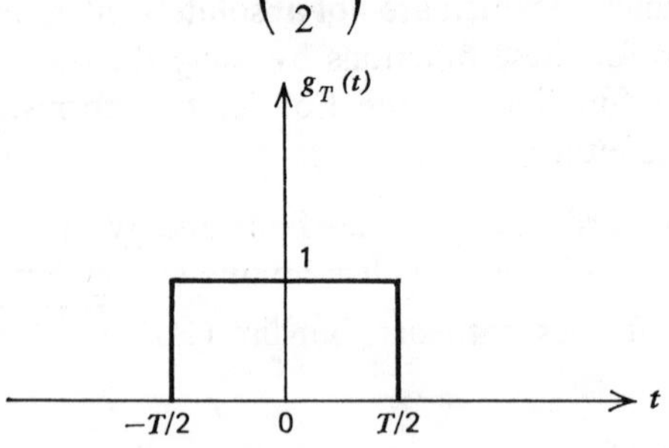

FIGURE 5.14

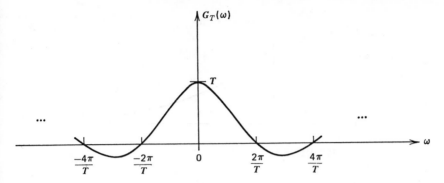

FIGURE 5.15

The amplitude spectrum is thus

$$|G_T(\omega)| = T \left| \text{sinc}\left(\frac{\omega T}{2}\right) \right| \tag{5.87}$$

and the phase spectrum is

$$\theta(\omega) = \begin{cases} 0, & \text{sinc}\left(\dfrac{\omega T}{2}\right) > 0 \\[3mm] \pi, & \text{sinc}\left(\dfrac{\omega T}{2}\right) < 0 \end{cases} \tag{5.88}$$

In this case, the spectrum $G_T(\omega)$ is a real number that is either positive or negative. The sign changes can be interpreted as phase changes of π radians. The plot of the spectrum of $G_T(\omega)$ is shown in Figure 5.15. The shape of the spectrum in Figure 5.15 is dependent on the shape of $g_T(t)$. However, a very general property of transform pairs can be illustrated using $g_T(t)$ and $G_T(\omega)$. We have shown for a periodic example (Example 5.7) that the majority of signal energy (power with periodic case) is contained between the first zero crossings of $G_T(\omega)$. The first zero crossing of $G_T(\omega)$ occurs in frequency at $f = 1/T$ Hz. As the pulse width T is decreased, this first zero crossing moves up in frequency. Conversely, as the pulse width T is increased, the first zero crossing moves toward the origin. This is a general property of all time-frequency transform pairs. The shorter the duration of a time signal, the more spread is its spectrum, and vice versa. We shall amplify this discussion when we consider the properties of the transform.

One-Sided Exponential Pulse

The exponential function is, as we have already seen, used again and again in the analysis of linear systems. Suppose we have the pulse shown in

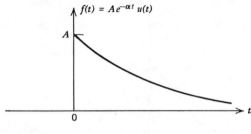

FIGURE 5.16

Figure 5.16. The Fourier transform of this function is

$$F(\omega) = \int_{-\infty}^{\infty} f(t)e^{-j\omega t}\, dt$$

$$= \int_{-\infty}^{\infty} Ae^{-\alpha t}u(t)e^{-j\omega t}\, dt$$

$$= A\int_{0}^{\infty} e^{-\alpha t}e^{-j\omega t}\, dt$$

$$= \frac{Ae^{-(\alpha+j\omega)t}}{-(\alpha + j\omega)}\bigg|_{0}^{\infty}$$

$$= \frac{A}{\alpha + j\omega} \tag{5.89}$$

The phase and amplitude spectra are shown in Figure 5.17. The analytic expressions for $|F(\omega)|$ and $\theta(\omega)$ are obtained by merely finding the magnitude and angle, respectively, of the complex function $F(\omega)$ in (5.89). These expressions are

$$|F(\omega)| = \frac{A}{(\alpha^2 + \omega^2)^{1/2}} \tag{5.90}$$

$$\theta(\omega) = \tan^{-1}\left(\frac{-\omega}{\alpha}\right) \tag{5.91}$$

Triangular Pulse

The triangular pulse of Figure 5.18 has the analytic expression given by

$$f(t) = \begin{cases} A\left(1 - \dfrac{|t|}{T}\right), & |t| < T \\ 0, & \text{otherwise} \end{cases} \tag{5.92}$$

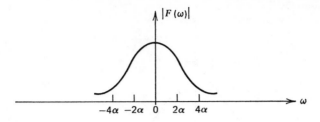

$$F(\omega) = \frac{A(\alpha - j\omega)}{\alpha^2 + \omega^2}$$

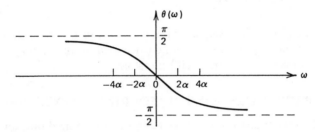

FIGURE 5.17

The Fourier integral of the time function can be calculated by use of the definition, as follows:

$$F(\omega) = \int_{-\infty}^{\infty} f(t)e^{-j\omega t}\, dt$$

$$= \int_{-T}^{0} \left(\frac{At}{T} + A\right)e^{-j\omega t}\, dt + \int_{0}^{T} \left(\frac{-At}{T} + A\right)e^{-j\omega t}\, dt$$

$$= \frac{A}{T} \left\{\left[\frac{te^{-j\omega t}}{-j\omega} - \frac{1}{(j\omega)^2} e^{-j\omega t}\bigg|_{-T}^{0}\right] + \frac{Ae^{-j\omega t}}{-j\omega}\bigg|_{-T}^{0}\right.$$

$$\left. + \frac{-A}{T} \left\{\left[\frac{te^{-j\omega t}}{-j\omega} - \frac{1}{(j\omega)^2} e^{-j\omega t}\bigg|_{0}^{T}\right] + \frac{Ae^{-j\omega t}}{-j\omega}\bigg|_{0}^{T}\right. \tag{5.93}$$

Simplifying the above expression (we leave the details to the reader) yields the transform as

$$F(\omega) = AT\, \text{sinc}^2 \left(\frac{\omega T}{2}\right) \tag{5.94}$$

In this case, the frequency spectrum is a real non-negative number for all ω

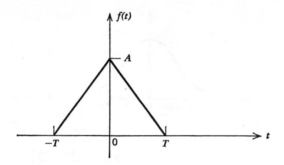

FIGURE 5.18

and so can be plotted in a single graph, as shown in Figure 5.19. The transforms of other energy functions can be calculated in like manner. Table 5.2 summarizes some of the more useful transform pairs.

5.7   PROPERTIES OF THE FOURIER TRANSFORM

The Fourier transform is, as we have already pointed out, an alternate and equivalent method of representing a function $f(t)$. These two descriptions are useful in engineering because often one description is easier to use in a particular application or one of the descriptions is more intuitive for a particular problem. This choice, of course, depends on one's own biases. Transforming between the two domains is relatively straightforward by means of the definitions. However, it is useful to study the effect in one domain caused by an operation in the other domain. Not only does this procedure allow one to transfer between the domains easily, but it also points out certain basic physical aspects of signals and systems that are not otherwise readily apparent (such as the inverse relationship between time duration

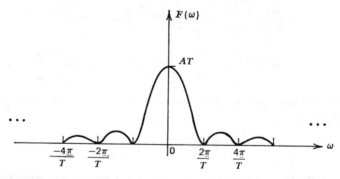

FIGURE 5.19

TABLE 5.2

Time Function, $f(t)$	Fourier Transform, $F(\omega)$
1. $e^{-at}u(t)$	$\dfrac{1}{a + j\omega}$
2. $te^{-at}u(t)$	$\left(\dfrac{1}{a + j\omega}\right)^2$
3. $g_T(t) = \begin{cases} 1, & \|t\| < \dfrac{T}{2} \\ 0, & \text{otherwise} \end{cases}$	$T \operatorname{sinc}\left(\dfrac{\omega T}{2}\right)$
4. $\begin{cases} A\left(1 - \dfrac{\|t\|}{T}\right), & \|t\| < T \\ 0, & \|t\| > T \end{cases}$	$AT \operatorname{sinc}^2\left(\dfrac{\omega T}{2}\right)$
5. $e^{-a\|t\|}$	$\dfrac{2a}{a^2 + \omega^2}$
6. $e^{-at} \sin \omega_0 t\, u(t)$	$\dfrac{\omega_0}{(a + j\omega)^2 + \omega_0^2}$
7. $e^{-at} \cos \omega_0 t\, u(t)$	$\dfrac{a + j\omega}{(a + j\omega)^2 + \omega_0^2}$
8. e^{-at^2}	$\dfrac{\pi}{a} e^{-\omega^2/4a}$
9. $\dfrac{t^{n-1}}{(n - 1)!} e^{-at}u(t)$	$\dfrac{1}{(j\omega + a)^n}$
10. $\dfrac{1}{a^2 + t^2}$	$\dfrac{\pi}{a} e^{-a\|\omega\|}$
11. $\dfrac{\cos bt}{a^2 + t^2}$	$\dfrac{\pi}{2a} [e^{-a\|\omega-b\|} + e^{-a\|\omega+b\|}]$
12. $\dfrac{\sin bt}{a^2 + t^2}$	$\dfrac{\pi}{2aj} [e^{-a\|\omega-b\|} - e^{-a\|\omega+b\|}]$
13. $\cos \omega_0 t \left[u\left(t + \dfrac{T}{2}\right) - u\left(t - \dfrac{T}{2}\right) \right]$	$\dfrac{T}{2}\left[\operatorname{sinc}\left(\dfrac{(\omega - \omega_0)}{2} T\right) + \operatorname{sinc}\left(\dfrac{(\omega + \omega_0)}{2} T\right) \right]$

and bandwidth). Thus one might ask, for example, what is the relationship between the transforms of a time function and the integral of this time function? What happens to the inverse of a frequency-domain function if the frequency-domain function is shifted in frequency? This section is concerned with answers to these kinds of questions.

The symmetry that exists between $f(t)$ and its transform $F(\omega)$ is a particularly powerful property which we shall exploit time and again. The symmetry is readily apparent from the defining equations repeated here.

$$f(t) = \frac{1}{2\pi} \int_{-\infty}^{\infty} F(\omega)e^{j\omega t} \, d\omega$$

$$F(\omega) = \int_{-\infty}^{\infty} f(t)e^{-j\omega t} \, dt \tag{5.95}$$

Actually, this symmetry could be made more complete by multiplying each integral by $1/\sqrt{2\pi}$ rather than multiplying by $1/2\pi$ and 1 as in (5.95). However, convention dictates that we use (5.95) to calculate $f(t)$ and $F(\omega)$. We again use the notation $f(t) \leftrightarrow F(\omega)$ to denote a transform pair. The reader should compare the properties of the Fourier transform with those of the Z-transform in Chapter 4. The properties are remarkably similar.

Symmetry

One of the features of the transform equations in (5.95) is their symmetry in the variables t and ω. This symmetry can be used to great advantage in extending our table of transform pairs. If

$$f(t) \leftrightarrow F(\omega)$$

then

$$F(t) \leftrightarrow 2\pi f(-\omega) \tag{5.96}$$

Proof: Because

$$f(t) = \frac{1}{2\pi} \int_{-\infty}^{\infty} F(\omega)e^{j\omega t} \, d\omega$$

then

$$2\pi f(-t) = \int_{-\infty}^{\infty} F(\omega')e^{-j\omega' t} \, d\omega'$$

where we have replaced the dummy variable ω by ω'. Now if we replace t by ω, we have

$$2\pi f(-\omega) = \int_{-\infty}^{\infty} F(\omega')e^{-j\omega'\omega} \, d\omega'$$

Finally, to obtain a more recognizable form, we replace ω' by t. Thus

$$2\pi f(-\omega) = \int_{-\infty}^{\infty} F(t)e^{-jt\omega} \, dt$$

$$= \mathscr{F}\{F(t)\}$$

Thus

$$F(t) \leftrightarrow 2\pi f(-\omega) \tag{5.97}$$

If $f(t)$ is an even function, $f(t) = f(-t)$, then (5.97) reduces to

$$\mathscr{F}\{F(t)\} = 2\pi f(\omega)$$

EXAMPLE 5.8. We can demonstrate the symmetry property using the triangular function

$$f(t) = \begin{cases} 1 - \dfrac{|t|}{A}, & |t| < A \\ 0, & |t| > A \end{cases}$$

The transform of this function is

$$F(\omega) = A \operatorname{sinc}^2 \left(\frac{\omega A}{2} \right)$$

The symmetry property states that the time function

$$F(t) = A \operatorname{sinc}^2 \left(\frac{tA}{2} \right)$$

has Fourier transform

$$2\pi f(-\omega) = \begin{cases} 2\pi \left(1 - \dfrac{|\omega|}{A} \right), & |\omega| < A \\ 0, & |\omega| > A \end{cases}$$

Figure 5.20 depicts the two transform pairs.

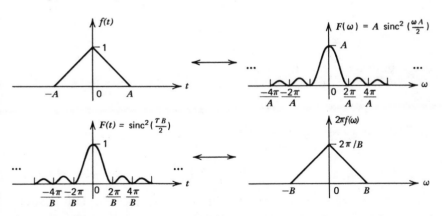

FIGURE 5.20

Linearity

The Fourier transform is a linear operation. That is, if

$$f_1(t) \leftrightarrow F_1(\omega)$$

$$f_2(t) \leftrightarrow F_2(\omega)$$

then

$$af_1(t) + bf_2(t) \leftrightarrow aF_1(\omega) + bF_2(\omega), \quad a \text{ and } b \text{ are constants} \quad (5.98)$$

The proof of the above is immediately evident because the transforms are integrals of the time functions and integration is a linear operation.

Scaling Property

If

$$f(t) \leftrightarrow F(\omega)$$

then for a real constant a,

$$f(at) \leftrightarrow \frac{1}{|a|} F\left(\frac{\omega}{a}\right) \quad (5.99)$$

Proof: Assume $a > 0$. Then the transform of $f(at)$ is

$$\mathscr{F}\{f(at)\} = \int_{-\infty}^{\infty} f(at)e^{-j\omega t}\, dt$$

Let $x = at$, so that $dx = a\, dt$. Substituting in the above, we obtain

$$\mathscr{F}\{f(at)\} = \int_{-\infty}^{\infty} f(x)e^{-(j\omega x/a)}\, \frac{dx}{a}$$

$$= \frac{1}{a} F\left(\frac{\omega}{a}\right)$$

If $a < 0$, then one can show that

$$\mathscr{F}\{f(at)\} = \frac{-1}{a} F\left(\frac{\omega}{a}\right)$$

combining these two results yields

$$f(at) \leftrightarrow \frac{1}{|a|} F\left(\frac{\omega}{a}\right)$$

The scaling property quantifies the time-duration to bandwidth relationship between a time function and its transform. If $|a| > 1$, then $f(at)$, as a function of t, is the function $f(t)$ with a time scale compressed by a factor of a.

Similarly, $F(\omega/a)$ represents the function $F(\omega)$ with a frequency scale, ω, expanded by the factor a. If $|a| < 1$, then $f(at)$ is an expansion of $f(t)$ and $F(\omega/a)$ is a compression of $F(\omega)$. Thus, as we compress the time duration of a signal, we expand the frequency spread of its spectrum. Compressing the time duration of a signal creates faster transitions, thereby requiring higher frequency components in the spectrum. Similarly, expanding the time duration of a signal means that transitions occur at more widely spaced intervals, which can be accomplished with lower frequency components in the spectrum.

Convolution

Convolution is a particularly powerful way of characterizing the input-output relationship of time-invariant linear systems. As we have seen in previous chapters, the convolution integral is not always simple to perform. The transform domain offers a convenient method of performing this convolution operation. There are two convolution theorems, one for the time domain and the other for the frequency domain.

Time Convolution. If

$$x(t) \leftrightarrow X(\omega)$$
$$h(t) \leftrightarrow H(\omega)$$

then

$$y(t) = \int_{-\infty}^{\infty} x(\tau)h(t - \tau)\, d\tau \leftrightarrow Y(\omega) = X(\omega)H(\omega) \qquad (5.100)$$

Proof:

$$\mathscr{F}\{y(t)\} = Y(\omega) = \int_{-\infty}^{\infty} e^{-j\omega t}\left[\int_{-\infty}^{\infty} x(\tau)h(t - \tau)\, d\tau\right] dt$$

$$= \int_{-\infty}^{\infty} x(\tau)\left[\int_{-\infty}^{\infty} h(t - \tau)e^{-j\omega t}\, dt\right] d\tau$$

Now let $a = t - \tau$. Then $da = dt$ and $t = a + \tau$. Thus

$$Y(\omega) = \int_{-\infty}^{\infty} x(\tau)\left[\int_{-\infty}^{\infty} h(a)e^{-j\omega(a+\tau)}\, da\right] d\tau$$

$$= \int_{-\infty}^{\infty} x(\tau)e^{-j\omega\tau}\, d\tau \int_{-\infty}^{\infty} h(a)e^{-j\omega a}\, da$$

$$= X(\omega)H(\omega)$$

The corresponding symmetric property involves the transform of a product of time functions.

Frequency Convolution.   If
$$f(t) \leftrightarrow F(\omega)$$
and
$$g(t) \leftrightarrow G(\omega)$$
then
$$f(t)g(t) \leftrightarrow \frac{1}{2\pi} F(\omega) * G(\omega) \tag{5.101}$$

Proof: Consider the inverse transform of $(F(\omega) * G(\omega))/2\pi$. We have

$$\mathscr{F}^{-1}\left(\frac{F(\omega) * G(\omega)}{2\pi}\right) = \left(\frac{1}{2\pi}\right)^2 \int_{-\infty}^{\infty} e^{j\omega t} \int_{-\infty}^{\infty} F(u)G(\omega - u) \, du \, d\omega$$

$$= \left(\frac{1}{2\pi}\right)^2 \int_{-\infty}^{\infty} F(u) \int_{-\infty}^{\infty} G(\omega - u)e^{j\omega t} \, d\omega \, du$$

Let $x = \omega - u$: then $dx = d\omega$ and $\omega = x + u$. Thus

$$\mathscr{F}^{-1}\left\{\frac{F(\omega) * G(\omega)}{2\pi}\right\} = \left(\frac{1}{2\pi}\right)^2 \int_{-\infty}^{\infty} F(u) \int_{-\infty}^{\infty} G(x)e^{j(x+u)t} \, dx \, du$$

$$= \frac{1}{2\pi} \int_{-\infty}^{\infty} F(u)e^{jut} \, du \, \frac{1}{2\pi} \int_{-\infty}^{\infty} G(x)e^{jxt} \, dx$$

$$= f(t) \cdot g(t)$$

Thus the convolution operation in one domain is transformed into a product operation in the other domain. This is, in many cases, a sufficient simplification to justify taking transforms to circumvent the convolution operation. In the study of linear systems this simplicity is probably the primary reason for the widespread use of transform methods as opposed to time-domain methods.

EXAMPLE 5.9.   In the following system, find the output voltage for an input voltage $x(t)$ of the form $te^{-at}u(t)$. Assume the constant $a = 1/RC$. From Example 2.7, we have the impulse response for the system

$$h(t) = \frac{1}{RC} e^{-t/RC}u(t)$$

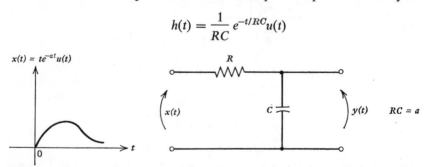

FIGURE 5.21

The output is given by

$$y(t) = x(t) * h(t)$$

In the transform domain, the above equation is

$$Y(\omega) = X(\omega)H(\omega)$$

$H(\omega) = \mathscr{F}[h(t)]$ is known as the system's transfer function or system function. From Table 5.2, we see that

$$H(\omega) = \mathscr{F}\left\{\frac{1}{RC}e^{-t/RC}u(t)\right\} = \frac{1}{RC}\left\{\frac{1}{\dfrac{1}{RC} + j\omega}\right\} = \frac{1}{1 + j\omega RC}$$

$$X(\omega) = \mathscr{F}\{te^{-at}u(t)\} = \frac{1}{(a + j\omega)^2} = \frac{1}{\left(\dfrac{1}{RC} + j\omega\right)^2} = \frac{(RC)^2}{(1 + j\omega RC)^2}$$

Thus

$$Y(\omega) = X(\omega)H(\omega)$$

$$= \left(\frac{1}{1 + j\omega RC}\right)\left(\frac{RC}{1 + j\omega RC}\right)^2 = \frac{(RC)^2}{(1 + j\omega RC)^3}$$

$$\therefore \quad y(t) = \mathscr{F}^{-1}\left\{\frac{(RC)^2}{(1 + j\omega RC)^3}\right\} = \mathscr{F}^{-1}\left\{\frac{1}{RC}\frac{1}{\left(\dfrac{1}{RC} + j\omega\right)^3}\right\}$$

$$= \frac{1}{RC}\frac{t^2 e^{-t/RC}}{2} u(t)$$

Time Shifting

If

$$f(t) \leftrightarrow F(\omega)$$

then

$$f(t - t_0) \leftrightarrow F(\omega)e^{-j\omega t_0} \tag{5.102}$$

Proof: The Fourier transform of $f(t - t_0)$ is by definition

$$\mathscr{F}\{f(t - t_0)\} = \int_{-\infty}^{\infty} f(t - t_0)e^{-j\omega t}\, dt$$

Let $x = t - t_0$: then $dx = dt$ and $t = t_0 + x$. Thus,

$$\mathscr{F}\{f(t - t_0)\} = \int_{-\infty}^{\infty} f(x)e^{-j\omega(t_0 + x)}\, dx$$

$$= e^{-j\omega t_0}F(\omega)$$

The function $f(t - t_0)$ is $f(t)$ delayed by t_0 seconds. The theorem states that the original spectrum is multiplied by $e^{-j\omega t_0}$. This multiplication does not affect the amplitude spectrum of the original spectrum. Each frequency component, however, is shifted in phase by an amount $-\omega t_0$. A little thought indicates that this result is reasonable, because a shift in time of t_0 corresponds to a phase shift of $-\omega t_0$ for a frequency component of ω rad/sec. For example, the double gate function of Fig. 5.22 has a spectrum

$$\mathscr{F}\{f(t)\} = e^{-j\omega D}T \text{ sinc} \left(\frac{\omega T}{2}\right) + e^{j\omega D}T \text{ sinc} \left(\frac{\omega T}{2}\right)$$

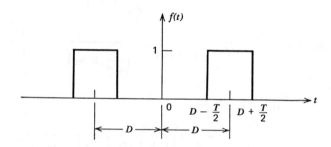

FIGURE 5.22

Frequency Shifting—Modulation

Frequency shifting or translation is an important operation in communication systems. This process is often known as modulation. If

$$f(t) \leftrightarrow F(\omega)$$

then

$$f(t)e^{j\omega_0 t} \leftrightarrow F(\omega - \omega_0) \tag{5.103}$$

Proof: The transform of $f(t)e^{j\omega_0 t}$ is by definition

$$\mathscr{F}\{f(t)e^{j\omega_0 t}\} = \int_{-\infty}^{\infty} f(t)e^{j\omega_0 t}e^{-j\omega t} \, dt$$

$$= \int_{-\infty}^{\infty} f(t)e^{-j(\omega - \omega_0)t} \, dt$$

$$= F(\omega - \omega_0)$$

that is,

$$f(t)e^{j\omega_0 t} \leftrightarrow F(\omega - \omega_0)$$

The expression in (5.103) is the mathematical basis for understanding modulation. Consider, for example, the multiplication of a time function

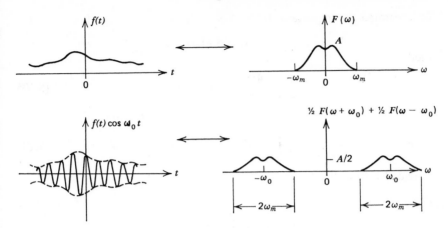

FIGURE 5.23

$f(t)$ by a sinusoid $\cos \omega_0 t$. The function $f(t)$ in this connection is known as the *modulating signal* and the sinusoid $\cos \omega_0 t$ is termed the *carrier* or *modulated signal*. Multiplying $f(t)$ by $\cos \omega_0 t$ shifts the spectrum of $f(t)$ by an amount ω_0. In symbols, the spectrum of $f(t) \cos \omega_0 t$ is

$$\mathscr{F}\{f(t) \cos \omega_0 t\} = \mathscr{F}\left\{f(t)\left[\frac{e^{j\omega_0 t} + e^{-j\omega_0 t}}{2}\right]\right\}$$

$$= \tfrac{1}{2}\mathscr{F}\{f(t)e^{j\omega_0 t}\} + \tfrac{1}{2}\mathscr{F}\{f(t)e^{-j\omega_0 t}\}$$

$$= \tfrac{1}{2}[F(\omega - \omega_0) + F(\omega + \omega_0)]$$

Thus, multiplying a time function by $\cos \omega_0 t$ shifts the original spectrum so that half the original spectrum is centered about ω_0 and the other half is centered about $-\omega_0$. Figure 5.23 depicts the various time functions and their corresponding spectra. Notice that we can use the frequency convolution to find the spectrum of $f(t) \cos \omega_0 t$, provided that we know the individual transforms. Thus

$$\mathscr{F}\{f(t) \cos \omega_0 t\} = \frac{1}{2\pi} \cdot \mathscr{F}\{f(t)\} * \mathscr{F}\{\cos \omega_0 t\} \qquad (5.104)$$

We shall return to (5.104) when we have discussed the transform of $\cos \omega_0 t$.

Time Differentiation and Integration

The Fourier transform can be used to solve linear differential equations. In this application, the transforms of time functions which are differentiated or integrated are important. If

$$f(t) \leftrightarrow F(\omega)$$

then

$$\frac{df(t)}{dt} \leftrightarrow j\omega F(\omega) \tag{5.105}$$

and

$$\int_{-\infty}^{t} f(t') \, dt' \leftrightarrow \frac{1}{j\omega} F(\omega) \tag{5.106}$$

provided $F(0) = 0$.

Proof:

$$f(t) = \frac{1}{2\pi} \int_{-\infty}^{\infty} F(\omega) e^{j\omega t} \, d\omega$$

Thus

$$\frac{df(t)}{dt} = \frac{1}{2\pi} \int_{-\infty}^{\infty} F(\omega) j\omega e^{j\omega t} \, d\omega$$

which implies that

$$\frac{df(t)}{dt} \leftrightarrow (j\omega) F(\omega)$$

We can extend this result to nth order derivatives by repeated differentiations within the integral. Thus

$$\frac{d^n f(t)}{dt^n} \leftrightarrow (j\omega)^n F(\omega) \tag{5.107}$$

Consider the function $g(t)$ defined by

$$g(t) = \int_{-\infty}^{t} f(t') \, dt'$$

Let $g(t)$ have Fourier transform $G(\omega)$. Now

$$\frac{dg(t)}{dt} = f(t)$$

and so, using (5.105), we see that

$$j\omega G(\omega) = G(\omega)$$

or

$$G(\omega) = \left(\frac{1}{j\omega}\right) F(\omega)$$

However, for $g(t)$ to have a transform $G(\omega)$, then, of course, $G(\omega)$ must exist. One condition (that is somewhat more restrictive than absolute integrability) is that

$$\lim_{t \to \infty} g(t) = 0$$

This means

$$\int_{-\infty}^{+\infty} f(t) \, dt = 0$$

which is equivalent to $F(0) = 0$ because

$$F(\omega)\big|_{\omega=0} = \int_{-\infty}^{\infty} f(t) \, dt$$

If $F(0) \neq 0$, then $g(t)$ is no longer an energy function and the transform of $g(t)$ includes an impulse function: i.e.,

$$\int_{-\infty}^{t} f(t') \, dt' \leftrightarrow \frac{1}{j\omega} F(\omega) + \pi F(0) \delta(\omega) \tag{5.108}$$

(see Section 5.9).

Thus we see that differentiation in the time domain corresponds to multiplication by $j\omega$ in the frequency domain. Similarly, integration in the time domain corresponds to division by $j\omega$ in the frequency domain.

Frequency Differentiation and Integration

If we differentiate the expression for $F(\omega)$, we can show that if

$$f(t) \leftrightarrow F(\omega)$$

then

$$-jtf(t) \leftrightarrow \frac{dF(\omega)}{d\omega} \tag{5.109}$$

Similarly, within an additive constant

$$\frac{f(t)}{-jt} \leftrightarrow \int F(\omega) \, d\omega \tag{5.110}$$

We leave the verification of these two properties as a problem for the reader.

Table 5.3 summarizes some of the important properties of the Fourier transform.

TABLE 5.3

1. Transformation	$f(t) \leftrightarrow F(\omega)$
2. Linearity	$a_1 f_1(t) + a_2 f_2(t) \leftrightarrow a_1 F_1(\omega) + a_2 F_2(\omega)$
3. Symmetry	$F(t) \leftrightarrow 2\pi f(-\omega)$
4. Scaling	$f(at) \leftrightarrow \dfrac{1}{\|a\|} F\left(\dfrac{\omega}{a}\right)$
5. Delay	$f(t - t_0) \leftrightarrow e^{-j\omega t_0} F(\omega)$
6. Modulation	$e^{j\omega_0 t} f(t) \leftrightarrow F(\omega - \omega_0)$
7. Convolution	$f_1(t) * f_2(t) \leftrightarrow F_1(\omega) F_2(\omega)$
8. Multiplication	$f_1(t) f_2(t) \leftrightarrow \dfrac{1}{2\pi} F_1(\omega) * F_2(\omega)$
9. Time Differentiation	$\dfrac{d^n}{dt^n} f(t) \leftrightarrow (j\omega)^n F(\omega)$
10. Time Integration	$\displaystyle\int_{-\infty}^{t} f(\tau)\, d\tau \leftrightarrow \dfrac{F(\omega)}{j\omega} + \pi F(0)\delta(\omega)$
11. Frequency Differentiation	$-jt f(t) \leftrightarrow \dfrac{dF(\omega)}{d\omega}$
12. Frequency Integration	$\dfrac{f(t)}{-jt} \leftrightarrow \displaystyle\int F(\omega')\, d\omega'$
13. Reversal	$f(-t) \leftrightarrow F(-\omega)$

5.8   THE ENERGY SPECTRUM

For a periodic function, we showed that the power in a time waveform can be associated with the power contained in each harmonic component of the signal. The same kind of results apply to nonperiodic signals represented by Fourier transforms. For nonperiodic energy signals, the energy over the interval $(-\infty, \infty)$ is finite, whereas the power (energy per unit time) is zero. Thus the energy spectrum (rather than the power spectrum) is the more useful concept for nonperiodic energy signals. The energy associated with $f(t)$ is defined as

$$E = \int_{-\infty}^{\infty} f^2(t)\, dt \tag{5.111}$$

Using the Fourier integral representation for $f(t)$: i.e.,

$$f(t) = \frac{1}{2\pi} \int_{-\infty}^{\infty} F(\omega) e^{j\omega t} \, d\omega$$

we can write the energy as

$$E = \int_{-\infty}^{\infty} f^2(t) \, dt = \int_{-\infty}^{\infty} f(t) \left(\frac{1}{2\pi} \int_{-\infty}^{\infty} F(\omega) e^{j\omega t} \, d\omega \right) dt$$

$$= \frac{1}{2\pi} \int_{-\infty}^{\infty} F(\omega) \left(\int_{-\infty}^{\infty} f(t) e^{j\omega t} \, dt \right) d\omega$$

$$= \frac{1}{2\pi} \int_{-\infty}^{\infty} F(\omega) F(-\omega) \, d\omega \qquad (5.112)$$

For real $f(t)$, $F(-\omega) = F^*(\omega)$ so that (5.112) becomes

$$E = \frac{1}{2\pi} \int_{-\infty}^{\infty} F(\omega) F^*(\omega) \, d\omega = \frac{1}{2\pi} \int_{-\infty}^{\infty} |F(\omega)|^2 \, d\omega$$

That is,

$$\int_{-\infty}^{\infty} f^2(t) \, dt = \frac{1}{2\pi} \int_{-\infty}^{\infty} |F(\omega)|^2 \, d\omega \qquad (5.113)$$

Equation (5.113) expresses the energy in $f(t)$ in terms of the continuous frequency spectrum of $f(t)$. The energy of $f(t)$ is given, as (5.109) indicates, by the area under the $|F(\omega)|^2/2\pi$ curve. The function $|F(\omega)|^2$ is a real and even function of ω, and so we can obtain the energy in $f(t)$ by integrating $|F(\omega)|^2/\pi$ from zero to infinity. That is,

$$E = \frac{1}{2\pi} \int_{-\infty}^{\infty} |F(\omega)|^2 \, d\omega = \frac{1}{\pi} \int_{0}^{\infty} |F(\omega)|^2 \, d\omega$$

$$= \int_{0}^{\infty} S(\omega) \, d\omega \qquad (5.114)$$

where $S(\omega) \stackrel{\Delta}{=} |F(\omega)|^2/\pi$ is called the *energy density spectrum* of $f(t)$. In our study of periodic functions, we associated certain quantities of power with each harmonic. In the case of energy signals, we associate energy with continuous bands of frequencies. The energy contained in the frequency band (ω_1, ω_2) is merely the area under $S(\omega)$ between ω_1 and ω_2. For example, if we have the gate function

$$g_T(t) = \begin{cases} 1, & |t| < \dfrac{T}{2} \\[2mm] 0, & |t| > \dfrac{T}{2} \end{cases} \qquad (5.115)$$

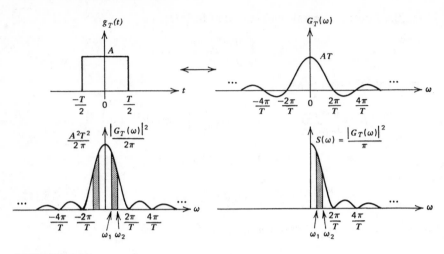

FIGURE 5.24

then the energy spectrum is shown in Figure 5.24. The energy in $g_T(t)$ in the frequency band (ω_1, ω_2) is the shaded area.

5.9 FOURIER TRANSFORMS OF POWER SIGNALS

We have thus far considered only energy signals: i.e., functions which possess finite energy over the interval $(-\infty, \infty)$. These energy functions are absolutely integrable and so satisfy a sufficient condition for the existence of $F(\omega)$. However, there are a number of functions which are very useful and are not absolutely integrable: i.e., they do not satisfy the condition

$$\int_{-\infty}^{\infty} |f(t)|\, dt < \infty \tag{5.116}$$

For example, sine waves or step functions do not satisfy (5.116). Many of these functions do, however, possess Fourier transforms if we allow the transforms to include impulse functions and, in some cases, higher-order singularity functions. The mathematical theory and justification for this process is known[7] but will not be covered here. We shall be content to obtain the transforms of these power signals by limiting processes. These limiting processes are straightforward, but they can be deceiving, because a wrong starting point often leads to an incorrect result. This class of signals is called

---

[7] Erdélyi, A., "*Operational Calculus and Generalized Functions,*" Holt, Rinehart, and Winston, New York, 1962.

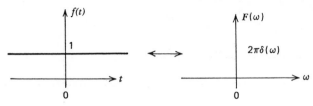

FIGURE 5.25

power signals because the signal energy is infinite over $(-\infty, \infty)$, but the power is finite: i.e.,

$$P = \lim_{T \to \infty} \frac{1}{T} \int_{-T/2}^{T/2} f^2(t) \, dt < \infty \qquad (5.117)$$

Impulse Function

The Fourier transform of the impulse function $\delta(t)$ is readily obtained by use of the defining relation for $\delta(t)$ given in (2.45):

$$F(\omega) = \mathcal{F}\{\delta(t)\} = \int_{-\infty}^{\infty} \delta(t) e^{-j\omega t} \, dt = 1 \qquad (5.118)$$

Thus we have the pair

$$\delta(t) \leftrightarrow 1 \qquad (5.119)$$

shown in Figure 5.25. Using the time-shift theorem we can obtain the transform of the shifted impulse $\delta(t - t_0)$ as

$$\delta(t - t_0) \leftrightarrow e^{-j\omega t_0} \qquad (5.120)$$

The transform of $\delta(t - t_0)$ has unit amplitude for all ω and a linear phase as shown in Figure 5.26. Using the symmetry property, we can derive the transform pairs

$$e^{j\omega_0 t} \leftrightarrow 2\pi \, \delta(\omega - \omega_0) \qquad (5.121)$$

$$1 \leftrightarrow 2\pi \, \delta(\omega) \qquad (5.122)$$

Thus the Fourier transform of the constant one is an impulse at the origin of area equal to 2π, as shown in Figure 5.27.

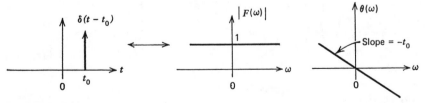

FIGURE 5.26

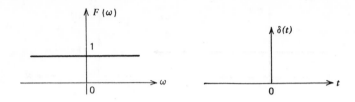

FIGURE 5.27

Sinusoidal Functions

We can use (5.121) to obtain the transforms of $\cos \omega_0 t$ and $\sin \omega_0 t$:

$$\cos \omega_0 t = \frac{e^{j\omega_0 t} + e^{-j\omega_0 t}}{2} \tag{5.123}$$

Using the transform pair $e^{j\omega_0 t} \leftrightarrow 2\pi\delta(\omega - \omega_0)$, we see that

$$\cos \omega_0 t \leftrightarrow \pi[\delta(\omega - \omega_0) + \delta(\omega + \omega_0)] \tag{5.124}$$

Similarly, we can write for the transform of $\sin \omega_0 t$

$$\sin \omega_0 t = \frac{e^{j\omega_0 t} - e^{-j\omega_0 t}}{2j} \leftrightarrow j\pi[\delta(\omega + \omega_0) - \delta(\omega - \omega_0)] \tag{5.125}$$

These transform pairs are depicted in Figure 5.28. Recall our discussion concerning frequency shifting. The calculation of the shifted spectrum involves a transform of the product $f(t) \cos \omega_0 t$. By the convolution theorem,

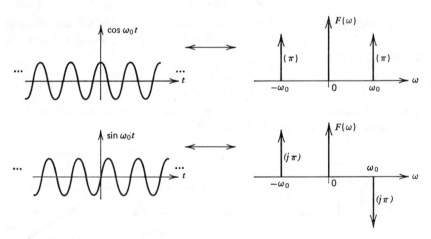

FIGURE 5.28

we can write $\mathscr{F}\{f(t)\cos\omega_0 t\}$ using (5.124) as

$$\mathscr{F}\{f(t)\cos\omega_0 t\} = F(\omega) * \tfrac{1}{2}[\delta(\omega + \omega_0) + \delta(\omega - \omega_0)]$$

$$= \frac{F(\omega + \omega_0) + F(\omega - \omega_0)}{2} \qquad (5.126)$$

The above convolution is, of course, very easy to perform, because convolution with an impulse merely reproduces the function $F(\omega)$ at the location of the impulse.

The Signum Function

The signum function, denoted by sgn (t), is defined as

$$\text{sgn } (t) = \begin{cases} -1, & t < 0 \\ 0, & t = 0 \\ 1, & t > 0 \end{cases} \qquad (5.127)$$

Its graph is shown in Figure 5.29. The transform of the signum function can be found by use of the time differentiation property. Recall that if

$$f(t) \leftrightarrow F(\omega)$$

then

$$f'(t) \leftrightarrow j\omega F(\omega)$$

Suppose we differentiate the signum function. Its derivative is $2\delta(t)$: i.e.,

$$\frac{d}{dt}\text{sgn } (t) = 2\,\delta(t)$$

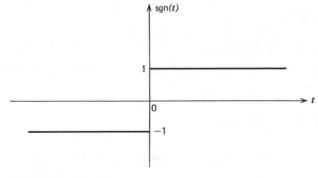

FIGURE 5.29

The transform of d/dt (sgn (t)) is

$$\frac{d}{dt} \text{sgn}\,(t) \leftrightarrow j\omega F(\omega) \tag{5.128}$$

where

$$j\omega\, F(\omega) = \mathscr{F}\{2\delta(t)\} = 2$$

Thus the transform of sgn (t) is

$$F(\omega) = \frac{2}{j\omega}$$

and so we have the transform pair

$$\text{sgn}\,(t) \leftrightarrow \frac{2}{j\omega} \tag{5.129}$$

A sketch of the spectrum for sgn (t) is shown in Figure 5.30.

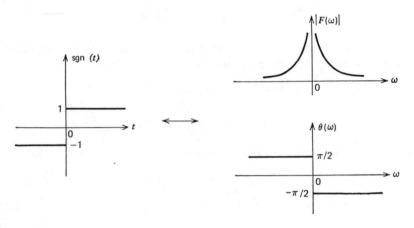

FIGURE 5.30

The Unit Step Function

The unit step function can be written in terms of the signum function as

$$u(t) = \tfrac{1}{2} + \tfrac{1}{2}\,\text{sgn}\,(t) \tag{5.130}$$

Thus, the transform of $u(t)$ is

$$\mathscr{F}\{u(t)\} = \mathscr{F}\{\tfrac{1}{2}\} + \tfrac{1}{2}\mathscr{F}\{\text{sgn}\,(t)\} = \pi\,\delta(\omega) + \frac{1}{j\omega} \tag{5.131}$$

In our discussion of the time integration property, we stated, in general, that the transform of integral of $f(t)$ includes an impulse in its transform: i.e.,

$$\int_{-\infty}^{t} f(t')\, dt' \leftrightarrow \frac{F(\omega)}{j\omega} + \pi\, F(0)\delta(\omega) \tag{5.132}$$

We can use (5.131) to show the above. Let

$$g(t) = \int_{-\infty}^{t} f(t')\, dt'$$

We can write $g(t)$ as a convolution of $f(t)$ and $u(t)$: i.e.,

$$g(t) = f(t) * u(t) = \int_{-\infty}^{\infty} f(t')u(t - t')\, dt'$$

$$= \int_{-\infty}^{t} f(t')\, dt' \tag{5.133}$$

Using the convolution theorem for time functions, we see that

$$G(\omega) = \mathcal{F}\{f(t) * u(t)\} = \mathcal{F}\{f(t)\} \cdot \mathcal{F}\{u(t)\}$$

$$= F(\omega) \cdot \left(\pi\, \delta(\omega) + \frac{1}{j\omega}\right)$$

$$= \pi\, F(\omega)\, \delta(\omega) + \frac{F(\omega)}{j\omega}$$

$$= \pi\, F(0)\, \delta(\omega) + \frac{F(\omega)}{j\omega} \tag{5.134}$$

Thus we have the transform pair of (5.132).

EXAMPLE 5.10. Once a few transform pairs are known, other pairs can be quickly derived through use of the properties of the Fourier transform. For example, consider the transform of the function $f(t)$ given by

$$f(t) = \cos \omega_0 t\, u(t) \tag{5.135}$$

The frequency convolution theorem can be used to obtain $F(\omega)$ as

$$F(\omega) = \mathcal{F}\{\cos \omega_0 t \cdot u(t)\} = \frac{1}{2\pi} \mathcal{F}\{\cos \omega_0 t\} * \mathcal{F}\{u(t)\}$$

$$= \frac{1}{2\pi} [\pi\, \delta(\omega - \omega_0) + \pi\, \delta(\omega + \omega_0)] * \left[\pi\, \delta(\omega) + \frac{1}{j\omega}\right]$$

$$= \frac{\pi}{2} [\delta(\omega - \omega_0) + \delta(\omega + \omega_0)] + \frac{1}{2}\left[\frac{1}{j(\omega - \omega_0)} + \frac{1}{j(\omega + \omega_0)}\right]$$

$$= \frac{\pi}{2} [\delta(\omega - \omega_0) + \delta(\omega + \omega_0)] + \frac{j\omega}{\omega_0^2 - \omega^2} \tag{5.136}$$

$|F(\omega)|$ is shown in Figure 5.31.

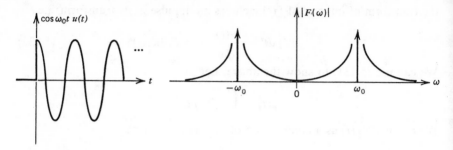

FIGURE 5.31

Periodic Functions

Periodic functions can, of course, be represented as a sum of complex exponentials; and because we can transform complex exponentials by means of (5.121), we should be able to represent a periodic function using the Fourier integral. Assume that $f(t)$ is periodic of period T. We can express $f(t)$ in terms of a Fourier series as

$$f(t) = \sum_{n=-\infty}^{\infty} F_n e^{jn\omega_0 t} \tag{5.137}$$

The Fourier transform of $f(t)$ is therefore

$$F(\omega) = \mathscr{F}\{f(t)\} = \mathscr{F}\left\{ \sum_{n=-\infty}^{\infty} F_n e^{jn\omega_0 t} \right\}$$

$$= \sum_{n=-\infty}^{\infty} F_n \mathscr{F}\{e^{jn\omega_0 t}\}$$

$$= 2\pi \sum_{n=-\infty}^{\infty} F_n \, \delta(\omega - n\omega_0) \tag{5.138}$$

where the F_n's are the Fourier coefficients associated with $f(t)$ and are given by

$$F_n = \frac{1}{T} \int_{-T/2}^{T/2} f(t) e^{-jn\omega_0 t} \, dt \tag{5.139}$$

Equation (5.138) states that the Fourier transform of a periodic function consists of impulses located at the harmonic frequencies of $f(t)$. The area associated with each impulse is equal to 2π times the Fourier coefficient obtained via the exponential Fourier series. Equation (5.138) is really nothing more than an alternate representation of the information contained

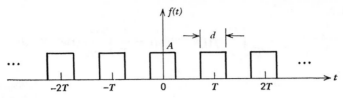

FIGURE 5.32

in the exponential Fourier series. This result should not really be a surprise to anyone familiar with the Fourier series representation.

EXAMPLE 5.11. Find the Fourier transform of a periodic on-off pulse train as shown in Figure 5.32. The Fourier series representation for this function has been found previously as

$$f(t) = \sum_{n=-\infty}^{\infty} F_n e^{jn\omega_0 t}$$

where the Fourier coefficients are

$$F_n = \frac{Ad}{T} \operatorname{sinc}\left(\frac{n\pi d}{T}\right)$$

The Fourier transform of $f(t)$ is therefore

$$\mathscr{F}\{f(t)\} = \frac{2\pi Ad}{T} \sum_{n=-\infty}^{\infty} \operatorname{sinc}\left(\frac{n\pi d}{T}\right) \delta(\omega - n\omega_0)$$

where $\omega_0 = 2\pi/T$. The transform of $f(t)$ consists of impulses located at $\omega = 0, \pm\omega_0, \pm2\omega_0, \ldots$. Each impulse has an area associated with it of value $(2\pi Ad/T) \operatorname{sinc}(n\pi d/T)$, where n is the number of the harmonic.

EXAMPLE 5.12. A special kind of periodic function is the unit impulse train shown in Figure 5.33. This function is useful in applications involving sampling of time waveforms. Because $f(t) = \sum_{k=-\infty}^{\infty} \delta(t - kT)$ is a periodic function, we can expand $f(t)$ in a Fourier series as

$$f(t) = \sum_{n=-\infty}^{\infty} F_n e^{jn\omega_0 t}$$

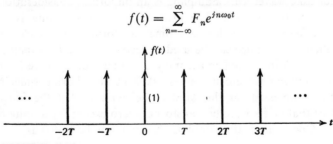

FIGURE 5.33

where

$$F_n = \frac{1}{T} \int_{-T/2}^{T/2} f(t) e^{-jn\omega_0 t} \, dt = \frac{1}{T} \int_{-T/2}^{T/2} \delta(t) e^{-jn\omega_0 t} \, dt$$

$$= \frac{1}{T} \tag{5.140}$$

Each of the F_n's is the same constant, $1/T$. Thus, the Fourier series representation of the unit impulse train is

$$f(t) = \frac{1}{T} \sum_{n=-\infty}^{\infty} e^{jn\omega_0 t} \tag{5.141}$$

If we now take the Fourier transform on both sides of (5.141), we obtain

$$F(\omega) = \mathscr{F}\left\{ \frac{1}{T} \sum_{n=-\infty}^{\infty} e^{jn\omega_0 t} \right\}$$

$$= \frac{2\pi}{T} \sum_{n=-\infty}^{\infty} \delta(\omega - n\omega_0), \qquad \omega_0 = \frac{2\pi}{T}$$

That is,

$$\sum_{k=-\infty}^{\infty} \delta(t - kT) \leftrightarrow \omega_0 \sum_{n=-\infty}^{\infty} \delta(\omega - n\omega_0) \tag{5.142}$$

A unit impulse train in the time domain has as a transform an impulse train in the frequency domain. The area associated with each transform in the frequency domain is ω_0 and the impulses are located at the harmonic frequencies $n\omega_0 = n2\pi/T$, $n = 0, \pm 1, \pm 2, \ldots$.

Table 5.4 summarizes many of the useful Fourier transforms for power signals.

5.10   SAMPLING OF TIME SIGNALS

One of the useful applications of transform techniques concerns the sampling of time waveforms. Sampling is an important consideration in the transmission of information. The idea involved in sampling is, roughly speaking, the following. If values of a time waveform are taken close enough together, then these values can be used to recover all the intervening values of the waveform with complete accuracy. In other words, there is an interdependence between waveform values taken at neighboring time instants. The condition we need for this kind of behavior is that the function be bandlimited: that is, its Fourier transform is zero except for a finite band of frequencies. One form of the sampling theorem may be stated as

TABLE 5.4

Time Functions, $f(t)$	Fourier Transform, $F(\omega)$
1. $k\delta(t)$	k
2. k	$2\pi k\delta(\omega)$
3. $u(t)$	$\pi\delta(\omega) + \dfrac{1}{j\omega}$
4. $\operatorname{sgn}(t)$	$2/j\omega$
5. $\cos \omega_0 t$	$\pi[\delta(\omega - \omega_0) + \delta(\omega + \omega_0)]$
6. $\sin \omega_0 t$	$j\pi[\delta(\omega + \omega_0) - \delta(\omega - \omega_0)]$
7. $e^{j\omega_0 t}$	$2\pi\delta(\omega - \omega_0)$
8. $t\,u(t)$	$j\pi\delta'(\omega) - \dfrac{1}{\omega^2}$
9. $\displaystyle\sum_{k=-\infty}^{\infty} \delta(t - kT)$	$\omega_0 \displaystyle\sum_{n=-\infty}^{\infty} \delta(\omega - n\omega_0), \quad \omega_0 = \dfrac{2\pi}{T}$
10. $\displaystyle\sum_{n=-\infty}^{\infty} F_n e^{jn\omega_0 t}$	$2\pi \displaystyle\sum_{n=-\infty}^{\infty} F_n \delta(\omega - n\omega_0)$
11. $\dfrac{d^n \delta(t)}{dt^n}$	$(j\omega)^n$
12. $|t|$	$\dfrac{-2}{\omega^2}$
13. t^n	$2\pi j^n \dfrac{d^n \delta(\omega)}{d\omega^n}$

A bandlimited signal which has no spectral components above ω_m rad/sec can be uniquely represented by its sampled values spaced at uniform intervals that are not more than $1/2f_m$ seconds apart (where $\omega_m = 2\pi f_m$).

The above statement of the sampling theorem is not the most general statement, because we require uniform spacing of the sampled values. It is this form, however, that is probably most widely used. If the Fourier transform of $f(t)$ is zero beyond a certain frequency $\omega_m = 2\pi f_m$, then the samples of $f(t)$ spaced no further than $T = 1/2f_m$ seconds apart are equivalent to knowing $f(t)$ at all time instants. For example, if $f(t)$ as shown in Figure 5.34, has a transform $F(\omega)$ as shown, then we need consider only the sampled values $f(nT)$, $n = 0, \pm 1, \pm 2, \ldots$ to have complete knowledge of $f(t)$ for every value of t.

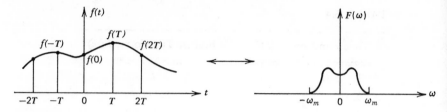

FIGURE 5.34

We can prove the sampling theorem with the help of the frequency convolution theorem. Consider a sampled version of $f(t)$ and denote it by $f_s(t)$. We can represent the sampled version of $f(t)$ by multiplying $f(t)$ by a unit impulse train with period T, equal to the sampling interval. That is,

$$f_s(t) = f(t) \cdot \sum_{n=-\infty}^{\infty} \delta(t - nT) \tag{5.143}$$

Now let us find the spectrum of $f_s(t)$. Using the frequency convolution theorem, we know that $F_s(\omega)$ is the convolution of $F(\omega)$ and the Fourier transform of the impulse train times $1/2\pi$. In symbols

$$F_s(\omega) = \frac{F(\omega) * \omega_0 \sum_{n=-\infty}^{\infty} \delta(\omega - n\omega_0)}{2\pi} \tag{5.144}$$

where $\omega_0 = 2\pi/T$. Thus, substituting for ω_0, we obtain the spectrum of the sampled values as

$$F_s(\omega) = \frac{1}{T} \sum_{n=-\infty}^{\infty} (F(\omega) * \delta(\omega - n\omega_0))$$

$$= \frac{1}{T} \sum_{n=-\infty}^{\infty} F(\omega - n\omega_0) \tag{5.145}$$

The right-hand side of (5.145) represents the function $F(\omega)$ repeating itself every ω_0 rad/sec. If the width of the band of nonzero frequencies in $F(\omega)$ is less than spacing between repeats of $F(\omega)$, then the repeated versions of $F(\omega)$ do not overlap. That is, $F(\omega)$ will repeat periodically in the frequency domain without overlap, provided that $\omega_0 \geq 2\omega_m$, which implies that

$$\frac{2\pi}{T} \geq 2(2\pi f_m)$$

or

$$T \leq \frac{1}{2f_m} \tag{5.146}$$

As long as we sample $f(t)$ at intervals not more than $1/2f_m$ seconds apart,

$F_s(\omega)$ will be a periodic replica of $F(\omega)$. This result can be also shown graphically as in Figure 5.35. The spectrum of $F_s(\omega)$ is the convolution of the impulse train $(\omega_0/2\pi) \sum_{n=-\infty}^{+\infty} \delta(\omega - n\omega_0)$ and $F(\omega)$. Because the impulse train $(1/T) \sum \delta(\omega - n\omega_0)$ is an even function of ω, the mirror image is the same as the original function. To perform the convolution, we shift the impulse train past $F(\omega)$. The impulses are separated by ω_0. Thus the reproduced versions of $F(\omega)$ that make up $F_s(\omega)$ are also spaced ω_0 rad/sec apart. This yields the $F_s(\omega)$ spectrum shown in Figure 5.35. Notice that the repeated spectrum is multiplied by $1/T$.

We can easily recover $F(\omega)$ (and so $f(t)$) from $F_s(\omega)$ by merely filtering out all frequency components above ω_m rad/sec. This can be accomplished by a low-pass filter which allows transmission of all frequencies below ω_m and attenuates all frequencies above ω_m. Such a filter characteristic is shown dotted in Figure 5.35.

The interval $T = 1/2f_m$ is called the *Nyquist interval*. If we sample $f(t)$ with an interval between samples larger than $1/2f_m$ sec, then the repeated versions of $F(\omega)$ that make up $F_s(\omega)$ overlap and $F(\omega)$ cannot be recovered from $F_s(\omega)$ without some error. This, of course, is what our intuition would indicate. If samples are spaced too far apart, it seems logical that the signal $f(t)$ could not be recovered from its samples.

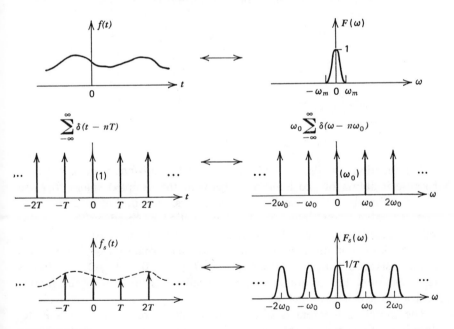

FIGURE 5.35

The process of recovering $f(t)$ from its sampled version $f_s(t)$ is accomplished by passing $f_s(t)$ through a low-pass filter. Mathematically, we can show this result using the theorem on time convolution. Let the sampling rate be $T = 1/2f_m$ and so $\omega_0 = 2\pi/T = 4\pi f_m = 2\omega_m$. The spectrum of $f_s(t)$ is given by (5.145).

$$F_s(\omega) = \frac{1}{T} \sum_{n=-\infty}^{\infty} F(\omega - n\omega_0) = \frac{1}{T} \sum_{n=-\infty}^{\infty} F(\omega - 2n\omega_m) \qquad (1.147)$$

The process of low-pass filtering is equivalent to multiplying $F_s(\omega)$ by a function of ω, which is 1 for $\omega < |\omega_m|$ and 0 otherwise: i.e., a gate function $G_{2\omega_m}(\omega)$.

$$G_{2\omega_m}(\omega) = \begin{cases} 1, & \omega < |\omega_m| \\ 0, & \omega > |\omega_m| \end{cases} \qquad (5.148)$$

Thus

$$\frac{F(\omega)}{T} = F_s(\omega)G_{2\omega_m}(\omega)$$

or

$$F(\omega) = TF_s(\omega)G_{2\omega_m}(\omega) \qquad (5.149)$$

The time convolution theorem applied to (5.149) yields

$$f(t) = Tf_s(t) * \frac{\omega_m}{\pi} \operatorname{sinc} \omega_m t$$

$$= f_s(t) * \operatorname{sinc} \omega_m t$$

$$= \sum_{n=-\infty}^{\infty} f(nT)\, \delta(t - nT) * \operatorname{sinc} \omega_m t$$

$$= \sum_{n=-\infty}^{\infty} f(nT) \operatorname{sinc} [\omega_m(t - nT)] \qquad (5.150)$$

The function $\operatorname{sinc} [\omega_m(t - nT)]$ is often called an *interpolation function*, because it allows one to interpolate between the sampled values $f(nT)$ to find $f(t)$ for all t. Graphically, the result is shown in Figure 5.36. Each sampled value is multiplied by the sinc function centered on the sampled value. These functions are then added to give the original waveform. Notice that (5.150) is nothing more than an orthogonal expansion of $f(t)$ in terms of the basis set of functions $\{\operatorname{sinc} [\omega_m(t - nT)]\}$ $n = 0, \pm1, \pm2, \ldots$. In this representation, the generalized Fourier coefficients are merely the sample values of the original function.

The sampling theorem is often used in the representation of a finite-duration signal, say on the interval $[0, T]$. Sampling at the Nyquist rate

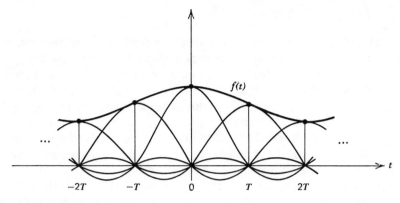

FIGURE 5.36

implies that one needs $2f_m T$ samples to represent a signal which is band-limited to f_m Hz. This use of the sampling theorem is, strictly speaking, erroneous, because a signal cannot be simultaneously both of finite duration and also bandlimited. The discontinuities at each end of the time signal cause the spectrum to exist for all frequencies. One way of showing this fact is as follows. Let $f(t)$ be of finite time duration, for example $f(t) = 0$ for $|t| > T/2$. We can write $f(t)$ as

$$f(t) = g_T(t) f(t) \tag{5.151}$$

where $g_T(t)$ is the gate function

$$g_T(t) = \begin{cases} 1, & |t| < \dfrac{T}{2} \\[2mm] 0, & |t| > \dfrac{T}{2} \end{cases} \tag{5.152}$$

The spectrum of $f(t)$ is therefore

$$F(\omega) = \frac{T}{2\pi} \operatorname{sinc}\left(\frac{\omega T}{2}\right) * F(\omega) \tag{5.153}$$

No matter what we assume for the spectrum of $F(\omega)$ on the right-hand side, $F(\omega)$ on the left-hand side of (5.153) exists for all ω because of the convolution with sinc $(\omega T/2)$. Thus, we conclude that if $f(t)$ is of finite duration, its spectrum exists for all ω. However, the error in representing a finite-duration signal on $[0, T]$ by $2f_m T$ samples is often minor and the loss of information using the samples is usually more than compensated for by the savings in complexity.

5.11   MODULATION

Modulation is the process of translating the frequency spectrum of a function. Modulation is used in communication systems to make the transmission process more efficient. For example, a time signal can be radiated effectively by an antenna only if the radiating antenna is of the order of one tenth or more of the wavelength of the frequencies comprising $f(t)$. If the signal to be transmitted is human speech, for instance, the maximum signal frequency is about 10,000 Hz. This frequency corresponds to a wavelength of 30,000 m. An efficient radiating antenna for human-speech signals would thus be on the order of 3000 m. If, however, we modulate or shift the frequency spectrum to a higher band of frequencies, we can reduce the antenna size accordingly. In actual practice, all radio and television signals are modulated for efficient transmission.

The understanding of the basic principles in modulation follows directly from the frequency convolution theorem. Suppose $f(t)$ is a signal with spectrum $F(\omega)$ centered about $\omega = 0$. If we want to shift the spectrum to a frequency ω_c, we need only to multiply $f(t)$ by $\cos \omega_c t$. That is, if

$$f(t) \leftrightarrow F(\omega)$$

then

$$f(t) \cos \omega_c t \leftrightarrow \frac{F(\omega + \omega_c) + F(\omega - \omega_c)}{2} \tag{5.154}$$

The signal $f(t)$ is called the *modulating signal*. The function $\cos \omega_c t$ is called the *carrier* or the *modulated signal*. Because multiplying $\cos \omega_c t$ by $f(t)$ varies the amplitude of the carrier signal, this modulation process is called *amplitude modulation*. The process of recovering $f(t)$ from the translated version is called *demodulation* or *detection*. Demodulation amounts to the inverse operation of modulation: i.e., frequency shifting down in frequency rather than up in frequency.

Schematically, this process is shown in Figure 5.37. The spectrum of $f(t)$ is shown. The modulated signal has a spectrum given by (5.154). To demodulate, we multiply $f(t) \cos \omega_c t$ by $\cos \omega_c t$ and then low-pass filter. Now the spectrum of $f(t) \cos \omega_c t \cdot \cos \omega_c t$ is the modulated spectrum convolved with the spectrum of $\cos \omega_c t$: i.e.,

$$\mathscr{F}\{f(t) \cos \omega_c t \cdot \cos \omega_c t\} = \frac{F(\omega + \omega_c) + F(\omega - \omega_c)}{2} * \frac{\delta(\omega + \omega_c) + \delta(\omega - \omega_c)}{2}$$

$$= \tfrac{1}{2}F(\omega) + \tfrac{1}{4}F(\omega + 2\omega_c) + \tfrac{1}{4}F(\omega - 2\omega_c) \tag{5.155}$$

The low-pass filter attenuates the frequency components centered about $\pm 2\omega_c$ and thus permits recovery of the original spectrum. The process of

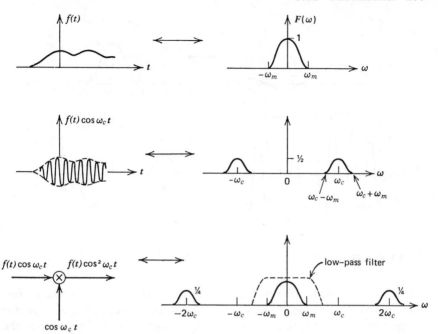

FIGURE 5.37

multiplying by $\cos \omega_c t$ at the receiver to recover the original spectrum is known as *coherent* or *synchronous* demodulation. This demodulation process depends on the fact that the one has a method of generating $\cos \omega_c t$ exactly. If the frequency of the cosine function at the receiver is not ω_c then one cannot recover the original signal without some error. It takes some expense to build very stable oscillators for generating the receiver carrier. Thus we often use other forms of demodulation which do not depend upon the generation of a local oscillator signal.

One method of circumventing the generation of a local oscillator is to transmit a large amount of a carrier signal. That is, we transmit $[A + f(t)] \cos \omega_c t$ instead of merely $f(t) \cos \omega_c t$. Here $A > |f(t)|_{\max}$. The spectrum of a typical time function might appear as shown in Figure 5.38. The amplitude-modulated signal and its spectrum are also shown. The impulses in $A(\omega)$ represent the carrier $A \cos \omega_c t$. The usual method of demodulating $a(t)$ to recover $f(t)$ is to use an envelope detector, as shown in Figure 5.39. The envelope detector has an output that follows the envelope of the modulated signal. On a peak voltage cycle, the capacitor charges to the peak of the modulated signal. As the modulated signal drops, the diode is cut off and the capacitor discharges through the resistance R at a rate depending on the

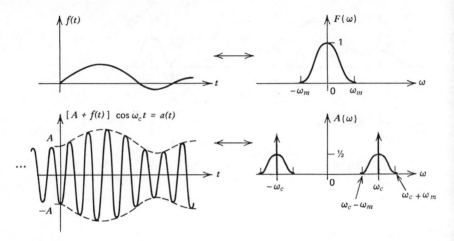

FIGURE 5.38

RC time constant. During the next cycle, the diode starts to conduct when the input voltage becomes greater than the capacitor voltage. The capacitor charges to the peak voltage of this new cycle and then discharges when the diode is cut off. During the cut-off period, the capacitor voltage does not change appreciably. The *RC* time constant is adjusted to follow the envelope of the modulated signal. The output voltage $y(t)$ contains a ripple frequency ω_c which can be smoothed by another low-pass filter. This type of demodulation is almost always used for detecting amplitude-modulation signals.

5.12 TRANSMISSION OF SIGNALS THROUGH LINEAR FILTERS

A system can be thought of as a method of processing an input signal $x(t)$ to form an output signal $y(t)$. The processing, of course, depends on the

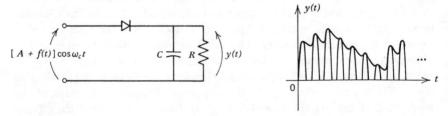

FIGURE 5.39

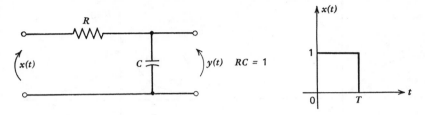

FIGURE 5.40

characteristic of system, which we can characterize by $h(t)$ or $H(\omega)$, the impulse response and its Fourier transform, respectively. The output $y(t)$ for any input $x(t)$ can be found by convolution: i.e.,

$$y(t) = x(t) * h(t) \tag{5.156}$$

In the transform domain, the above is

$$Y(\omega) = X(\omega) H(\omega) \tag{5.157}$$

The system thus modifies or filters the spectrum of the input. In other words, $H(\omega)$ changes the relative importance of the frequencies contained in the input signal both in amplitude and in phase. The filtering of the input spectrum depends on the transfer function $H(\omega)$. That is, $H(\omega)$ weights the various frequency components of $x(t)$. This filtering process is most easily interpreted in the frequency domain.

For example, consider the low-pass filter approximation provided by the RC network of Figure 5.40. The impulse response of this system is (see Example 2.15)

$$h(t) = e^{-t}u(t) \tag{5.158}$$

Thus the transfer function $H(\omega)$ is

$$H(\omega) = \mathscr{F}\{h(t)\} = \int_0^\infty e^{-t}e^{-j\omega t}\, dt = \frac{1}{1 + j\omega} \tag{5.159}$$

This system has a filter characteristic shown in Figure 5.41, where both $|H(\omega)|$ and $\theta(\omega)$ are plotted. If we excite this system with the pulse input shown in Figure 5.40, we can obtain the output spectrum of $y(t)$ from (5.157). Thus

$$|Y(\omega)| = |X(\omega)H(\omega)| = |X(\omega)|\, |H(\omega)| \tag{5.160}$$

and

$$\theta_y(\omega) = \theta_x(\omega) + \theta_h(\omega) \tag{5.161}$$

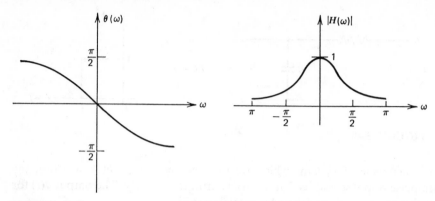

FIGURE 5.41

The input amplitude spectrum is thus filtered or shaped by the form of $|H(\omega)|$. Likewise, the output phase is obtained by adding the phase characteristic of the system to the phase characteristic of the input. Figure 5.42 depicts the input and output amplitude spectra. The system in this example attenuates the higher-frequency components in $x(t)$ and thus approximates a low-pass filter. The output time function $y(t)$ is a distorted version of the input pulse, as shown in Figure 5.42. The attenuation of the high frequencies contained in $x(t)$ means that the output has smoother corners, because it is the high frequencies in $x(t)$ that combine to produce its sharp transitions. Because the system attenuates all high-frequency components, the output voltage cannot change as rapidly as the input voltage.

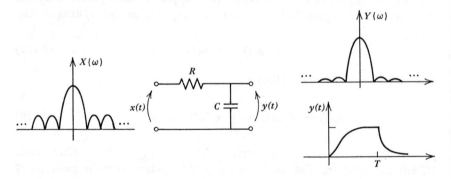

FIGURE 5.42

A Distortionless Filter

The preceding example was one in which the system distorted the input signal as it passed through the system. In certain applications, this effect is

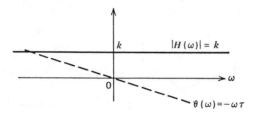

FIGURE 5.43

precisely what is desired. However, in other applications, we may desire distortionless transmission. In distortionless transmission, we permit a scale change of the input waveform (a gain change) and a time delay of the input signal, provided that the shape or form of the input signal is unchanged. Thus, if $x(t)$ is the input signal, a distortionless system has as an output $kx(t - \tau)$, where k is the gain change and τ the time delay. We can now deduce the frequency characteristic of a distortionless system. Let the Fourier transform of $x(t)$ be $X(\omega)$. By the time-shifting theorem,

$$\mathscr{F}\{kx(t - \tau)\} = k\, X(\omega)\, e^{-j\omega\tau} \qquad (5.162)$$

we have the relationship between input and output as

$$X(\omega)\, H(\omega) = k\, X(\omega)\, e^{-j\omega\tau} \qquad (5.163)$$

As we compare the right-hand and left-hand sides of (5.163), it is clear that

$$H(\omega) = ke^{-j\omega\tau} \qquad (5.164)$$

The magnitude of the transfer function is $|H(\omega)| = k$, which is independent of ω. The phase shift of this transfer function is $-\omega\tau$. A plot of $H(\omega)$ is shown in Figure 5.43. A filter is said to be *amplitude distorted* if $|H(\omega)|$ is not a constant, and *phase distorted* if $\theta(\omega)$ is not linear.

Linear Phase Systems

Consider a system which has no phase distortion: i.e., the phase shift of the filter is linear. A typical transfer function is shown in Figure 5.44. The analytic form of a linear phase system is

$$H(\omega) = |H(\omega)|\, e^{-j\omega\tau} \qquad (5.165)$$

The impulse response of this system is

$$h(t) = \mathscr{F}^{-1}\{|H(\omega)|\, e^{-j\omega\tau}\}$$

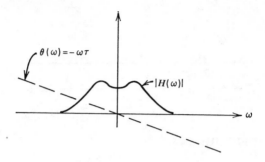

FIGURE 5.44

$$= \frac{1}{2\pi} \int_{-\infty}^{\infty} |H(\omega)| \, e^{-j\omega\tau} e^{j\omega t} \, d\omega$$

$$= \frac{1}{\pi} \int_{0}^{\infty} |H(\omega)| \cos \left[\omega(t - \tau) \right] \, d\omega \qquad (5.166)$$

This impulse response is symmetrical about τ, because

$$h(t + \tau) = h(\tau - t)$$

Also, the maximum value of $h(t)$ is reached at $t = \tau$.

$$h_{max} = h(\tau) = \frac{1}{\pi} \int_{0}^{\infty} |H(\omega)| \, d\omega \qquad (5.167)$$

Any other value of t can only decrease the value of the integrand in (5.166), because $\cos \omega t$ takes on its maximum value at $t = 0$ for all ω. We conclude from these two properties that $h(t)$ has the general form sketched in Figure 5.45. The impulse response is, in general, nonzero for $t < 0$.

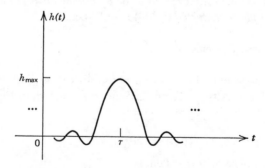

FIGURE 5.45

Ideal Filters

An ideal filter is a system which transmits without distortion all the frequencies in a certain band. The amplitude spectrum over the band is a constant and the phase spectrum over the band is linear. Figure 5.46 depicts the filter characteristics for the ideal low-pass and band-pass filters.

The ideal low-pass filter thus is defined by the transfer $H(\omega)$ given by

$$H(\omega) = \begin{cases} e^{-j\omega t_0}, & \omega < |\omega_m| \\ 0, & \omega > |\omega_m| \end{cases} \qquad (5.168)$$

The frequency ω_m is often called the cut-off frequency of the filter. We can find the impulse response of this system by taking the inverse Fourier transform of (5.168). Thus

$$\begin{aligned} h(t) &= \mathscr{F}^{-1}\{H(\omega)\} \\ &= \frac{1}{2\pi} \int_{-\infty}^{\infty} H(\omega)e^{j\omega t}\, d\omega \\ &= \frac{1}{2\pi} \int_{-\omega_m}^{\omega_m} e^{j\omega(t-t_0)}\, d\omega = \frac{1}{2\pi j(t-t_0)}\, e^{j\omega(t-t_0)}\Big|_{-\omega_m}^{\omega_m} \\ &= \frac{\omega_m}{\pi}\, \text{sinc}\,[\omega_m(t-t_0)] \end{aligned} \qquad (5.169)$$

A sketch of the impulse response is shown in Figure 5.46. We conclude from the impulse-response function that the peak value of the response, ω_m/π, is proportional to the cut-off frequency. The width of the main pulse is $2\pi/\omega_m$.

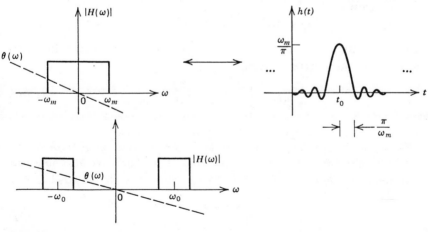

FIGURE 5.46

It is convenient to characterize this width as the *effective duration* of the output pulse T_d. Notice that as $\omega_m \to \infty$ the filter becomes an all-pass filter, $T_d \to 0$, and the output response peak $\to \infty$. In other words, the output response approaches the input, an impulse. Again, the impulse response is not causal, of course, because of the ideal characteristic of the amplitude spectrum. If we limit the frequency spectrum of $h(t)$ to be nonzero for a finite band of frequencies, then $h(t)$ must exist for all time.

The step response $g(t)$ of this ideal low-pass filter can be obtained by integrating the impulse response, as we discussed in Chapter 2. Thus

$$g(t) = \int_{-\infty}^{t} h(t')\, dt' = \frac{1}{\pi} \int_{-\infty}^{t} \frac{\sin\,(\omega_m(t' - t_0))}{t' - t_0}\, dt'$$

In the above integral, let $x = \omega_m(t' - t_0)$. Then we have

$$g(t) = \frac{1}{\pi} \int_{-\infty}^{\omega_m(t-t_0)} \frac{\sin x}{x}\, dx$$

$$= \frac{1}{\pi} \int_{-\infty}^{0} \frac{\sin x}{x}\, dx + \frac{1}{\pi} \int_{0}^{\omega_m(t-t_0)} \frac{\sin x}{x}\, dx \qquad (5.170)$$

The integral $\int_0^y (\sin x)/x\, dx$ is a tabulated function known as the sine-integral function and denoted by

$$Si(y) \triangleq \int_0^y \text{sinc } x\, dx$$

From the properties of the sinc function, we can deduce the following:

(1) $Si(y)$ is an odd function: i.e., $Si(y) = -Si(-y)$
(2) $Si(0) = 0$
(3) $Si(\infty) = \pi/2,\ Si(-\infty) = -\pi/2$

A sketch of $Si(y)$ is shown in Figure 5.47. Using the sine-integral function, the step response can be expressed as

$$g(t) = \frac{1}{2} + \frac{1}{\pi} Si[\omega_m(t - t_0)] \qquad (5.171)$$

A sketch of the step response $g(t)$ is shown in Figure 5.48.

We again observe the distortion in the step input as it passes through this system, resulting from the limited pass band of the filter. We also note that the response is nonzero before $t = 0$. As $\omega_m \to \infty$, the response $g(t)$ becomes

$$g(t) = \begin{cases} \frac{1}{2} + Si(-\infty) = \frac{1}{2} - \frac{1}{2} = 0, & t < t_0 \\ \frac{1}{2} + Si(\infty) = \frac{1}{2} + \frac{1}{2} = 1, & t > t_0 \end{cases} \qquad (5.172)$$

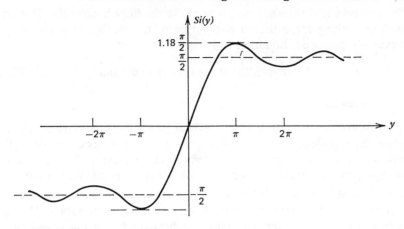

FIGURE 5.47

i.e., $g(t)$ approaches a delayed unit step $u(t - t_0)$, as it should. Notice that the abrupt rise of the input step corresponds to the more gradual rise of the output in the region denoted by t_r. The rise time or buildup of the response of $g(t)$ can be related to the cut-off frequency ω_m. Define the rise time t_r as the interval between the intercepts of the tangent at $t = 0$ with the lines $g(t) = 0$ and $g(t) = 1$. From Figure 5.48, we conclude that

$$\frac{dg(t)}{dt}\bigg|_{t=t_0} = \frac{1}{t_r} = \frac{d}{dt}\left(\frac{1}{\pi} Si[\omega_m(t - t_0)]\bigg|_{t=t_0}\right) = \frac{\omega_m}{\pi}$$

Hence

$$t_r = \frac{\pi}{\omega_m} \tag{5.173}$$

or

$$\omega_m t_r = \pi \tag{5.174}$$

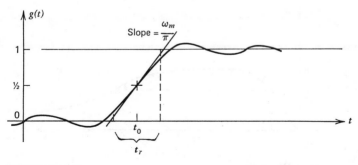

FIGURE 5.48

The rise time t_r is inversely proportional to the filter bandwidth. This statement is nothing more than a paraphrase of the scaling property we have previously discussed. Equation (5.174) indicates that

$$(\text{bandwidth}) \times (\text{rise time}) = \text{constant} \tag{5.175}$$

Bandwidth

Bandwidth is a number we attach to an amplitude spectrum to characterize the significant band of frequencies in a signal spectrum or transfer function. There are several ways of defining the bandwidth for a system. For example, one definition that is often used is that the bandwidth of the system $H(\omega)$ is that interval of frequencies for which $|H(\omega)|$ remains within $1/\sqrt{2}$ (within 3 db) of its maxium value. If the amplitude spectrum is $|H(\omega)|$ as shown in Figure 5.49, the bandwidth by this definition is $(\omega_2 - \omega_1)$ rad/sec.

Another approach is to define the bandwidth of a spectrum $H(\omega)$ by the following:

$$2Bw = \frac{1}{|H(\omega)|_{\max}} \int_{-\infty}^{\infty} |H(\omega)| \, d\omega \tag{5.176}$$

Equation (5.176) is equivalent to replacing the amplitude spectrum by a rectangle with area equal to the area under the curve $|H(\omega)|$ and height equal to $|H(\omega)|_{\max}$. This definition of bandwidth is shown schematically in Figure 5.50.

Energy Spectra Through Linear Systems

The energy spectrum of a signal $f(t)$ is

$$S(\omega) = \frac{1}{\pi} |F(\omega)|^2$$

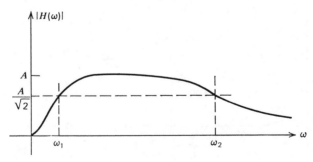

FIGURE 5.49

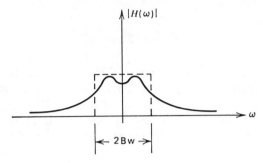

FIGURE 5.50

The energy spectrum of the output $y(t)$ of a linear system with input $f(t)$ is therefore

$$S_{out}(\omega) = \frac{1}{\pi} |Y(\omega)|^2$$

$$= \frac{1}{\pi} |F(\omega) H(\omega)|^2 = \frac{1}{\pi} |F(\omega)|^2 |H(\omega)|^2$$

$$= S_{in}(\omega) |H(\omega)|^2 \qquad (5.177)$$

where $H(\omega)$ is the transfer function of the linear system. Notice that the output energy spectrum is independent of the phase characteristics of both the input and the system.

5.13 NUMERICAL CALCULATION OF FOURIER TRANSFORMS—DISCRETE FOURIER TRANSFORMS

The problem of computing a Fourier transform numerically for a given time function is important if one wishes to use digital computers in computing Fourier transforms.

Suppose that we approximate the Fourier integral of $f(t)$ by a sum: i.e.,

$$F(\omega) = \int_{-\infty}^{\infty} f(t)e^{-j\omega t}\, dt$$

$$\cong \sum_{n=-\infty}^{\infty} f(n\,\Delta t)e^{-j\omega n\Delta t}\, \Delta t = \tilde{F}(\omega) \qquad (5.178)$$

This approximation is equivalent to Simpson's trapezoidal rule. Denote the approximating sum in (5.178) by $\tilde{F}(\omega)$. How close is $\tilde{F}(\omega)$ to $F(\omega)$ and what factors determine the goodness of the approximation?

We note from (5.178) that $\tilde{F}(\omega)$ is periodic in ω of period $\Omega = 2\pi/\Delta t$

because

$$\tilde{F}(\omega) = \tilde{F}\left(\omega + \frac{2\pi}{\Delta t}\right)$$

In general, of course, $F(\omega)$ is not periodic. However, the approximation $\tilde{F}(\omega)$ is always periodic. The reason $\tilde{F}(\omega)$ is periodic is that our approximation is based on sampling the original function $f(t)$. Whenever a function is sampled periodically, the transform of the sampled function is periodic. In symbols, if we denote by $f_s(t)$ the sampled version of $f(t)$, we can write

$$f_s(t) = f(t) \sum_{n=-\infty}^{\infty} \delta(t - n\,\Delta t) \tag{5.179}$$

The spectrum of $f_s(t)$, by the frequency convolution property, is

$$F_s(\omega) = \mathscr{F}\{f_s(t)\} = \frac{\mathscr{F}\{f(t)\} * \mathscr{F}\left\{\sum_n \delta(t - n\,\Delta t)\right\}}{2\pi}$$

$$= \frac{F(\omega) * \omega_0 \sum_n \delta(\omega - n\omega_0)}{2\pi}, \qquad \omega_0 = \frac{2\pi}{\Delta t}$$

$$= \frac{1}{\Delta t} \sum_{n=-\infty}^{\infty} F(\omega - n\omega_0) \tag{5.180}$$

We can also write the transform of (5.179) as

$$\mathscr{F}\{f_s(t)\} = \mathscr{F}\left\{\sum_{n=-\infty}^{\infty} f(n\,\Delta t)\,\delta(t - n\,\Delta t)\right\}$$

$$= \sum_{n=-\infty}^{\infty} f(n\,\Delta t)\mathscr{F}\{\delta(t - n\,\Delta t)\}$$

$$= \sum_{n=-\infty}^{\infty} f(n\,\Delta t)e^{-j\omega n\Delta t} \tag{5.181}$$

Because (5.180) and (5.181) are equal, we have

$$\sum_{n=-\infty}^{\infty} f(n\,\Delta t)e^{-j\omega n\Delta t} = \frac{1}{\Delta t} \sum_{n=-\infty}^{\infty} F(\omega - n\omega_0), \qquad \omega_0 = \frac{2\pi}{\Delta t} \tag{5.182}$$

From (5.182), we see that $\tilde{F}(\omega)$ is

$$\tilde{F}(\omega) = \sum_{n=-\infty}^{\infty} f(n\,\Delta t)e^{-j\omega n\Delta t}\,\Delta t = \sum_{n=-\infty}^{\infty} F(\omega - n\omega_0) \tag{5.183}$$

Thus the sum in (5.178) does not give $F(\omega)$, but rather an aliased version of $F(\omega)$. This relationship is illustrated in Figure 5.51.

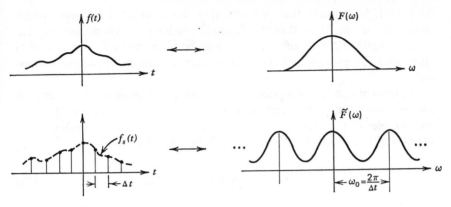

FIGURE 5.51

How do we obtain a good approximation of $F(\omega)$? From (5.183) and Figure 5.49, we see that as ω_0 is increased, $\tilde{F}(\omega)$ approaches $F(\omega)$, at least for $|\omega| < \omega_0/2$. However, increasing ω_0 means decreasing Δt, the sampling interval. Intuitively, this is just what we would expect. To better approximate the integral in (5.178), we should take smaller steps in forming the approximating sum. Thus Δt should be chosen small enough so that $\tilde{F}(\omega)$ is sufficiently close to $F(\omega)$. Of course, if we do not know $F(\omega)$, we may have to try successively smaller values of Δt until results remain essentially unchanged as Δt is further reduced.

As a further practical note, we observe that the sum in (5.178) cannot be taken over an infinite number of samples of $f(t)$ if we are actually to perform the computations. We must assume that $f(t)$ is essentially of finite time duration and, having picked Δt, sum over all the nonzero values of $f(n\,\Delta t)$. The approximation $\tilde{F}(\omega)$ then becomes

$$\tilde{F}(\omega) = \sum_{n=n_1}^{n_2} f(n\,\Delta t)e^{-j\omega n\Delta t}\,\Delta t \qquad (5.184)$$

There is an analogous problem of calculating the inverse transform numerically. That is, given $F(\omega)$, find the corresponding time function. In this case, the approximating sum is

$$\tilde{f}(t) = \frac{1}{2\pi}\sum_{m=-\infty}^{\infty} F(m\,\Delta\omega)e^{jm\Delta\omega t}\,\Delta\omega \qquad (5.185)$$

where $\Delta\omega$ is the sampling period in ω. Using arguments exactly like those for $\tilde{F}(\omega)$, we find that $\tilde{f}(t)$, is periodic of period $T = 2\pi/\Delta\omega$ and that $\tilde{f}(t)$ in (5.185) is an aliased version of $f(t)$: i.e.,

$$\tilde{f}(t) = \sum_{k} f(t - kT), \qquad T = \frac{2\pi}{\Delta\omega} \qquad (5.186)$$

As before, to improve the approximation, we want T to be large (which means $\Delta\omega$ is small) so that the repeated versions of $f(t)$ do not overlap greatly. Again, the sum of (5.185) actually will be over a finite range of m although large enough to include all significant portions of the given transform.

The preceding discussion points to one motivation for developing a discrete Fourier transform pair relating time samples of a given function $\{f(n\,\Delta t)\}$ and samples of the spectrum $\{F(k\,\Delta\omega)\}$. Suppose now that $\{f(n\,\Delta t)\}$ are N samples of a time function $f(t)$. The sampled time function is

$$f_s(t) = \sum_{n=0}^{N-1} f(n\,\Delta t)\,\delta(t - n\,\Delta t) \tag{5.187}$$

The spectrum is of $f_s(t)$ is thus

$$F_s(\omega) = \sum_{n=0}^{N-1} f(n\,\Delta t)e^{-j\omega n\Delta t} \tag{5.188}$$

If we now sample $F_s(\omega)$ N times, we obtain an expression for $F_s(m\,\Delta\omega)$ as

$$F_s(m\,\Delta\omega) = \sum_{n=0}^{N-1} f(n\,\Delta t)e^{-jm\Delta\omega n\Delta t}, \qquad m = 0, 1, \ldots N-1 \tag{5.189}$$

Thus, given a sequence of samples $\{f(n\,\Delta t)\}$, we obtain the sampled spectrum for these samples $F_s(m\,\Delta\omega)$, as given in (5.189). The sampling interval in frequency is chosen as

$$\Delta\omega = \frac{2\pi}{N\,\Delta t} \tag{5.190}$$

The sample values $\{f(n\,\Delta t)\}$ can be obtained from the sampled spectrum $F_s(m\,\Delta\omega)$ by multiplying (5.189) by $\exp\{jm\,\Delta\omega l\,\Delta t\}$ and summing over m. Equation (5.189) becomes

$$\sum_{m=0}^{N-1} F_s(m\,\Delta\omega)e^{jm\Delta\omega l\Delta t} = \sum_{m=0}^{N-1}\sum_{n=0}^{N-1} f(n\,\Delta t)e^{-jm\Delta\omega n\Delta t}e^{jm\Delta\omega l\Delta t}$$

$$= \sum_{m=0}^{N-1}\sum_{n=0}^{N-1} f(n\,\Delta t)e^{j\Delta\omega\Delta t m(l-n)} \tag{5.191}$$

Recall that

$$\sum_{k=0}^{N} \alpha^k = \begin{cases} \dfrac{1 - \alpha^{N+1}}{1 - \alpha}, & \alpha \neq 1 \\[2mm] N + 1, & \alpha = 1 \end{cases}$$

Thus

$$\sum_{m=0}^{N-1} e^{j\Delta\omega\Delta t m(l-n)} = \frac{1 - e^{j\Delta\omega\Delta t(l-n)N}}{1 - e^{j\Delta\omega\Delta t(l-n)}}$$

$$= \begin{cases} 0, & l \neq n \\ N, & l = n \end{cases} \qquad (5.192)$$

Using (5.192), we can evaluate the right-hand side of (5.191). Interchanging the sum over m and n, we have

$$\sum_{n=0}^{N-1} f(n \Delta t) \sum_{m=0}^{N-1} e^{j\Delta\omega\Delta t m(l-n)} = \sum_{m=0}^{N-1} F_s(m \Delta\omega)e^{jm\Delta\omega l\Delta t}$$

The inner sum yields, by (5.192), the value N for $l = n$ and zero otherwise. Thus only a single term in the outer sum is nonzero, namely for $l = n$. Thus (5.191) can be written as

$$\sum_{m=0}^{N-1} F_s(m \Delta\omega)e^{jm\Delta\omega l\Delta t} = f(l \Delta t) \cdot N \qquad (5.193)$$

Thus we have the pair of formulas given in (5.189) and (5.193) relating $F_s(m \Delta\omega)$ and $f(l \Delta t)$.

$$F_s(m \Delta\omega) = \sum_{n=0}^{N-1} f(n \Delta t)e^{-jm\Delta\omega n\Delta t} \qquad (5.194)$$

$$f(l \Delta t) = \frac{1}{N} \sum_{m=0}^{N-1} F_s(m \Delta\omega)e^{jm\Delta\omega l\Delta t} \qquad (5.195)$$

Equations (5.194) and (5.195) are a discrete Fourier transform pair. Equation (5.194) is the *discrete Fourier transform* (DFT) of the sequence $\{f(n \Delta t)\}$. Because these formulas are both discrete and involve only sequences of numbers, they can easily be computed by means of digital computers. In fact, there are very efficient means for computing the sums of (5.194) and (5.195).

Notice that in a straightforward evaluation of these sums, we must perform N multiplications and additions for each value of $F_s(m \Delta\omega)$ or $f(l \Delta t)$ we wish to calculate. Thus, to calculate the total sampled spectrum $\{F_s(m \Delta\omega)\}$, we would need to perform N^2 operations. However, if we arrange the computations in a particular way, we can obtain the same results with the number of operations approximately $N \log_2 N$. This procedure is known as the *Fast Fourier Transform* (FFT) algorithm. For example, if $N = 256$, the FFT allows one to compute in $N \log_2 N = \frac{1}{32}$ of the time required by a straightforward calculation. The FFT algorithm is described in detail by Cooley et al.[8] together with a sample FORTRAN program.

[8] Cooley, J. W., P. A. Lewis, and P. D. Welch, "The Fast Fourier Transform and Its Applications," *IEEE Transactions on Education* E-12(1), 27–34 (March 1969).

Thus far, we have assumed $f(t)$ to be of finite time duration. Suppose it is not. We can proceed in two ways.

For one method, multiply $f(t)$ by a gating function which is 1 on the time interval of interest and 0 elsewhere. That is, multiply $f(t)$ by $g_D(t)$, where

$$g_D(t) = \begin{cases} 1, & |t| \le D/2 \\ \\ 0, & |t| > D/2 \end{cases}$$

Now form a periodic function $f_p(t)$ by extending $f(t)g_D(t)$ and sample $f_p(t)$ every Δt sec to obtain the samples $\{f_p(n\,\Delta t)\}$, $n = 0, \pm 1, \pm 2, \ldots, \pm N$. The spectrum of $f(t)g_D(t)$ is

$$\mathscr{F}\{f(t)g_D(t)\} = \frac{F(\omega) * D\sin c\,(\omega D/2)}{2\pi} \triangleq F_2(\omega)$$

If we apply the DFT to the samples $\{f(n\,\Delta T)\}$, we obtain

$$F_m \triangleq F(m\,\Delta\omega) = \sum_{n=0}^{N-1} f(n\,\Delta t)e^{-jm\Delta\omega n\Delta t}$$

$$= \sum_{n=0}^{N-1} f_n e^{-j(m2\pi n\Delta t/N\Delta t)}$$

$$= \sum_{n=0}^{N-1} f_n e^{-j(mn2\pi/N)} \tag{5.196}$$

where $f_n \triangleq f(n\,\Delta t)$ and $F_m \triangleq F(m\,\Delta\omega)$. The numbers $F_0, F_{\pm 1}, F_{\pm 2}, \ldots$, are samples of the aliased form of $F_2(\omega) = F(\omega) * G_D(\omega)/2\pi$: ie, if $\tilde{F}(\omega)$ is defined as

$$\tilde{F}(\omega) = \sum_{n=-\infty}^{\infty} F_2\left(\omega - n\frac{2\pi}{\Delta t}\right) \tag{5.197}$$

Then the numbers F_m are samples of $\tilde{F}(\omega)$, with

$$F_m = \frac{1}{\Delta t}\,\tilde{F}(m\,\Delta\omega) \tag{5.198}$$

The factor $(1/\Delta t)$ must be included to scale the spectrum $\tilde{F}(m\,\Delta\omega)$ correctly, as we discussed previously in Section 5.10 (see Figure 5.35). Using (5.197), we can write (5.198) as

$$F_m = \frac{1}{\Delta t}\,\tilde{F}(m\,\Delta\omega)$$

$$= \frac{1}{\Delta t}\sum_{n=-\infty}^{\infty} F_2\left(m\,\Delta\omega - n\frac{2\pi}{\Delta t}\right)$$

$$= \frac{1}{\Delta t}\sum_{n=-\infty}^{\infty} F_2\left((m - nN)\frac{2\pi}{N\,\Delta t}\right) \tag{5.199}$$

The sample values of the spectrum $\{F_m\}$ are a good approximation to the samples of $F(\omega) = \mathscr{F}\{f(t)\}$ provided that

(1) $G_D(\omega)$ is close to an impulse, which implies that $g_D(t)$ is relatively wide and that $F_2(\omega)$ above is a good approximation to $F(\omega)$.

(2) The wound-up or aliased version of $F(\omega) * G_D(\omega)$ is approximately equal to $F(\omega)$. This means that the repeated versions of $F(\omega)$ must be separated far enough in ω, which implies that the sampling interval Δt must be small.

Figure 5.52 depicts the relationship between the time functions and their spectra in this process just described.

The second method for applying the DFT sum to calculate a sampled Fourier transform involves first forming an aliased version of the time function $f(t)$. If $f_a(t)$ denotes the aliased version of $f(t)$, then

$$f_a(t) = \sum_{k=-\infty}^{\infty} f(t - kP) \tag{5.200}$$

where P is the separation between repeated versions of $f(t)$. The transform of $f_a(t)$ is

$$
\begin{aligned}
F_a(\omega) &= \mathscr{F}\{f_a(t)\} = \mathscr{F}\left\{ \sum_{k=-\infty}^{\infty} f(t - kP) \right\} \\
&= \mathscr{F}\left\{ f(t) * \sum_{k=-\infty}^{\infty} \delta(t - kP) \right\} \\
&= F(\omega) \cdot \Delta\omega \sum_{k=-\infty}^{\infty} \delta(\omega - k\,\Delta\omega)
\end{aligned}
\tag{5.201}
$$

where $\Delta\omega = 2\pi/P$. If we apply the DFT sum to the samples of $f_a(t)$, we obtain

$$F_m \triangleq \tilde{F}_a(m\,\Delta\omega) = \sum_{n=0}^{N-1} f_n e^{-j(nm2\pi/N)}$$

The numbers F_0, $F_{\pm 1}$, $F_{\pm 2} \ldots$, are samples of the aliased spectrum $\tilde{F}_a(\omega)$, where

$$\tilde{F}_a(\omega) = \frac{1}{\Delta t} \sum_{n=0}^{N-1} F_a\left(\omega - \frac{n2\pi}{\Delta t} \right) \tag{5.202}$$

and Δt is the time between samples of the sequence $\{f_n\}$. The numbers F_m are thus

$$\tilde{F}_m = \tilde{F}_a(m\,\Delta\omega) = \frac{1}{\Delta t} \sum_{n=0}^{N-1} F_a\left(m\,\Delta\omega - \frac{n2\pi}{\Delta t} \right) \tag{5.203}$$

where $\Delta\omega = 2\pi/N\,\Delta t$ is the sampling interval in ω. These numbers will form a good approximation to the samples of $F(\omega)$, provided that

(1) The aliasing period P in the time domain is large enough so that $f_a(t)$ closely approximates $f(t)$,

(2) The sampling period Δt is small enough so that the aliased version of $F_a(\omega)$, denoted $\tilde{F}_a(\omega)$, is close to $\mathscr{F}\{f_a(t)\} = F_a(\omega)$. This second method is shown schematically in Figure 5.53.

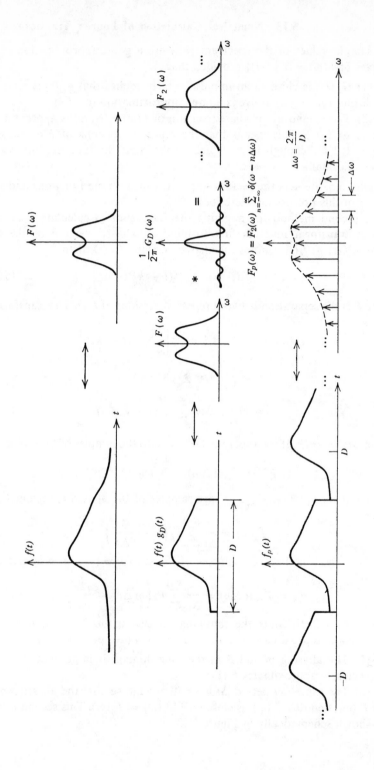

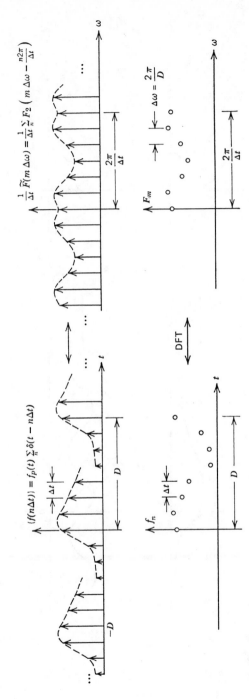

FIGURE 5.52

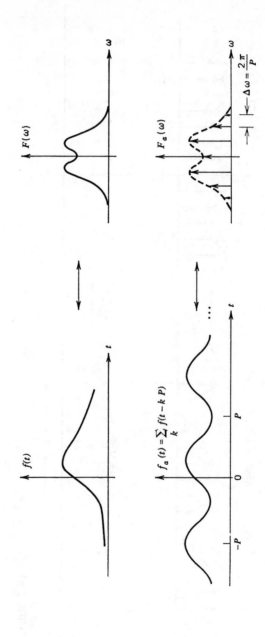

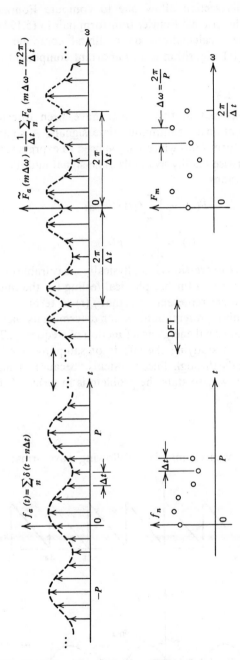

FIGURE 5.53

The results of this section allow one to compute Fourier transforms numerically. Using the discrete Fourier transform pair in (5.194) and (5.195), we can perform these calculations on a digital computer. For efficient computations, the FFT algorithm is an important component of the overall computation.

5.14   SUMMARY

The topics discussed in this chapter have been directed primarily toward the representation of continuous time signals. A signal $f(t)$ has a unique Fourier transform $F(\omega)$ which can be used to express various properties of $f(t)$ not otherwise readily available. The signal and its transform are related by the expressions

$$F(\omega) = \int_{-\infty}^{\infty} f(t)e^{-j\omega t}\,dt$$

$$f(t) = \frac{1}{2\pi} \int_{-\infty}^{\infty} F(\omega)e^{j\omega t}\,d\omega$$

The variable ω in these expressions is a physically measurable quantity. Thus, by knowing $F(\omega)$, we gain a further physical feeling for the time signal $f(t)$.

The Fourier transform represents a signal $f(t)$ in terms of its harmonic structure, and from this representation, we can determine its energy spectrum: i.e., the distribution of signal energy as a function of frequency. The spectrum of a signal is useful in studying the effects of sampling, modulation, and transmission of signals through linear systems, because in many physical problems, the natural way to state the problem is in terms of the frequency domain.

PROBLEMS

1. Find the frequency spectrum of the following waveforms:

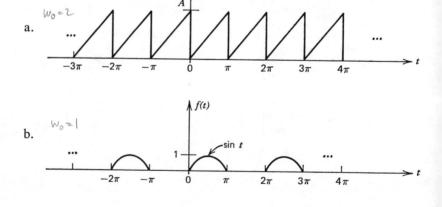

2. Define the energy of a signal $f(t)$ on the interval $[t_1, t_2]$ as

$$\text{Energy} = \int_{t_1}^{t_2} f^2(t') \, dt'$$

a. What is the energy of the sum of two functions $f_1(t)$ and $f_2(t)$ on $[t_1, t_2]$?

b. What is the energy of $f_1(t) + f_2(t)$ if $f_1(t)$ and $f_2(t)$ are orthogonal on $[t_1, t_2]$?

3. Suppose an electric circuit is excited by a voltage $v(t)$ given by

$$v(t) = V_0 + \sum_{n=1}^{\infty} V_n \cos(n\omega_0 t + \theta_n)$$

The corresponding steady-state current is $i(t)$, given by

$$i(t) = I_0 + \sum_{n=1}^{\infty} I_n \cos(n\omega_0 t + \phi_n)$$

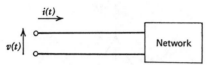

as shown. Define the input power at the input terminals as

$$P = \frac{1}{T} \int_{-T/2}^{T/2} v(t) \, i(t) \, dt, \qquad T = \frac{2\pi}{\omega_0}$$

Show that the input power can also be written as

$$P = V_0 I_0 + \sum_{n=1}^{\infty} \frac{V_n I_n}{2} \cos(\theta_n - \phi_n)$$

4. A square-wave voltage $v(t)$ whose waveform is shown below is applied to a series RL circuit. Find the first five harmonics in the response current $i(t)$.

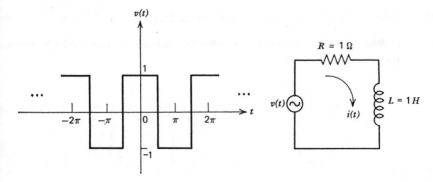

5. Show that the response of a linear time-invariant system to an exponential function $e^{j\omega t}$ is also an exponential function of the same frequency ω and proportional to the input. In symbols, if $G(\cdot)$ represents the transformation that the system performs on the input, then $G(e^{j\omega t}) = k(\omega)e^{j\omega t}$. (In general k is a complex function of ω.)

6. A periodic function $f(t)$ is known to have only the first ten harmonics (the remaining harmonics all have zero coefficients). Show that such a signal is uniquely specified by any $(2 \cdot 10 + 1) = 21$ sample values in a single period. In other words, show that the set of samples $\{f(t_1), f(t_2), \ldots, f(t_{21})\}$, $t_i \in [t_0, t_0 + T]$, can precisely specify $f(t)$ on $[t_0, t_0 + T]$.

7. A periodic waveform as shown below is to be approximated by sinusoids How many sinusoids (added together) are needed to approximate $f(t)$ so that the approximating series is within 1% of the total energy of $f(t)$ over a single period?

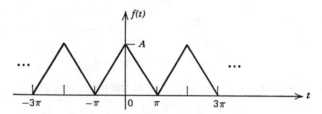

8. Sam Mooge has been experimenting with various waveforms for use in his electronic music machine. He is now searching for a signal which is amplitude limited between the values A and $-A$ and has maximum power at a particular frequency ω_0. What waveform should Sam use?
   a. Prove your answer for the class of all even or odd functions.
   b. Prove your answer in general.

9. The following differential equation is a model for a linear system with input $x(t)$ and output $y(t)$.

$$\dot{y}(t) + y(t) = x(t)$$

If the input $x(t)$ is the square waveform shown, then find the third harmonic in the output.

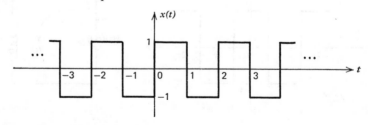

10. The following set of samples taken from the waveform $f(t)$ at the Nyquist interval defines a bandlimited function, $\{\ldots, 0, 0, 10, 30, 50, 50, -40, -10, 0, \ldots\}$. Give an exact expression for $f(t)$. Estimate the maximum value of this waveform as best you can.

11. Suppose a function $f(t)$ is continuous and bandlimited to W rad/sec. Show that $f(t)$ can be represented as

$$f(t) = \frac{\omega}{\pi} \left[f(t) * \text{sinc}\,(\omega t) \right], \qquad \text{for all} \quad \omega > W$$

12. A communications channel has a bandwidth of 200 kHz. Pulses are to be transmitted over this channel with information coded by varying the amplitude of the pulses. If the pulses all have width δ sec and separation between centers of T sec, what values of δ and T should be chosen to transmit as much information per unit time as possible?

13. Sketch the output waveform frequency spectrum for the following system. Is this system linear?

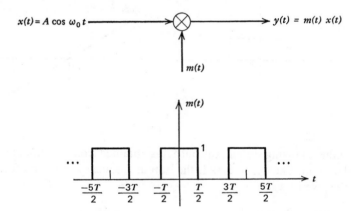

14. Sketch the phase spectrum between $[-2\pi, 2\pi]$ rad/sec of the time function shown, consisting of 2 pulses whose center values are separated by D seconds.

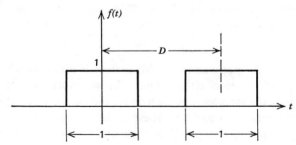

15. Find the time function $g(t)$ for the frequency spectrum $G(\omega)$ as shown. Use properties of the Fourier transform to help determine the time function.

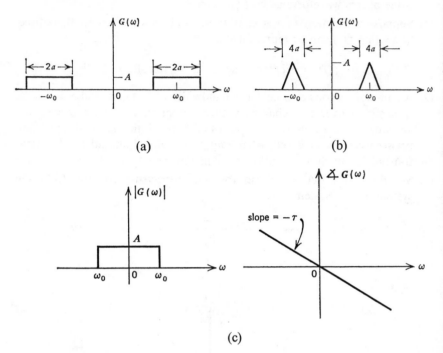

(a)

(b)

(c)

16. In the following system, the output is obtained from the input by squaring the input. Sketch, in detail, the output frequency spectrum for an input $x(t)$ given by

a. $x(t) = \text{sinc}\,(Wt)$

b. $x(t) = \sum_{n=80}^{90} \cos\,(n\omega_0 t), \qquad \omega_0 = 10^3\ \text{rad/sec}$

$$x(t) \longrightarrow \boxed{\text{System}} \longrightarrow y(t) = x^2\,(t)$$

17. A pulse code modulation (PCM) system is a system which takes input analog information $x(t)$, samples the waveform at the Nyquist rate, quantizes the samples, and then codes the quantized sampled values by a binary code of 0's and 1's. These coded samples are then transmitted. For example, if the quantization operation performed on the sampled values divides the possible sampled values into 8 regions, the binary code for each sample would be chosen from one of the following: {000, 001, 101, 011, 100, 101, 110, 111}. Suppose the transmission channel has a bandwidth of W Hz and the input signal is bandlimited to f_m Hz, where $W > f_m$. What is the finest quantization one can obtain if the input signal is to be transmitted in the actual duration of the signal?

18. Sketch the frequency spectrum of $y(t)$ for the following modulation system where $f(t) = 2 \cos 10t + 4 \cos 20t$ and $m(t) = \cos 200t$.

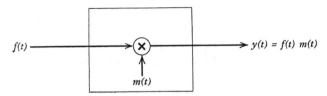

19. The impulse response of the following system is desired. The only equipment available to obtain this response is an oscilloscope and a pulse generator with a pulse width that can be varied between 1 msec and 1 sec. Can the system's impulse response be found by use of this equipment? Explain.

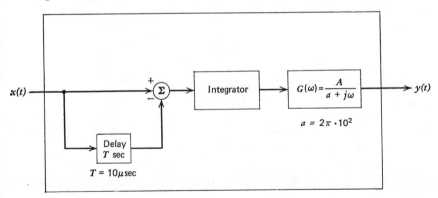

20. The waveform $f(t)$ modulates the pulse train $p_T(t)$ as shown below. Consider two cases. In case a, the modulated pulses in $m_a(t)$ have constant amplitudes. In case b, the pulses in $m_b(t)$ follow $f(t)$.
   (1) Express the time waveforms $m_a(t)$ and $m_b(t)$ in terms of $f(t)$, $p(t)$, and the time impulse train $\delta_T(t) = \sum_{n=-\infty}^{\infty} \delta(t - nT)$, using the operations of multiplication and convolution.
   (2) Find the spectra of $m_a(t)$ and $m_b(t)$ in terms of the transforms $F(\omega)$ and $P(\omega)$. Sketch $M_a(\omega)$ and $M_b(\omega)$ for $F(\omega)$ as shown and $\tau = T/4$, and compare.

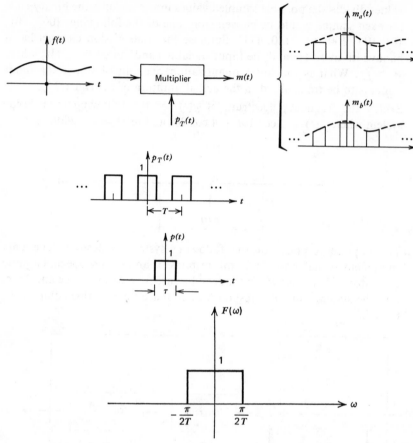

21. A function $f(t)$ has the Fourier transform shown below. Carefully sketch the Fourier transform of $f^2(t)$.

$$F(\omega) = \delta(\omega) + \begin{cases} 1, & 2 \le |\omega| \le 4 \\ 0, & \text{otherwise} \end{cases}$$

22. Use the energy theorem: i.e.,

$$\int_{-\infty}^{\infty} f^2(t)\, dt = \frac{1}{2\pi} \int_{-\infty}^{\infty} |F(\omega)|^2\, d\omega$$

to find the value of

a. $\displaystyle\int_{-\infty}^{\infty} \operatorname{sinc}^2(t)\, dt$

b. $\displaystyle\int_{-\infty}^{\infty} \frac{dx}{(1 + x^2)^2}$

23. A full-wave rectifier is a system whose output $y(t)$ to an input $x(t)$ is $y(t) = |x(t)|$. Suppose $a(t) \sin \omega_0 t$, $a(t) > 0$, is the input to a full-wave rectifier. Show that the output spectrum is

$$Y(\omega) = -\frac{2}{\pi} \sum_{n=-\infty}^{\infty} \frac{A(\omega - 2n\omega_0)}{4n^2 - 1}, \qquad A(\omega) = \mathscr{F}\{a(t)\}.$$

SUGGESTED READINGS

Bracewell, R. M., *The Fourier Integral and Its Applications*, McGraw-Hill, New York, 1965.
This book is written at a senior-graduate level. A very complete and detailed discussion of the Fourier transform with many interesting and varied examples. A very good reference book. The book's notation is somewhat difficult at first.

Lathi, B. P., *Signals, Systems, and Communications*, Wiley, New York, 1967.
Chapters 3 and 4 present Fourier Series and Fourier transforms at a level comparable to the present text. The book is written more for the engineer than for the mathematician.

Papoulis, A., *The Fourier Integral and Its Applications*, McGraw-Hill, New York, 1962.
This book is also written at a senior-graduate level. It is more mathematical in approach than Bracewell. Another good reference on Fourier transforms.

6

The Laplace Transform

The Fourier transform of a function $f(t)$ is a representation of the function as a continuous sum of exponential functions of the form $e^{j\omega t}$, where ω is a real frequency. This interpretation of the Fourier transform gives one a physical feeling for an otherwise abstract mathematical transformation. The Laplace transform can be viewed as an extension of the Fourier transform in which we represent a function $f(t)$ by a continuous sum of exponential functions of the form e^{st}, where $s = \sigma + j\omega$ is a complex frequency. From this viewpoint, the Fourier transform is a special case in which $s = j\omega$. Recall that the Fourier transform pair for $f(t)$ is

$$F(\omega) = \int_{-\infty}^{\infty} f(t)e^{-j\omega t}\, dt \tag{6.1}$$

$$f(t) = \frac{1}{2\pi}\int_{-\infty}^{\infty} F(\omega)e^{j\omega t}\, d\omega \tag{6.2}$$

Now suppose we define a function $a(t)$ as $f(t)$ times the function $e^{-\sigma t}$, where σ is a real number: i.e., $a(t) = f(t)e^{-\sigma t}$. The Fourier transform of $a(t)$ is

$$\mathscr{F}\{a(t)\} = \int_{-\infty}^{\infty} a(t)e^{-j\omega t}\, dt = \int_{-\infty}^{\infty} f(t)e^{-\sigma t}e^{-j\omega t}\, dt$$

$$= \int_{-\infty}^{\infty} f(t)e^{-(\sigma+j\omega)t}\, dt \tag{6.3}$$

Equation (6.3) can thus be written as

$$\hat{F}(\sigma + j\omega) = \int_{-\infty}^{\infty} f(t)e^{-(\sigma+j\omega)t}\, dt \tag{6.4}$$

306

and $f(t)$ can be expressed as

$$f(t) = \frac{1}{2\pi} \int_{-\infty}^{\infty} \hat{F}(\sigma + j\omega)e^{(\sigma + j\omega)t} \, d\omega \qquad (6.5)$$

Now let s be the complex frequency $\sigma + j\omega$: i.e., $s = \sigma + j\omega$. Then $d\omega = (1/j) \, ds$ and (6.4) and (6.5) become

$$\hat{F}(s) = \int_{-\infty}^{\infty} f(t)e^{-st} \, dt \qquad (6.6)$$

$$f(t) = \frac{1}{2\pi j} \int_{\sigma - j\infty}^{\sigma + j\infty} \hat{F}(s)e^{st} \, ds \qquad (6.7)$$

The limits on the integral of (6.7) result because of the substitution $s = \sigma + j\omega$. Equations (6.6) and (6.7) constitute the *complex Fourier transform pair* or the *bilateral Laplace transform pair* or the *two-sided Laplace transform pair*. We shall use the latter designation with the notation

$$F(s) = \mathcal{L}_b\{f(t)\}$$

$$f(t) = \mathcal{L}_b^{-1}\{F(s)\}$$

The two-sided Laplace transform of $f(t)$ can be obtained from the Fourier transform of $f(t)$ by merely substituting s for $j\omega$ provided $f(t)$ is absolutely integrable. For functions which possess Fourier transforms only in the limit; i.e., power functions, one must evaluate the Laplace transform directly using (6.6). Notice that in this development we have multiplied a given function $f(t)$ by $e^{-\sigma t}$, where σ is any real number. Thus the convergence of the resulting Fourier integral is greatly enhanced by the so-called convergence factor $e^{-\sigma t}$. This means that the Laplace transform exists for many functions for which there is no Fourier transform. This is one of the principal advantages of the Laplace transform: the ability to transform functions which are not ab-solutely integrable.

6.1   CONVERGENCE OF THE LAPLACE TRANSFORM

To appreciate the generality possessed by the two-sided Laplace trans-form, one must understand the convergence properties of the integral (6.6). The two-sided Laplace transform exists if

$$F(s) = \int_{-\infty}^{\infty} f(t)e^{-st} \, dt \qquad (6.8)$$

is finite. Therefore $F(s)$ is guaranteed to exist if

$$\int_{-\infty}^{\infty} |f(t)e^{-st}| \, dt = \int_{-\infty}^{\infty} |f(t)| \, e^{-\sigma t} \, dt$$

is finite. Suppose there exists a real positive number $\mathscr{R}$ so that for some real α and β we know that

$$|f(t)| < \begin{cases} \mathscr{R}e^{\alpha t}, & t > 0 \\ \mathscr{R}e^{\beta t}, & t < 0 \end{cases} \tag{6.9}$$

Then $F(s)$ converges for

$$\alpha < \text{Re}\,(s) < \beta$$

To see this conclusion, write (6.8) as

$$F(s) = \int_{-\infty}^{0} f(t)e^{-st}\,dt + \int_{0}^{\infty} f(t)e^{-st}\,dt$$

Using the inequalities of (6.9), we see that

$$F(s) < \int_{-\infty}^{0} \mathscr{R}e^{(\beta-s)t}\,dt + \int_{0}^{\infty} \mathscr{R}e^{(\alpha-s)t}\,dt$$

$$< \mathscr{R}\left[\frac{1}{\beta - s}\, e^{(\beta-s)t} \Big|_{-\infty}^{0} + \frac{1}{a - s}\, e^{(\alpha-s)t} \Big|_{0}^{\infty} \right] \tag{6.10}$$

In (6.10) the first integral (negative time portion of $f(t)$) converges for $\text{Re}\,(s) < \beta$. Similarly, the second integral (positive time portion of $f(t)$) converges for $\text{Re}\,(s) > \alpha$. Thus both integrals, and therefore $F(s)$, converge in the common region, $\alpha < \text{Re}\,(s) < \beta$. Notice that the left-hand limit α of the convergence region is governed by the behavior of the positive-time portion of $f(t)$, while the right-hand limit β depends on the behavior of the negative-time portion of $f(t)$.

Because $F(s)$ converges in the strip $\alpha < \sigma < \beta$, there are no singularities or poles[1] of $F(s)$ in this region. Now if $f(t)$ is nonzero for $(0, \infty)$ only, then the left-hand limit α alone defines the convergence region, and all the poles of $F(s)$ must be to the left of α. Likewise, if $f(t)$ is nonzero for $(-\infty, 0)$ only, then β alone defines the convergence region. In this case, all poles of $F(s)$ must lie to the right of β. In other words, poles to the left of the convergence region arise from the positive time portion of $f(t)$ and poles to the right of the convergence region arise from the negative time portion of $f(t)$. This simple fact is the key to finding inverse transforms for two-sided Laplace transforms. Whenever the region of convergence of $F(s)$ includes the $j\omega$ axis, then one can obtain the Fourier transform from $F(s)$ simply by substituting $j\omega$ for s and vice versa. If the region of convergence does not include the $j\omega$ axis, then the function $f(t)$ is not absolutely integrable and so does not possess a Fourier transform in the strict sense. Figure 6.1 depicts several time functions and their corresponding regions of convergence.

---

[1] A pole of $F(s)$ is a value of s such that $|F(s)| \to \infty$. For example, if $F(s) = k(s + c)/(s - a)(s + b)$, then a and $-b$ are poles of $F(s)$.

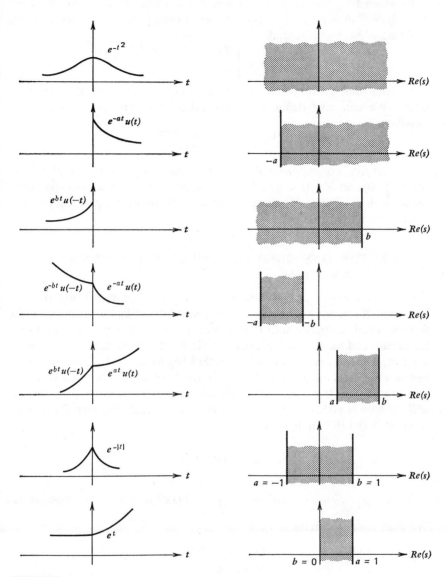

FIGURE 6.1

In summary, the two-sided Laplace transform exists, provided that $\int_{-\infty}^{\infty} |f(t)|e^{-\sigma t}\, dt$ is finite. This is equivalent to requiring the existence of real numbers $\mathscr{R}$, α, and β such that

$$|f(t)| < \begin{cases} \mathscr{R}e^{\alpha t}, & t > 0 \\ \mathscr{R}e^{\beta t}, & t < 0 \end{cases}$$

A function that satisfies the above inequalities is said to be of exponential order. We can also define an exponential-order function as one which satisfies

$$\lim_{t \to \infty} e^{\alpha t} f(t) = 0, \qquad \lim_{t \to -\infty} e^{\beta t} f(t) = 0$$

If a signal $f(t)$ is of exponential order, $f(t)$ does not grow more rapidly than $\mathscr{R}e^{\alpha t}(t > 0)$ and $\mathscr{R}e^{\beta t}(t < 0)$. There are functions that are not of exponential order, such as e^{t^3} or t^t. However, it is difficult to envision a physical problem where they might arise.

6.2 THE ONE-SIDED OR UNILATERAL LAPLACE TRANSFORM

In the majority of applications involving Laplace transforms, the time functions of interest are causal: i.e., $f(t) = 0$ for $t < 0$. For example, if the forcing function applied to a causal system is zero for negative values of t, the response of the system is also causal. Generally, in the analysis of physical systems, one can assume an excitation that begins at $t = 0$. Thus, one often restricts the class of functions for use in the Laplace transform to functions nonzero only for $0 < t < \infty$. The resulting Laplace transform is termed the *unilateral* or *one-sided Laplace transform*. In this case, the transform equations (6.6) and (6.7) reduce to

$$F(s) = \int_0^\infty f(t)e^{-st}\, dt \tag{6.11}$$

$$f(t) = \frac{1}{2\pi j} \int_{\sigma - j\infty}^{\sigma + j\infty} F(s)e^{st}\, ds \tag{6.12}$$

We shall use the shorthand notation

$$F(s) = \mathscr{L}\{f(t)\}$$
$$f(t) = \mathscr{L}^{-1}\{F(s)\}$$

for the one-sided Laplace transform pair. The relationship of the two-sided to the one-sided Laplace transform is analogous to what occurs in the one and two-sided Z-transform. For example, if one is dealing with time functions nonzero on $(-\infty, \infty)$, then the region of convergence must be known in order to obtain a unique inverse. This same information is needed to obtain unique inverses in the case of two-sided Z-transforms (see Section 4.5). The properties of the one-sided and two-sided Laplace transforms differ only slightly, in much the same way as we have already seen for the case of Z-transforms.

6.3 PROPERTIES OF THE LAPLACE TRANSFORM

The properties of the two-sided Laplace transform are so similar to those of the Fourier transform of Chapter 5 that we merely summarize them here. Notice that the region of convergence is also added so as to obtain a unique transform pair.

(1) Linearity

$$af_1(t) + bf_2(t) \leftrightarrow aF_1(s) + bF_2(s), \qquad \max(\alpha_1, \alpha_2) < \sigma < \min(\beta_1, \beta_2)$$

(2) Scaling

$$f(at) \leftrightarrow \frac{1}{|a|} F\left(\frac{s}{a}\right), \qquad |\alpha|\, a < \sigma < |a|\, \beta$$

(3) Time shift

$$f(t - \tau) \leftrightarrow F(s)e^{-\tau s}, \qquad \alpha < \sigma < \beta$$

(4) Frequency shift

$$e^{-at}f(t) \leftrightarrow F(s + a), \qquad \alpha - \operatorname{Re}(a) < \sigma < \beta - \operatorname{Re}(a)$$

(5) Time convolution

$$f_1(t) * f_2(t) \leftrightarrow F_1(s)F_2(s), \qquad \text{same as (1)}$$

(6) Frequency convolution

$$f_1(t)f_2(t) \leftrightarrow \frac{1}{2\pi j} \int_{c-j\infty}^{c+j\infty} F_1(u) \qquad \begin{cases} \alpha_1 + \alpha_2 < \sigma < \beta_1 + \beta_2 \\ \alpha_1 < c < \beta_1 \end{cases}$$
$$\times F_2(s - u)\, du,$$

(7) Time differentiation

$$\frac{df(t)}{dt} \leftrightarrow sF(s), \qquad \text{same as (3)}$$

(8) Time integration

$$\int_{-\infty}^{t} f(u)\, du \leftrightarrow \frac{F(s)}{s}, \qquad \max(\alpha, 0) < \sigma < \beta$$

$$\int_{t}^{\infty} f(u)\, du \leftrightarrow \frac{F(s)}{s}, \qquad \alpha < \sigma < \min(\beta, 0)$$

(9) Frequency differentiation

$$(-t)^n f(t) \leftrightarrow \frac{d^n F(s)}{ds^n}, \qquad \text{same as (3)}$$

These properties for the two-sided Laplace transform hold for the one-sided transform with the exception of properties (7) and (8). There are also initial and final value theorems for one-sided transforms. The time-differentiation and time-integration properties are very important properties for the one-sided transform for the analysis of transient phenomona in linear systems. The time-convolution theorem again plays a central role in linear system analysis.

(10) One-sided time-differentiation

If
$$f(t) \leftrightarrow F(s)$$
then
$$f'(t) \leftrightarrow sF(s) - f(0) \qquad (6.13)$$
and, in general, we have

$$\frac{d^n f(t)}{dt^n} \leftrightarrow s^n F(s) - s^{n-1}f(0) - s^{n-2}f'(0) - \cdots - f^{(n-1)}(0) \qquad (6.14)$$

Proof: By definition,

$$\mathscr{L}\left\{\frac{df(t)}{dt}\right\} = \int_0^\infty \frac{df(t)}{dt} e^{-st} \, dt$$

Integrate by parts to obtain

$$\mathscr{L}\{f'(t)\} = f(t)e^{-st}\big|_0^\infty + s\int_0^\infty f(t)e^{-st} \, dt$$

$$= f(t)e^{-st}\big|^{t=\infty} - f(0) + sF(s)$$

Now because $F(s)$ exists, it follows that $f(t)e^{-st}$ evaluated at $t = \infty$ is zero. Thus,

$$\mathscr{L}\{f'(t)\} = sF(s) - f(0)$$

The general case is found by repeated application of the above process. The value of $f(0), f'(0), \ldots$ which is used must be consistent with the definition of the integral in (6.11). If $f(t)$ is discontinuous at the origin, $f'(t)$ will possess an impulse equal in area to the height of the discontinuity. If this impulse is to be included in the Laplace transform of $f(t)$, then $f(0), f'(0), \ldots$ are evaluated at 0^+, and the one-sided Laplace integral of (6.11) has a lower limit of 0^+. One can also exclude these discontinuities at the origin provided $f(0)$, $f'(0), \ldots$ are evaluated at 0^- and the one-sided Laplace integral has a lower limit of 0^-. In other words, we can use either 0^- or 0^+, in (6.11) and (6.14), provided that we are consistent in our choice of "$t = 0$." To illustrate, consider the following example.

EXAMPLE 6.1.   Find the transform of $f'(t)$, where $f(t) = e^{-at}u(t)$ as shown in Figure 6.2.

(1) Suppose we consider $t = 0$ as being $t = 0^-$. Then, using the defining integral, we have

$$\mathcal{L}\{f'(t)\} = \int_{0^-}^{\infty} f'(t)e^{-st}\,dt = \int_{0^-}^{\infty} [e^{-at}\,\delta(t) - ae^{-at}u(t)]e^{-st}\,dt$$

$$= \int_{0^-}^{\infty} (e^{-at}e^{-st}\,\delta(t) - ae^{-(a+s)t})\,dt$$

$$= 1 + \frac{ae^{-(a+s)t}}{a+s}\Big|_{0^-}^{\infty} = 1 - \frac{a}{a+s} = \frac{s}{s+a}$$

Because we are using 0^- as $t = 0$, $f(0^-) = 0$. Thus, (6.13) yields

$$\mathcal{L}\{f'(t)\} = sF(s) - f(0^-) = sF(s)$$

where

$$F(s) = \int_{0}^{\infty} e^{-at}e^{-st}\,dt = \frac{e^{-(a+s)t}}{-(s+a)}\Big|_{0}^{\infty} = \frac{1}{s+a}$$

$$\therefore\ \ \mathcal{L}\{f'(t)\} = s\,F(s) = \frac{s}{s+a}$$

which agrees with the expression obtained by direct calculation.
(2) Suppose we consider $t = 0$ as being $t = 0^+$. The defining integral is therefore

$$\mathcal{L}\{f'(t)\} = \int_{0^+}^{\infty} f'(t)e^{-st}\,dt = \int_{0^+}^{\infty} [e^{-(a+s)t}\,\delta(t) - ae^{-(a+s)t}]\,dt$$

$$= 0 + \frac{ae^{-(a+s)t}}{s+a}\Big|_{0^+}^{\infty} = \frac{-a}{s+a}$$

If the lower limit is 0^+ in (6.13), then $f(0^+) = 1$. Thus we have

$$\mathcal{L}\{f'(t)\} = sF(s) - f(0^+) = s\left(\frac{1}{s+a}\right) - 1 = \frac{-a}{s+a}$$

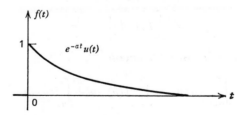

FIGURE 6.2

which again agrees with the direct calculation. Thus, the Laplace transform depends on whether we use 0^- or 0^+, but (6.13) and (6.14) give results consistent with the defining integral.

This property of the one-sided Laplace transform furnishes one with an extremely effective method of solving transient problems in linear systems. The initial conditions of a system can easily be incorporated into the solution by use of (6.13) or (6.14).

(11) One-sided time integration.

If

$$f(t) \leftrightarrow F(s)$$

then

$$\int_0^t f(u)\, du \leftrightarrow \frac{F(s)}{s} \tag{6.15}$$

and

$$\int_{-\infty}^t f(u)\, du \leftrightarrow \frac{F(s)}{s} + \frac{f^{(-1)}(0)}{s} \tag{6.16}$$

where

$$f^{(-1)}(0) \triangleq \int_{-\infty}^t f(u)\, du \,\bigg|_{t=0}$$

Proof: By definition,

$$\mathscr{L}\left\{ \int_0^t f(u)\, du \right\} = \int_0^\infty \left(\int_0^t f(u)\, du \right) e^{-st}\, dt$$

Integrating by parts we obtain

$$\mathscr{L}\left\{ \int_0^t f(u)\, du \right\} = \frac{-e^{-st}}{s} \int_0^t f(u)\, du \,\bigg|_0^\infty + \frac{1}{s} \int_0^\infty f(t) e^{-st}\, dt$$

If $f(t)$ is of exponential order, so then is its integral. Thus the term $e^{-st} \int_0^t f(u)\, du$ evaluated at $t = \infty$ is zero for values of s in the convergence region. Thus

$$\mathscr{L}\left\{ \int_0^t f(u)\, du \right\} = \frac{F(s)}{s}$$

If the lower limit is $-\infty$, then we obtain

$$\int_{-\infty}^t f(u)\, du = \int_{-\infty}^0 f(u)\, du + \int_0^t f(u)\, du$$

$$= f^{(-1)}(0) u(t) + \int_0^t f(u)\, du$$

Now

$$\mathscr{L}\{u(t)\} = \int_0^\infty e^{-st}\,dt = \frac{e^{-st}}{-s}\bigg|_0^\infty = \frac{1}{s}$$

Thus

$$\mathscr{L}\left\{\int_{-\infty}^t f(u)\,du\right\} = \frac{f^{(-1)}(0)}{s} + \frac{F(s)}{s}$$

This property of the Laplace transform is useful in transforming equations containing integrals and derivatives, as the next example demonstrates.

EXAMPLE 6.2. In the following *RLC* network, find the current resulting from the voltage source $v(t)$. The switch closes at $t = 0$. The inductor has an initial current of $i(0)$ and the capacitor an initial charge of $q(0)$. Using Kirchhoff's voltage law, we see that

$$v(t) = R\,i(t) + L\frac{di(t)}{dt} + \frac{1}{C}\int_{-\infty}^t i(t')\,dt' \qquad (6.17)$$

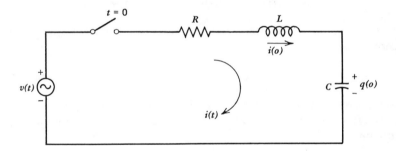

FIGURE 6.3

Usually, we solve (6.17) by differentiating it to remove the integral and obtain a differential equation: however, using (6.16) directly, we can take Laplace transforms on both sides of (6.17) to obtain

$$V(s) = RI(s) + L(sI(s) - i(0)) + \frac{1}{C}\left(\frac{I(s)}{s} + \frac{i^{(-1)}(0)}{s}\right)$$

Now $i(0)$ is merely the current in the inductor at $t = 0$. The term

$$\frac{i^{(-1)}(0)}{C} = \frac{\int_{-\infty}^0 i(t')dt'}{C} = \frac{q(0)}{C} = v(0)$$

is the voltage across the capacitor at $t = 0$. Thus, solving for $I(s)$, we have

$$I(s) = \frac{V(s) - Li(0) + \dfrac{v(0)}{s}}{R + Ls + \dfrac{1}{Cs}}$$

Given the parameter values, we can take the inverse transform of $I(s)$ to obtain $i(t)$. The value of $t = 0$ in this problem is just after the switch closes at $t = 0^+$.

(12) Initial value.

The initial-value property permits one to calculate $f(0)$ directly from the transform $F(s)$ without the need of inverting the transform. It is important to understand which value of $f(0)$ this theorem gives when $f(t)$ is discontinuous at the origin. Now (6.13) states

$$\mathscr{L}\{f'(t)\} = \int_0^\infty f'(t)e^{-st}\,dt = sF(s) - f(0) \tag{6.18}$$

If $f(t)$ is continuous at the origin, $f'(t)$ is finite and as $s \to \infty$, the integral goes to zero. Thus

$$\lim_{s \to \infty} sF(s) = f(0) \tag{6.19}$$

If $f(t)$ is discontinuous at the origin, then $f'(t)$ contains an impulse term namely $[f(0^+) - f(0^-)]\,\delta(t)$. Thus, if 0^- is taken as the lower limit in (6.18), we obtain

$$\lim_{s \to \infty} \int_{0^-}^\infty f'(t)e^{-st}\,dt = f(0^+) - f(0^-) = \lim_{s \to \infty} [sF(s) - f(0^-)]$$

Therefore,

$$\lim_{s \to \infty} sF(s) = f(0^+) - f(0^-) + f(0^-)$$

$$= f(0^+) \tag{6.20}$$

If 0^+ is taken as the lower limit, then as $s \to \infty$ (6.18) is

$$\lim_{s \to \infty} \int_{0^+}^\infty f'(t)e^{-st}\,dt = 0$$

so that

$$0 = \lim_{s \to \infty} [sF(s) - f(0^+)]$$

which again yields

$$\lim_{s \to \infty} sF(s) = f(0^+) \tag{6.21}$$

Thus the initial value is always the limit of $f(t)$ as $t \to 0^+$, provided that it exists, independent of the lower limit, 0^- or 0^+, used in the defining equation for the Laplace transform.

EXAMPLE 6.3. In the proof of the property (11), we calculated the Laplace transform of the unit step function $u(t)$ as $1/s$. The initial value of the unit step by (6.21) is therefore

$$\lim_{s \to \infty} sF(s) = \lim_{s \to \infty} s\left(\frac{1}{s}\right) = 1$$

which agrees with the correct value of $u(0^+)$.

(13) Final Value.

The value which $f(t)$ approaches as t becomes large may also be found directly from its transform $F(s)$. Using (6.13) again, we take the limit as $s \to 0$.

$$\lim_{s \to 0} \int_0^\infty f'(t)e^{-st}\, dt = \lim_{s \to 0} [sF(s) - f(0)] \tag{6.22}$$

If $sF(s)$ has all its singularities in the left half of the s plane, then $\lim_{s \to 0} sF(s)$ exists. Then (6.22) can be written

$$\int_0^\infty f'(t)\, dt = \lim_{s \to 0} [sF(s) - f(0)]$$

or

$$\lim_{t \to \infty} f(t) - f(0) = \lim_{s \to 0} sF(s) - f(0)$$

so that

$$\lim_{t \to \infty} f(t) = \lim_{s \to 0} sF(s)$$

A simple pole in $F(s)$ at the origin is permitted, but otherwise all other poles of $F(s)$ must be strictly in the left half of the s plane.

EXAMPLE 6.4. Find the final value of the time function which corresponds to the transform (one-sided)

$$F(s) = \frac{1}{s + a}, \qquad a > 0$$

The singularity of $sF(s)$ is a pole at $s = -a$ (which is in the left half of the s plane). Thus

$$\lim_{s \to 0} sF(s) = \lim_{s \to 0} \left(\frac{s}{s + a} \right) = 0$$

This expression is correct from our knowledge of the time function $f(t) = e^{-at}u(t)$, which approaches zero as $t \to \infty$.

6.4   LAPLACE TRANSFORMS OF SIMPLE FUNCTIONS

It is seldom necessary to evaluate a Laplace transform by integration. Usually with systems that have a small number of natural modes, one need employ only a small number of basic transforms and knowledge of their properties. We have already calculated the key pair, namely

$$e^{-at}u(t) \leftrightarrow \frac{1}{s + a} \tag{6.23}$$

For $a = 0$, we have the unit step transform pair

$$u(t) \leftrightarrow \frac{1}{s} \tag{6.24}$$

If $a = +j\omega$ or $-j\omega$, then

$$e^{j\omega t}u(t) \leftrightarrow \frac{1}{s + j\omega} \tag{6.25}$$

$$e^{-j\omega t}u(t) \leftrightarrow \frac{1}{s - j\omega} \tag{6.26}$$

Adding the last two pairs yields a transform pair for the truncated cosine function.

$$\cos \omega t\, u(t) \leftrightarrow \frac{1}{2} \frac{1}{s + j\omega} + \frac{1}{2} \frac{1}{s - j\omega}$$

$$\leftrightarrow \frac{s}{s^2 + \omega^2} \tag{6.27}$$

Subtracting (6.26) from (6.25) yields a transform pair for the truncated sine function.

$$\sin \omega t\, u(t) \leftrightarrow \frac{1}{2j} \frac{1}{s - j\omega} - \frac{1}{2j} \frac{1}{s + j\omega}$$

$$\leftrightarrow \frac{\omega}{s^2 + \omega^2} \tag{6.28}$$

The time-integration property applied to the unit step function $u(t)$ yields a ramp function $tu(t)$ with the transform

$$tu(t) \leftrightarrow \frac{1}{s^2} \tag{6.29}$$

The time-derivative property applied to the unit step function $u(t)$ yields an impulse function $\delta(t)$ with the transform

$$\delta(t) \leftrightarrow 1 \tag{6.30}$$

One can apply the time-derivative property to the impulse function $\delta(t)$ to obtain $\delta'(t)$ with the transform

$$\delta'(t) \leftrightarrow s \tag{6.31}$$

Suppose that $a = \sigma \pm j\omega$ in (6.23). The resulting transform pair is

$$e^{-(\sigma \pm j\omega)t}u(t) \leftrightarrow \frac{1}{s + (\sigma \pm j\omega)} \tag{6.32}$$

If we break (6.32) into real and imaginary parts, we obtain two transform pairs.

$$e^{-\sigma t}\cos \omega t\, u(t) \leftrightarrow \frac{s + \sigma}{(s + \sigma)^2 + \omega^2} \tag{6.33}$$

$$e^{-\sigma t}\sin \omega t\, u(t) \leftrightarrow \frac{\omega}{(s + \sigma)^2 + \omega^2} \tag{6.34}$$

Table 6.1 summarizes some of the more useful transform pairs primarily for one-sided transforms.

6.5 INVERSION OF THE LAPLACE TRANSFORM

In order to use the Laplace transform effectively in system analysis, one must be able to carry out the transformation from the s domain to the t domain easily. This inverse transformation can be accomplished in several ways. The most direct method is to use the defining equation for $f(t)$: i.e.,

$$f(t) = \frac{1}{2\pi j} \int_{\sigma - j\infty}^{\sigma + j\infty} F(s)e^{st}\, ds \tag{6.35}$$

Evaluation of (6.35) requires an understanding of complex variables. The integration is usually calculated by means of a line integral in the s plane. It is not always necessary to use (6.35) to find $f(t)$, and for certain classes of systems, other methods are easier.

TABLE 6.1
Useful Transform Pairs

$f(t)$	$F(s)$	Convergence Region
1. $e^{-at}u(t)$	$\dfrac{1}{s+a}$	$-\operatorname{Re}(a) < \operatorname{Re}(s)$
2. $u(t)$	$\dfrac{1}{s}$	$0 < \operatorname{Re}(s)$
3. $tu(t)$	$\dfrac{1}{s^2}$	$0 < \operatorname{Re}(s)$
4. $t^n u(t)$	$n!/s^{n+1}$	$0 < \operatorname{Re}(s)$
5. $\delta(t)$	1	all s
6. $\delta'(t)$	s	all s
7. $\operatorname{sgn} t$	$\dfrac{2}{s}$	$\operatorname{Re}(s) = 0$
8. $-u(-t)$	$\dfrac{1}{s}$	$\operatorname{Re}(s) < 0$
9. $te^{-at}u(t)$	$\dfrac{1}{(s+a)^2}$	$-\operatorname{Re}(a) < \operatorname{Re}(s)$
10. $t^n e^{-at}u(t)$	$\dfrac{n!}{(s+a)^{n+1}}$	$-\operatorname{Re}(a) < \operatorname{Re}(s)$
11. $e^{-a|t|}$	$\dfrac{2a}{a^2 - s^2}$	$-\operatorname{Re}(a) < \operatorname{Re}(s) < \operatorname{Re}(a)$
12. $(1 - e^{-at})u(t)$	$\dfrac{a}{s(s+a)}$	$\max(0, -\operatorname{Re}(a)) < \operatorname{Re}(s)$
13. $\cos \omega t\, u(t)$	$\dfrac{s}{s^2 + \omega^2}$	$0 < \operatorname{Re}(s)$
14. $\sin \omega t\, u(t)$	$\dfrac{\omega}{s^2 + \omega^2}$	$0 < \operatorname{Re}(s)$
15. $e^{-\sigma t}\cos \omega t\, u(t)$	$\dfrac{s + \sigma}{(s+\sigma)^2 + \omega^2}$	$-\sigma < \operatorname{Re}(s)$
16. $e^{-\sigma t}\sin \omega t\, u(t)$	$\dfrac{\omega}{(s+\sigma)^2 + \omega^2}$	$-\sigma < \operatorname{Re}(s)$
17. $\begin{cases} 1 - |t|, & |t| < 1 \\ 0, & |t| > 1 \end{cases}$	$\left(\dfrac{\sinh s/2}{s/2}\right)^2$	all s
18. $\displaystyle\sum_{n=0}^{\infty} \delta(t - nT)$	$\dfrac{1}{1 - e^{-sT}}$	all s

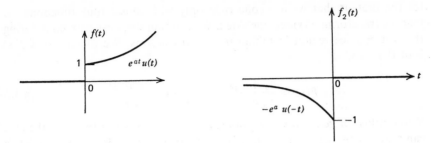

FIGURE 6.4

As with the Z-transform, if one has $F(s)$ and its region of convergence, then the corresponding time function $f(t)$ is unique. If $F(s)$ is a one-sided transform, then knowledge of the region of convergence is not needed to obtain a unique inverse. This situation is analogous to the one that occurs for one-sided and two-sided Z-transforms. The following example demonstrates these ideas.

EXAMPLE 6.5.   Consider the two time functions $f_1(t) = e^{at}u(t)$ and $f_2(t) = -e^{at}u(-t)$, shown in Figure 6.4. The corresponding Laplace transforms are

$$F_1(s) = \int_{-\infty}^{\infty} e^{at}u(t)e^{-st}\,dt = \int_0^{\infty} e^{at}e^{-st}\,dt$$

$$= \frac{-1}{s-a}\, e^{-(s-a)t}\,\Big|_0^{\infty} = \frac{1}{s-a}, \qquad \sigma > a$$

$$F_2(s) = \int_{-\infty}^{\infty} -e^{at}u(-t)e^{-st}\,dt = \int_{-\infty}^{0} -e^{at}e^{-st}\,dt$$

$$= \frac{1}{s-a}\, e^{-(s-a)t}\,\Big|_{-\infty}^{0} = \frac{1}{s-a}, \qquad \sigma < a$$

The Laplace transform of both $f_1(t)$ and $f_2(t)$ is $1/(s-a)$. However, the regions of convergence are different. If one allows time functions to be nonzero for both positive and negative t, then one must specify the region of convergence of $F(s)$ in order to specify a unique time function.

Partial Fraction Expansions

Linear time-invariant systems with lumped parameters generally lead to transforms which are rational functions of s: that is, transforms that are ratios of polynomials in s. Rational functions of this form can be represented as a sum of simpler fractions whose inverse transforms are tabulated. Assume for the present that we are concerned only with causal time functions, so that the transforms $F(s)$ are one sided. We can thus dispense with considering the convergence region in taking inverse transforms. We assume that $F(s)$ is of the form

$$F(s) = \frac{a_0 + a_1 s + a_2 s^2 + \cdots + a_m s^m}{b_0 + b_1 s + b_2 s^2 + \cdots + b_n s^n} \tag{6.36}$$

Without loss of generality, we assume $m < n$. If this is not the case, then we can always write $F(s)$ as the sum of a polynomial $Q(s)$ of degree $m - n$ plus a ratio of polynomials with the numerator degree one less than the denominator degree.

In (6.36), to find a partial fraction expansion, we first find the roots of the denominator polynomial, $p_1, p_2, \ldots, p_n$. These roots are the poles or singularities of $F(s)$. We write $F(s)$ as

$$F(s) = \frac{a_0 + a_1 s + a_2 s^2 + \cdots + a_m s^m}{b_n(s - p_1)(s - p_2) \cdots (s - p_n)}$$

$$= \frac{c_1}{s - p_1} + \frac{c_2}{s - p_2} + \cdots + \frac{c_n}{s - p_n} \tag{6.37}$$

The coefficients c_i in (6.37) can be found for the case of nonrepeated poles as

$$c_i = (s - p_i)F(s)\big|_{s=p_i} \tag{6.38}$$

Equation (6.38) is obtained by multiplying both sides of (6.37) by $(s - p_i)$ and then evaluating the resulting equation at $s = p_i$. In the case of multiple roots, say p_i repeated r times, the expansion of (6.37) must include terms

$$\frac{c_{i1}}{s - p_i} + \frac{c_{i2}}{(s - p_i)^2} + \cdots + \frac{c_{ir}}{(s - p_i)^r}$$

The coefficients c_{i_k} are evaluated by multiplying both sides of (6.37) by $(s - p_i)^r$, differentiating $(r - k)$ times, and then evaluating the resultant equation at $s = p_i$. Thus

$$c_{ir} = (s - p_i)^r F(s)\big|_{s=p_i}$$

$$c_{ir-1} = \frac{d}{ds}\{(s - p_i)^r F(s)\}\big|_{s=p_i}$$

$$\cdots$$

$$c_{ir-k} = \frac{1}{k!}\frac{d^k}{ds^k}\{(s - p_i)^r F(s)\}\big|_{s=p_i} \tag{6.39}$$

$$\cdots$$

$$c_{i1} = \frac{1}{(r-1)!}\frac{d^{r-1}}{ds^{r-1}}\{(s - p_i)^r F(s)\}\big|_{s=p_i}$$

The general form, assuming that the pole p_i is repeated r_i times, for $i = 1, 2, \ldots, k$ and $r_1 + r_2 + \cdots + r_k = n$, is

$$F(s) = \frac{c_1}{s - p_1} + \cdots + \frac{c_{i_1}}{s - p_i}$$

$$+ \frac{c_{i_2}}{(s - p_i)^2} + \cdots + \frac{c_{i r_i}}{(s - p_i)^{r_i}} + \cdots + \frac{c_k}{(s - p_k)^{r_k}} \qquad (6.40)$$

The above expansions can always be checked by recombining the terms in the partial fraction expansion. These expansions are valid for complex roots also. However, because complex roots always occur as a complex conjugate pair (for real $f(t)$), they can be expressed as a single quadratic factor. Quadratic factors of the denominator polynomial can be separated either as a single entity or as two simple poles. Suppose there is a complex conjugate pair of poles in the denominator polynomial. Three alternate representations for these roots are

$$(s + \alpha + j\omega_0)(s + \alpha - j\omega_0) = (s + \alpha)^2 + \omega_0^2$$
$$= s^2 + 2\alpha s + (\alpha^2 + \omega_0^2)$$
$$= s^2 + as + b \qquad (6.41)$$

The corresponding partial fraction expansion can take any of the following forms, depending on the representation used in (6.41).

$$F(s) = \frac{A(s)}{B(s)} = \frac{a_0 + a_1 s + a_2 s^2 + \cdots + a_m s^m}{b_0 + b_1 s + b_2 s^2 + \cdots + b_n s^n} = \frac{A(s)}{B_1(s)(s^2 + as + b)}$$

$$= \frac{\tilde{a}_1 s + \tilde{a}_2}{s^2 + as + b} + \frac{A_1(s)}{B_1(s)}$$

$$= \frac{\tilde{b}_1 s + \tilde{b}_2}{(s + \alpha)^2 + \omega_0^2} + \frac{A_1(s)}{B_1(s)}$$

$$= \frac{\tilde{c}_1 + j\tilde{c}_2}{s + \alpha + j\omega_0} + \frac{\tilde{c}_1 - j\tilde{c}_2}{s + \alpha - j\omega_0} + \frac{A_1(s)}{B_1(s)} \qquad (6.42)$$

The constants $\tilde{a}$, $\tilde{b}$, and $\tilde{c}$ in (6.42) are all real numbers. Usually, the latter two forms in (6.42) are the easiest to use. Notice that in the last form in (6.42), the two factors are complex conjugates, and so only one factor need be evaluated. The other can then be written down immediately. The following examples demonstrate finding inverse transforms by partial fraction expansions.

EXAMPLE 6.6. Assume that $f(t)$ is a causal function and that $F(s)$ is given by $F(s) = (s + 3)/(s^3 + 3s^2 + 6s + 4)$. Because $f(t)$ is causal, we need not worry about the convergence region of $F(s)$ to obtain a unique inverse. We factor the denominator polynomial by guessing at a root, say a, and then dividing $s^3 + 3s^2 + 6s + 4$ by $s - a$. We find quickly that $s = -1$ is a root, and so

$$F(s) = \frac{s + 3}{[(s + 1)^2 + 3](s + 1)} = \frac{A}{s + 1} + \frac{Bs + C}{s^2 + 2s + 4} \quad (6.43)$$

We evaluate A by multiplying both sides of (6.43) by $(s + 1)$, then setting $s = -1$:

$$A = (s + 1)F(s)\big|_{s=-1} = \frac{(-1) + 3}{((-1) + 1)^2 + 3} = \frac{2}{3}$$

The values of B and C can now be found by clearing (6.43) of fractions and equating the coefficients of powers of s.

$$s + 3 = \tfrac{2}{3}s^2 + \tfrac{4}{3}s + \tfrac{8}{3} + Bs^2 + Bs + Cs + C$$

Now we set
$$\tfrac{2}{3} + B = 0, \qquad B = -\tfrac{2}{3}$$
and
$$\tfrac{4}{3} + B + C = 1, \qquad C = \tfrac{1}{3}$$

Therefore we have that

$$F(s) = \frac{2}{3}\frac{1}{s + 1} + \frac{-\tfrac{2}{3}s + \tfrac{1}{3}}{s^2 + 2s + 4}$$

$$= \frac{2}{3}\frac{1}{s + 1} - \frac{2}{3}\frac{s - \tfrac{1}{2}}{(s + 1)^2 + 3}$$

$$= \frac{2}{3}\frac{1}{s + 1} - \frac{2}{3}\frac{s + 1}{(s + 1)^2 + 3} + \frac{1}{3}\frac{3}{(s + 1)^2 + 3} \quad (6.44)$$

Equation (6.44) is now in a form which can be found immediately in Table 6.1. Thus

$$f(t) = (\tfrac{2}{3}e^{-t} - \tfrac{2}{3}e^{-t}\cos 3t + \tfrac{1}{3}e^{-t}\sin 3t)u(t)$$

EXAMPLE 6.7. Assume again that $f(t)$ is causal and that its transform is given by

$$F(s) = \frac{2s^2 + 3s + 3}{(s + 1)(s + 3)^3}$$

In this case, $F(s)$ has a first-order pole at $s = -1$ and third-order pole at $s = -3$. We thus expand $F(s)$ as

$$F(s) = \frac{2s^2 + 3s + 3}{(s + 1)(s + 3)^3} = \frac{A}{s + 1} + \frac{C_1}{s + 3} + \frac{C_2}{(s + 3)^2} + \frac{C_3}{(s + 3)^3} \quad (6.45)$$

We can evaluate coefficient A as in Example 6.6:

$$A = (s + 1)F(s)\big|_{s=-1} = \tfrac{1}{4}$$

Coefficient C_3 is found by multiplying (6.45) by $(s + 3)^3$ and then evaluating at $s = -3$.

$$C_3 = (s + 3)^3 F(s)\big|_{s=-3} = -6$$

To find C_2 and C_1, we have to differentiate. Thus

$$C_2 = \frac{d}{ds}\{(s + 3)^3 F(s)\}\bigg|_{s=-3} = \frac{d}{ds}\left\{\frac{2s^2 + 3s + 3}{s + 1}\right\}\bigg|_{s=-3}$$

$$= \left(\frac{4s + 3}{s + 1} - \frac{2s^2 + 3s + 3}{(s + 1)^2}\right)\bigg|_{s=-3} = \frac{3}{2}$$

Also,

$$C_1 = \frac{1}{2!}\frac{d^2}{ds^2}\{(s + 3)^3 F(s)\}\bigg|_{s=-3} = \frac{1}{2}\frac{d}{ds}\left\{\frac{4s + 3}{s + 1} - \frac{2s^2 + 3s + 3}{(s + 1)^2}\right\}\bigg|_{s=-3}$$

$$= \frac{1}{2}\left\{\frac{4}{s + 1} - \frac{4s + 3}{(s + 1)^2} - \frac{4s + 3}{(s + 1)^2} + \frac{2(2s^2 + 3s + 3)}{(s + 1)^3}\right\}\bigg|_{s=-3} = -\frac{1}{8}$$

and so

$$F(s) = \frac{\frac{1}{4}}{s + 1} + \frac{-\frac{1}{8}}{s + 3} + \frac{\frac{3}{2}}{(s + 3)^2} + \frac{-6}{(s + 3)^3}$$

Using Table 6.1, we obtain the inverse transform as

$$f(t) = (\tfrac{1}{4}e^{-t} - \tfrac{1}{8}e^{-3t} + \tfrac{3}{2}te^{-3t} - 3t^2e^{-3t})u(t)$$

Inversion of Two-Sided Laplace Transforms

We have to this point considered finding inverse transforms only for causal functions. Suppose now that $f(t)$ is noncausal. In order to find $f(t)$ from $F(s)$ we must know the region of convergence for $F(s)$. The location of the poles of $F(s)$ with respect to the convergence region determines whether a given singularity refers to a positive or negative time portion of $f(t)$. If, for example, a pole of $F(s)$ lies to the right of the convergence region, then by our discussion in Section 6.1, this pole gives rise to a nonzero time function for $t < 0$. Similarly, poles to the left of the convergence region give rise to a positive time portion of $f(t)$.

To illustrate the procedure suppose that $F(s)$ is given by

$$F(s) = \frac{2s}{(s + 1)(s + 2)}, \qquad -2 < \text{Re}(s) < -1$$

In this case, we see that the pole at $s = -1$ lies to the right of the convergence region and the pole at $s = -2$ lies to the left of the convergence region. Thus, the pole at $s = -1$ gives rise to a time function nonzero for

$t < 0$ and the pole at $s = -2$ gives rise to a time function nonzero for $t > 0$. We now expand $F(s)$ in partial fractions to obtain

$$F(s) = \frac{2s}{(s+1)(s+2)} = \frac{-2}{s+1} + \frac{4}{s+2}, \qquad -2 < \text{Re}\,(s) < -1$$

Having identified which poles correspond to negative time portions and positive time portions of $f(t)$, we can proceed with inverting $F(s)$. The term $4/(s+2)$ yields, from Table 6.1,

$$\frac{4}{s+2} \leftrightarrow 4e^{-2t}u(t)$$

The term $-2/(s+1)$ corresponds to a negative time portion of $f(t)$. Using the reasoning of Example 6.1, we find that $f(t) = e^{at}u(-t)$ has transform $-a/(s-a)$. Thus,

$$\frac{-2}{s+1} \leftrightarrow 2e^{-t}u(-t)$$

The inverse transform of $F(s)$ is thus

$$f(t) = 2e^{-t}u(-t) + 4e^{-2t}u(t)$$

The inverse transform of many two-sided Laplace transforms can be found using the procedure outlined above. First, expand $F(s)$ in a partial fraction expansion. Identify which poles lie to the right and the left of the convergence region. For poles to the left of the convergence region, invert each term to obtain a time function which is nonzero for $t > 0$. For poles to the right of the convergence region, invert each term to obtain a time function which is nonzero for $t < 0$. The key transform pairs are

$$\frac{t^n e^{-at}u(t)}{n!} \leftrightarrow \frac{1}{(s+a)^n}, \qquad \text{Re}\,(s) > a \tag{6.46}$$

$$\frac{(-t)^n e^{-at}u(-t)}{n!} \leftrightarrow \frac{1}{(s+a)^n}, \qquad \text{Re}\,(s) < a \tag{6.47}$$

Another method of finding the inverse of two-sided Laplace transforms makes use of tables for one-sided transforms. The underlying idea is to express the Laplace transform as the sum of two one-sided transforms.

Consider a function $f(t)$ nonzero for both positive and negative t, as shown in Figure 6.5. We can always decompose $f(t)$ into a sum of two functions, $f_1(t)$ nonzero for $t > 0$ and $f_2(t)$ nonzero for $t < 0$. Thus

$$f(t) = f_1(t)u(t) + f_2(t)u(-t)$$

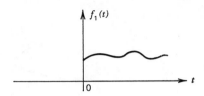

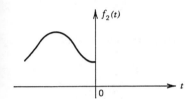

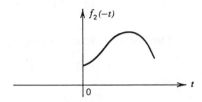

FIGURE 6.5

Suppose we now write the Laplace transform of $f(t)$:

$$\mathscr{L}_b\{f(t)\} = \int_{-\infty}^{\infty} f(t)e^{-st}\,dt = \int_{-\infty}^{0} f_2(t)e^{-st}\,dt + \int_{0}^{\infty} f_1(t)e^{-st}\,dt \quad (6.48)$$

The first integral in (6.48) can be put in standard one-sided form by substituting $t' = -t$. Thus

$$F(s) = \int_{0}^{\infty} f_2(t')e^{st'}\,dt' + \int_{0}^{\infty} f_1(t)e^{-st}\,dt$$

$$= F_2(-s) + F_1(s) \quad (6.49)$$

where $F_1(s) = \mathscr{L}\{f_1(t)\}$
$\quad\quad\ F_2(s) = \mathscr{L}\{f_2(-t)\}$

The function $f_2(-t)$ is the mirror image of $f_2(t)$ about the vertical axis at $t = 0$. We now have a method for obtaining $F(s)$ using tables for one-sided transforms. Merely break $f(t)$ into its positive and negative time portions. Take the mirror reflection of the negative time portion $f_2(-t)$ and find its one-sided Laplace transform $F_2(s)$. Now replace s by $-s$ for the negative time portion and add $F_2(-s)$ to the one-sided Laplace transform for the positive time portion.

EXAMPLE 6.8. Find the Laplace transform of $f(t)$, where $f(t) = e^{-\alpha t}u(t) + e^{\beta t}u(-t)$, as shown in Figure 6.6.

Clearly

$$f_1(t) = e^{-\alpha}\,u(t)$$
$$f_2(t) = e^{\beta t}u(-t)$$

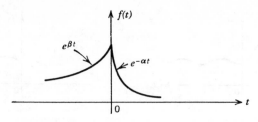

FIGURE 6.6

The mirror reflection of $f_2(t)$ about the vertical axis at $t = 0$ is

and so
$$f_2(-t) = e^{-\beta t}u(t)$$

$$F_2(s) = \mathscr{L}\{f_2(-t)\} = \frac{1}{s + \beta}, \qquad -\beta < \text{Re}\,(s)$$

Thus, $F_2(-s) = 1/(-s + \beta)$, $\text{Re}\,(s) < \beta$. The positive time portion $f_1(t)$ has a transform

$$F_1(s) = \frac{1}{s + \alpha}, \qquad -\alpha < \text{Re}\,(s)$$

The two-sided transform of $F(s)$ is therefore

$$\mathscr{L}_b\{f(t)\} = F_1(s) + F_2(-s) = \frac{1}{s + \alpha} + \frac{1}{-s + \beta} = \frac{(\alpha + \beta)}{(s + \alpha)(-s + \beta)},$$

$$-\alpha < \sigma < \beta$$

EXAMPLE 6.9. Find the inverse Laplace transform of $F(s) = (2s + 3)/(s + 1)(s + 2)$ if the convergence regions are:

(1) $-2 < \text{Re}\,(s) < -1$
(2) $\text{Re}\,(s) > -1$
(3) $\text{Re}\,(s) < -2$.

We first break $F(s)$ into partial fractions.

$$F(s) = \frac{2s + 3}{(s + 1)(s + 2)} = \frac{A}{s + 1} + \frac{B}{s + 2}$$

Evaluating A and B, we obtain

$$A = (s + 1)F(s)\big|_{s=-1} = \frac{2s + 3}{s + 2}\bigg|_{s=-1} = 1$$

$$B = (s + 2)F(s)\big|_{s=-2} = \frac{2s + 3}{s + 1}\bigg|_{s=-2} = 1$$

and so

$$F(s) = \frac{1}{s+1} + \frac{1}{s+2}$$

(1) For the convergence region $-2 < \text{Re } (s) < -1$ we see that the pole at $s = -1$ lies to the right of the convergence region and the pole at $s = -2$ lies to the left of the convergence region. The first term, therefore, gives rise to a time function nonzero for negative t and the second term gives rise to a time function nonzero for positive t. In the notation of (6.49), we have that

$$F_1(s) = \frac{1}{s+2}, \qquad F_2(-s) = \frac{1}{s+1}$$

The positive time portion of $f(t)$ is therefore

$$f_1(t) = \mathscr{L}^{-1}\{F_1(s)\} = e^{-2t}u(t)$$

The mirror reflection of the negative time portion of $f(t)$ is the inverse transform of $F_2(s) = 1/(-s+1)$. Thus

$$f_2(-t) = \mathscr{L}^{-1}\{F_2(s)\} = \mathscr{L}^{-1}\left\{\frac{1}{-s+1}\right\} = -e^t\, u(t)$$

and so

$$f_2(t) = -e^{-t}\, u(-t)$$

Thus,

$$f(t) = -e^{-t}\, u(-t) + e^{-2t}\, u(t)$$

(2) For the convergence region $\text{Re } (s) > -1$, we see that both poles in $F(s)$ lie to the left of the convergence region. Therefore $f(t)$ must be non-zero for positive t only: i.e., $f(t)$ is causal. Thus, $F_2(-s) = 0$ and so

$$f(t) = (e^{-t} + e^{-2t})u(t)$$

(3) For the convergence region $\text{Re } (s) < -2$, both poles in $F(s)$ lie to the right of the convergence region. Thus $f(t)$ is nonzero for negative t only. Therefore, in the notation of (6.49),

$$F_1(s) = 0, \qquad F_2(-s) = \frac{1}{s+1} + \frac{1}{s+2}$$

and so

$$f_2(-t) = \mathscr{L}^{-1}\{F_2(s)\} = \mathscr{L}^{-1}\left\{\frac{1}{-s+1} + \frac{1}{-s+2}\right\}$$

$$= -e^t\, u(t) - e^{2t}\, u(t)$$

This relation implies that

$$f(t) = f_2(t) = (-e^{-t} - e^{-2t})u(-t)$$

This example demonstrates how the "same" Laplace transform $F(s)$ can represent various time functions, depending on the region of convergence. The inversion process merely involves identifying poles to the left and right of the convergence region. By our previous discussions, poles to the left of the convergence region give rise to positive time functions and those to the right of the convergence region give rise to negative time functions. By converting that part of $F(s)$ with singularities to the right of the convergence region into a positive time function (replacing s by $-s$) we are able to use one-sided transform tables to evaluate two-sided transforms. In physical applications, the region of convergence is usually not essential, because the correct region of convergence can be obtained by physical reasoning. Time functions which approach $\pm\infty$ as $t \to \pm\infty$ do not exist physically. Thus, it is a fairly easy matter to consider all the possible convergence regions and rule out those which give rise to time functions that approach $\pm\infty$ as $t \to \pm\infty$. Only one "practical" time function will remain.

To summarize, the process of inverting two-sided Laplace transforms using one-sided transform pairs consists of these steps:

(1) Expand $F(s)$ in a partial fraction expansion.

(2) Identify those poles to the right of the convergence region. These terms give rise to the negative time portion of $f(t)$ and are denoted as $F_2(-s)$.

(3) Replace $-s$ by s in $F_2(-s)$. Invert $F_2(s)$ to obtain the mirror image of the negative time portion of $f(t)$, denoted as $f_2(-t)$.

(4) Finally, replace $-t$ by t to obtain $f_2(t)$. This term is added to the positive time portion of $f(t)$, $f_1(t)$, to obtain the complete time function.

6.6   APPLICATIONS OF LAPLACE TRANSFORMS— DIFFERENTIAL EQUATIONS

In Chapter 1, the basic models developed to describe linear systems were linear differential or difference equations with constant coefficients. If we have a linear time-invariant system with input $x(t)$ and output $y(t)$, the basic model that we have used to describe the system is

$$b_n \frac{d^n}{dt^n} y(t) + b_{n-1} \frac{d^{n-1}}{dt^{n-1}} y(t) + \cdots + b_0 y(t)$$

$$= a_m \frac{d^m}{dt^m} x(t) + a_{m-1} \frac{d^{m-1}}{dt^{m-1}} x(t) + \cdots + a_0 x(t) \quad (6.50)$$

We can solve this differential equation by using the Laplace transform. Multiplying both sides of (6.50) by e^{-st} and integrating each term from $-\infty$ to ∞, we obtain

$$b_n \int_{-\infty}^{\infty} y^{(n)}(t)e^{-st}\,dt + \cdots + b_0 \int_{-\infty}^{\infty} y(t)e^{-st}\,dt$$

$$= a_m \int_{-\infty}^{\infty} x^{(m)}(t)e^{-st}\,dt + \cdots + a_0 \int_{-\infty}^{\infty} x(t)e^{-st}\,dt \quad (6.51)$$

Denote the Laplace transforms for $y(t)$ and $x(t)$ by $Y(s)$ and $X(s)$, respectively. Equation (6.51) can then be written, using the Laplace transform for the derivative, as

$$b_n Y(s)s^n + \cdots + b_0 Y(s) = a_m X(s)s^m + \cdots + a_0 X(s)$$

Solving for $Y(s)$, we have

$$Y(s) = \left(\frac{a_m s^m + a_{m-1}s^{m-1} + \cdots + a_0}{b_n s^n + b_{n-1}s^{n-1} + \cdots + b_0}\right)X(s) \quad (6.52)$$

Equation (6.52) is an algebraic equation in s which can readily be solved for $Y(s)$. $Y(s)$ can then be expanded in partial fractions and the inverse transform taken to find $y(t)$. Equation (6.52) is a conceptually useful equation also. The system is characterized by the ratio of polynomials in s: i.e., $(a_m s^m + \cdots + a_0)/(b_n s^n + \cdots + b_0)$. We term this ratio of polynomials the system transfer function or system function $H(s)$:

$$H(s) = \frac{a_m s^m + a_{m-1}s^{m-1} + \cdots + a_0}{b_n s^n + b_{n-1}s^{n-1} + \cdots + b_0} \quad (6.53)$$

Thus the transform of the output of a linear time-invariant system is always given by multiplying $H(s)$ by the Laplace transform of the input $X(s)$: i.e.,

$$Y(s) = H(s) \cdot X(s)$$

By definition, the output of a linear system when the input is an impulse function is the impulse response of the system. Recall that the Laplace transform of an impulse is 1. Thus, for an impulse input, the output $y(t)$ is $h(t)$, the impulse response, and

$$Y(s) = H(s) \cdot 1 \quad (6.54)$$

Equation (6.54) merely states that $H(s)$ is the Laplace transform of the impulse response of the system, i.e.,

$$H(s) = \mathscr{L}_b\{h(t)\}$$

Notice that in going from (6.51) to (6.52), we assumed all initial conditions as zero. This assumption is always made in finding the system transfer function $H(s)$.

One of the advantages of the transform method is the automatic inclusion of initial conditions into the solution process. To demonstrate this property, consider the next example.

EXAMPLE 6.10.   Consider the simple RLC series network shown in Figure 6.7. There is an initial current i_L flowing in the inductor and initial voltage v_c across the capacitor. The switch is closed at $t = 0$. The differential equation for the system is

$$e(t) = R\, i(t) + L\frac{di(t)}{dt} + \frac{1}{C}\int_{-\infty}^{t} i(t')\, dt' \qquad (6.55)$$

Transforming this equation, we obtain

$$E(s) = R\, I(s) + L(s\, I(s) + i_L) + \frac{1}{C}\left(\frac{I(s)}{s} - \frac{v_c}{s}\right)$$

Notice that the signs on the initial conditions depend on the assumed direction for $i(t)$ relative to the flow of the current in L and the voltage across C. Rearranging terms yields

$$I(s)\left[R + Ls + \frac{1}{Cs}\right] = E(s) - Li_L + \frac{v_c}{s} \qquad (6.56)$$

In (6.56), the bracketed terms are characteristic of the system, while the terms $E(s) - Li_L + (v_c/s)$ are characteristic of the input. The initial energy storage terms v_c and i_L act as forcing functions on the system (in conjunction with $E(s)$). We find $i(t)$ by finding the inverse transform of $I(s)$ given by

$$I(s) = \frac{E(s) - Li_L + \dfrac{v_c}{s}}{R + Ls + \dfrac{1}{Cs}}$$

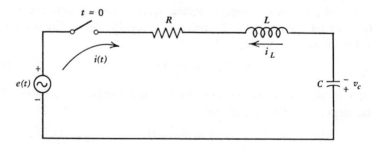

FIGURE 6.7

The Laplace transform is used almost exclusively to solve differential equations like (6.50) because the initial conditions are easily incorporated into the solution. However, the Fourier transform can be used in the same way. The main differences lie in the restricted class of functions which can be Fourier transformed and in the way initial conditions must be introduced. The initial conditions, when the Fourier transform is used, are lumped with the input forcing function in the following manner. Assume that we have system described by the differential equation

$$a_3 \frac{d^3 y(t)}{dt^3} + a_2 \frac{d^2 y(t)}{dt^2} + a_1 \frac{dy(t)}{dt} + a_0 y(t) = x(t), \qquad t > 0 \quad (6.57)$$

In order to obtain a complete solution to (6.57), we need three initial conditions, such as $y(0)$, $y'(0)$, $y''(0)$. Assuming a nonzero solution for $t > 0$, we define the solution to be $y_1(t)$, where

$$y_1(t) = y(t) u(t)$$

thus we have that

$$\frac{dy_1(t)}{dt} = y'(t) u(t) + y(t) \delta(t)$$

$$= y'(t) u(t) + y(0) \delta(t)$$

Similarly

$$\frac{d^2 y_1(t)}{dt^2} = y''(t) u(t) + y'(0) \delta(t) + y(0) \delta'(t)$$

$$\frac{d^3 y_1(t)}{dt^3} = y'''(t) u(t) + y''(0) \delta(t) + y'(0) \delta'(t) + y(0) \delta''(t)$$

Substituting in (6.57), we write the original differential equation as

$$a_3 \frac{d^3 y_1(t)}{dt^3} + a_2 \frac{d^2 y_1(t)}{dt^2} + a_1 \frac{dy_1(t)}{dt} + a_0 y_1(t)$$

$$= x(t) + [a_1 y(0) + a_2 y'(0) + a_3 y''(0)] \delta(t)$$

$$+ [a_2 y(0) + a_3 y'(0)] \delta'(t) + a_3 y(0) \delta''(t) \quad (6.58)$$

This process has placed the initial conditions of the differential equation at time $t = 0$ on the right-hand side as part of the forcing function. We can now take the Fourier transform of (6.58) and proceed in much the same manner as in the Laplace transform process.

EXAMPLE 6.11.   A system is described by the differential equation

$$y'(t) + a y(t) = e^{-\alpha t} u(t)$$

with initial condition $y(0) = c$. We rewrite the differential equation to include the initial condition c on the right-hand side. Thus

$$y'(t) + a\,y(t) = e^{-\alpha t}\,u(t) + c\,\delta(t)$$

Taking Fourier transforms on both sides yields

$$j\omega\,Y(\omega) + a\,Y(\omega) = \frac{1}{\alpha + j\omega} + c$$

Solving for $Y(\omega)$, we obtain

$$Y(\omega) = \frac{c}{a + j\omega} + \frac{1}{(a + j\omega)(\alpha + j\omega)}$$

$$= \frac{c}{a + j\omega} + \frac{A}{a + j\omega} + \frac{B}{\alpha + j\omega}$$

where

$$A = \frac{1}{\alpha + j\omega}\bigg|_{j\omega=-a} = \frac{1}{\alpha - a}$$

$$B = \frac{1}{a + j\omega}\bigg|_{j\omega=-\alpha} = \frac{1}{a - \alpha}$$

and so

$$Y(\omega) = \frac{c}{a + j\omega} + \frac{1}{\alpha - a}\left(\frac{1}{a + j\omega} - \frac{1}{\alpha + j\omega}\right)$$

Therefore

$$y(t) = \left\{ce^{-at} + \frac{1}{\alpha - a}(e^{-at} - e^{-\alpha t})\right\}u(t)$$

This problem can also be solved by using the Laplace transforms. Transforming the original differential equation, we have

$$sY(s) - c + aY(s) = \frac{1}{s + \alpha}$$

Solving for $Y(s)$, we obtain

$$Y(s) = \frac{1}{(s + \alpha)(s + a)} + \frac{c}{s + a}$$

$$= \frac{1}{\alpha - a}\left(\frac{1}{s + a} - \frac{1}{s + \alpha}\right) + \frac{c}{s + a}$$

The inverse transform is therefore

$$y(t) = \left\{ ce^{-at} + \frac{1}{\alpha - a} (e^{-at} - e^{-\alpha t}) \right\} u(t)$$

which is the same as the result obtained using the Fourier transform.

6.7 STABILITY IN THE *s* DOMAIN

One of the important questions that often must be answered in the analysis of a system concerns the stability of the system. In Chapter 3, we defined a stable system as one which had a bounded output for a bounded input. Obviously, this concept of stability is important if we are going to operate a system for any period of time. Stability, as we showed in Chapter 3, is a characteristic of the system and does not depend on the input to the system. In the state variable formulation, we determined stability based on knowledge of the eigenvalues of the system matrix **A**. The eigenvalues of the matrix **A** allow us to determine the response of the natural modes of the system and thus determine stability. Stability can also be determined in the frequency or *s* domain.

The transfer function of a system $H(s)$ is the Laplace transform of the impulse response $h(t)$. We have seen that either $H(s)$ or $h(t)$ characterizes the system in the sense that we can find the output to any input given $H(s)$ or $h(t)$. Thus, because the stability of a system is characteristic of the system itself, independent of the input, we should be able to use $H(s)$ to determine stability.

Assume that $H(s)$ is the transfer function of a causal time-invariant system and is the ratio of two polynomials in *s*, as in (6.53). The transformed output is

$$Y(s) = \left(\frac{a_m s^m + a_{m-1} s^{m-1} + \cdots + a_0}{b_n s^n + b_{n-1} s^{n-1} + \cdots + b_0} \right) X(s) \tag{6.59}$$

Consider the time response of (6.59). If we make a partial fraction expansion of (6.59), the denominator polynomial of $H(s)$, $b_n s^n + \cdots + b_0$, can factor into a variety of terms.

1. Simple poles of the form $c/(s + a)$. This form corresponds to a simple pole at $s = -a$. If *a* is positive, the pole is in the left half of the *s* plane, as shown in Figure 6.8. The corresponding time response is $ce^{-at} u(t)$, and as *t* increases, the time response dies away to zero. If *a* is negative, so that the pole is in the right half of the *s* plane, then the time response increases without bound as *t* increases. Thus, a stable system must have real-valued poles of $H(s)$ in the left half of the *s* plane. Repeated simple poles also give

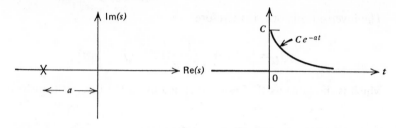

FIGURE 6.8

rise to exponentially damped time responses for poles in the left half of the s plane.

2. Complex conjugate poles of the form $c/[(s + \alpha)^2 + \omega^2]$. This term can be factored into two terms

$$c_1/(s + \alpha + j\omega) + c_1^*/(s + \alpha - j\omega)$$

This term thus represents a pair of complex conjugate poles. If α is positive, the poles are in the left half of the s-plane as shown in Figure 6.9. The corresponding time response is of the form $(ce^{-\alpha t}/\omega) \sin \omega t\, u(t)$, as shown in Figure 6.9. If α is negative, thereby placing the poles in the right half of the s plane, then the time function is $(ce^{|\alpha|t}/\omega) \sin \omega t\, u(t)$. Again we see that left-plane poles have time functions which die away as t increases and right plane poles correspond to time functions which increase without bound as t increases.

3. Complex conjugate poles of the form $c/(s^2 + \omega_0^2)$. Terms of this form represent complex conjugate poles on the $j\omega$ axis. The corresponding time function is $(c/\omega_0) \sin \omega_0 t$. In this case, there is no exponential damping, and so the response does not die away as t increases. It may appear at first glance that the time response does not increase as t increases. However, if the system is forced with a sinusoid of the same frequency ω_0, then a double complex conjugate pair of poles results, and $Y(s)$ has a term of the form $[1/(s^2 + \omega_0^2)]^2$.

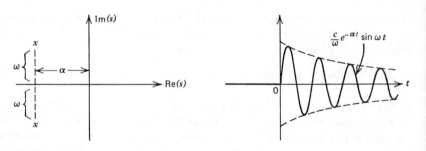

FIGURE 6.9

This term gives rise to a time response

$$(1/2\omega_0^3)\,(\sin \omega_0 t - \omega_0 t \cos \omega_0 t)$$

which increases without bound as t increases. Physically, we are exciting a natural resonance of the system with an input at precisely the resonant frequency. Because there is no loss ($\alpha = 0$) associated with this mode of the system, the output grows without bound. This is often called a marginally stable system. The same kind of considerations apply to a simple pole at the origin. This term gives rise to a time function which is a step. If this system is forced with a step, then a ramp output will result. Notice that if $H(s)$ has repeated poles on the $j\omega$ axis, terms occur of the form $[1/(s^2 + \omega^2)^2]$ which lead to time functions growing without bound.

To summarize, a causal, time-invariant lumped parameter system with transfer function $H(s)$ is stable if

(1) All poles of $H(s)$ are in the left half of the s plane
(2) The degree of the denominator polynomial of $H(s)$ is greater than or equal to the degree of the numerator polynomial.

The last consideration is required in order to rule out terms like s^{m-n}, a differentiator of degree $m - n$. If an input like $\sin \omega t$ were used to excite the system, then $y(t)$ would include a term $\omega^{m-n} \sin \omega t$. This term could be made as large as desired by increasing the input frequency ω.

EXAMPLE 6.12. How is the stability of the system in Figure 6.10 affected by the amplifier gain g? This system is known as a feedback system. It has the very desirable property that the total response is not substantially affected by parameter changes within the feedback loop. The overall transfer function of the system is $H(s) = Y(s)/X(s)$ and can be calculated as follows. The output of the amplifier (in the transform domain) is $g\,Y(s)$; thus, the input to the box containing $G(s)$ is

$$E(s) = X(s) - g\,Y(s)$$

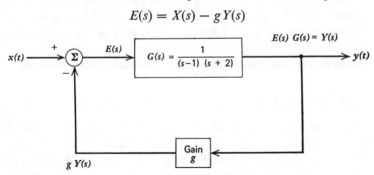

FIGURE 6.10

The output of the box containing $G(s)$ is, of course, $Y(s) = E(s) G(s)$. Therefore

$$
\begin{aligned}
Y(s) &= E(s) G(s) \\
&= [X(s) - g Y(s)] G(s) \\
&= X(s) G(s) - g Y(s) G(s)
\end{aligned}
$$

To find $H(s)$ we form the ratio $Y(s)/X(s)$. Thus

$$
X(s) G(s) = Y(s) + g Y(s) G(s) = Y(s)[1 + gG(s)]
$$

and so

$$
H(s) = \frac{Y(s)}{X(s)} = \frac{G(s)}{1 + gG(s)}
$$

Stability of $H(s)$ depends only on the zeros of the denominator $1 + gG(s)$.

$$
1 + gG(s) = 1 + g \frac{1}{(s - 1)(s + 2)} = \frac{(s - 1)(s + 2) + g}{(s - 1)(s + 2)}
$$

The zeros of the denominator are thus determined by the roots of

$$
(s + 2)(s - 1) + g = 0
$$

or

$$
s^2 + s - 2 + g = 0
$$

The roots are thus

$$
s_1, s_2 = -\frac{1}{2} \pm \sqrt{\frac{1 - 4(-2 + g)}{4}} = -\frac{1}{2} \pm \sqrt{\tfrac{9}{4} - g}
$$

For $g < \frac{9}{4}$, the roots s_1 and s_2 are real. For $g > \frac{9}{4}$, the roots become complex conjugates. For $g = \frac{9}{4}$ there is a double root on the real axis of the s plane at $\text{Re}\,(s) = -\frac{1}{2}$. If we were to plot a locus of the roots in the s plane as g varies, we would obtain the graph of Figure 6.11. As we increase the gain of this feedback system, we actually stabilize an initially unstable system. The opposite phenomenon can also occur: i.e., increasing the gain can make a stable feedback system unstable.

6.8 NONCAUSAL SYSTEMS AND INPUTS

In the analysis of most systems, the assumption is made that both the input $x(t)$ and the impulse response function $h(t)$ are causal functions. A noncausal impulse response implies that there exists some output before an impulse is applied. This interpretation thus leads one to term noncausal impulse responses as physically nonrealizable. Suppose that $h(t)$ is as shown in

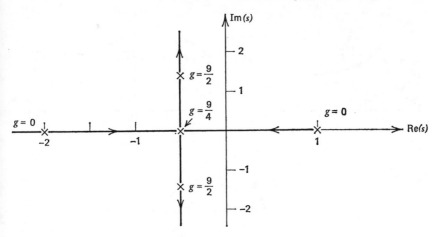

FIGURE 6.11

Figure 6.12. If we wish to find the output of this system at some time t resulting from an input $x(t)$, we can convolve $h(t)$ and $x(t)$ as shown graphically in Figure 6.12. Figure 6.12 shows why $h(t)$ is called unrealizable. To obtain the output at time t, we must weight future inputs of the input $x(t)$. If the processing is to occur in real time, this approach, of course, is impossible. However, there are many applications in which we can use noncausal impulse responses provided that

(1) we can store the data $x(t)$ on a tape or in some other fashion or;
(2) we are willing to accept a time delay in obtaining the output $y(t)$.

There are many applications in which one processes stored data. For example, stored data is often used in seismic processing or in processing signals from space probes. In these cases, there is no reason not to use noncausal impulse responses to process the input data $x(t)$. In other cases, if we are willing to accept a time delay, of say t_0, in when we obtain the output, we can approximate noncausal impulse responses as closely as we wish. Thus, suppose we wish to approximate the impulse response of Figure 6.12. If we were to use a translated version of $h(t)$, as shown in Figure 6.13, truncated at $t = 0$, the output at time t would be obtained as shown in Figure 6.13. Notice that the actual output at time t is a good approximation to the noncausal output at time $(t - t_0)$. In other words, if we are willing to accept a delay of t_0, we can obtain a good approximation to the response of a noncausal system. This discussion thus serves to justify the use of noncausal impulse-response functions in practical applications. The two-sided Laplace transform is the appropriate transform tool to use for these noncausal systems. The

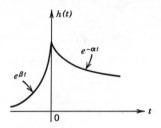

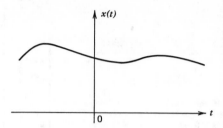

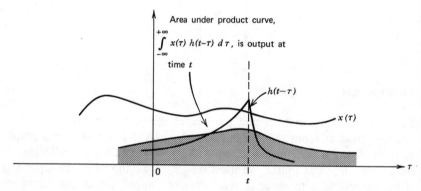

FIGURE 6.12

Fourier transform can also be used to analyze such systems if we restrict our interest to functions which are absolutely integrable on $(-\infty, \infty)$, as the next example demonstrates.

EXAMPLE 6.13.   A system with impulse response $h(t)$ given by

$$h(t) = e^{-t} u(t) + e^{+2t} u(-t)$$

is excited by a signal $x(t)$ given by

$$x(t) = e^{-2t} u(t)$$

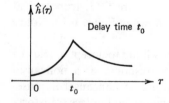

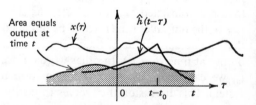

FIGURE 6.13

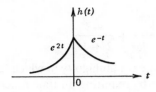

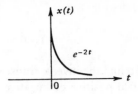

FIGURE 6.14

what is the output response? The input signal and impulse response are shown in Figure 6.14.

The output $y(t)$ is given by

$$y(t) = x(t) * h(t)$$

which in the Fourier transform domain is

$$Y(\omega) = X(\omega) H(\omega)$$

The system function $H(\omega)$ is

$$H(\omega) = \mathscr{F}\{h(t)\} = \int_{-\infty}^{\infty} [e^{2t}u(-t) + e^{-t}u(t)]e^{-j\omega t}\, dt$$

$$= \int_{-\infty}^{0} e^{2t}e^{-j\omega t}\, dt + \int_{0}^{\infty} e^{-t}e^{-j\omega t}\, dt$$

$$= \frac{1}{2 - j\omega} e^{(2-j\omega)t} \bigg|_{-\infty}^{0} + \frac{-1}{1 + j\omega} e^{-(1+j\omega)t} \bigg|_{0}^{\infty}$$

$$= \frac{1}{2 - j\omega} + \frac{1}{1 + j\omega} = \frac{3}{(1 + j\omega)(2 - j\omega)}$$

Similarly, $X(\omega)$ is

$$X(\omega) = \frac{1}{2 + j\omega}$$

The transformed output is therefore

$$Y(\omega) = X(\omega)H(\omega) = \frac{3}{(1 + j\omega)(2 - j\omega)(2 + j\omega)}$$

$$= \frac{A}{1 + j\omega} + \frac{B}{2 - j\omega} + \frac{C}{2 + j\omega}$$

where

$$A = \frac{3}{(2 - j\omega)(2 + j\omega)} \bigg|_{j\omega=-1} = 1$$

$$B = \frac{3}{(1 + j\omega)(2 + j\omega)} \bigg|_{j\omega=2} = \frac{1}{4}$$

$$C = \frac{3}{(1 + j\omega)(2 - j\omega)} \bigg|_{j\omega=-2} = -\frac{3}{4}$$

and so

$$Y(\omega) = \frac{1}{1 + j\omega} + \frac{\frac{1}{4}}{2 - j\omega} + \frac{-\frac{3}{4}}{2 + j\omega}$$

To find the inverse transform, we eliminate the ambiguity in which terms yield positive or negative time functions by requiring all time functions to be absolutely integrable. Because the term $(\frac{1}{4})/(2 - j\omega)$ has an inverse transform $\frac{1}{4}e^{2t}$, which approaches $+\infty$ as $t \to +\infty$ and 0 as $t \to -\infty$, we identify this term as a negative time function. The inverse transform is, by this process

$$y(t) = \tfrac{1}{4}e^{2t}\,u(-t) + (e^{-t} - \tfrac{3}{4}e^{-2t})\,u(t)$$

Suppose we now work this problem using the two-sided Laplace transform. The transformed output is

$$Y(s) = X(s)H(s)$$

where

$$X(s) = \mathscr{L}_b\{x(t)\} = \frac{1}{s + 2}, \qquad \sigma > -2$$

$$H(s) = \mathscr{L}_b\{h(t)\} = \frac{1}{s + 1} + \frac{1}{-s + 2}, \qquad -1 < \sigma < 2$$

$$= \frac{3}{(s + 1)(-s + 2)}$$

Thus

$$Y(s) = H(s)X(s) = \frac{3}{(s + 1)(-s + 2)(s + 2)}$$

$$= \frac{A}{s + 1} + \frac{B}{-s + 2} + \frac{C}{s + 2}$$

where

$$A = 1, \qquad B = \tfrac{1}{4}, \qquad C = -\tfrac{3}{4}$$

Thus

$$Y(s) = \frac{1}{s + 1} + \frac{\frac{1}{4}}{-s + 2} + \frac{-\frac{3}{4}}{s + 2}, \qquad -1 < \sigma < 2$$

The poles at $s = -1$ and $s = -2$ lie to the left of the convergence region $-1 < \text{Re}\,(s) < 2$ and so correspond to positive time functions. Similarly, the pole at $s = 2$ lies to the right of the convergence region and so corresponds to a negative time function. One could also reason physically as we did in the case of the Fourier transform to determine which terms in the partial fraction expansion of $Y(s)$ correspond to positive and negative time functions. The inverse transform is again

$$y(t) = \tfrac{1}{4}e^{2t}\,u(-t) + (e^{-t} - \tfrac{3}{4}e^{-2t})\,u(t)$$

6.9 TRANSIENT AND STEADY-STATE RESPONSE OF A LINEAR SYSTEM

The complete response of a system to any forcing function is composed of a transient portion which is characteristic of the system and a steady-state portion which depends on both the system and the forcing function. These ideas were first introduced in Chapter 1 in conjunction with the time-domain solution of differential equations. These concepts can also be illustrated in the frequency domain.

Suppose that we apply $x(t)$ to a system with impulse response $h(t)$. The output $y(t)$ in the transform domain is

$$Y(s) = X(s)\,H(s)$$

Let the transfer function of the system be

$$H(s) = \frac{N_1(s)}{D_1(s)} = \frac{N_1(s)}{(s - p_1)(s - p_2) \cdots (s - p_n)}$$

and the transformed input be

$$X(s) = \frac{N_2(s)}{D_2(s)} = \frac{N_2(s)}{(s - q_1)(s - q_2) \cdots (s - q_m)}$$

Then

$$Y(s) = \frac{N_1(s)\,N_2(s)}{(s - p_1) \cdots (s - p_n)(s - q_1) \cdots (s - q_m)}$$

We can expand $Y(s)$ in partial fractions as

$$Y(s) = \frac{c_1}{s - p_1} + \cdots + \frac{c_n}{s - p_n} + \frac{k_1}{s - q_1} + \cdots + \frac{k_m}{s - q_m}$$

and obtain the inverse transform

$$y(t) = \sum_{i=1}^{n} c_i e^{-p_i t} + \sum_{i=1}^{m} k_i e^{-q_i t} \tag{6.60}$$

The sum $\sum_{i=1}^{n} c_i e^{-p_i t}$ in (6.60) is the transient response of $y(t)$. Notice that these terms result from the singularities of $H(s)$. (In general, the coefficients c_i depend on $H(s)$ and $X(s)$.) The transient response is a linear combination of terms oscillating at the natural frequencies of the system. In other words, the transient response results from the natural modes of the system. The sum $\sum_{i=1}^{m} k_i e^{-q_i t}$ is called the steady-state portion of the response $y(t)$. These terms arise from the singularities in $X(s)$, and the coefficients k_i depend on both $H(s)$ and $X(s)$. In stable systems, the terms in the first sum die away with increasing t, while the steady-state portion may contain terms that are nonzero indefinitely. The impulse response $h(t)$ is obtained for an impulse input which in the s domain is an input of $X(s) = 1$. Thus, $h(t)$ is a particular linear combination of the natural modes of the system. The transient and steady-state portions of $y(t)$ correspond to what we termed the homogeneous and nonhomogeneous solutions, respectively, for the differential equation describing the system.

EXAMPLE 6.14.   A voltage of $10 \cos 4t\, u(t)$ is applied to the network shown in Figure 6.15. What is the output voltage $v_0(t)$? The transfer function of this system is $H(s)$, given by

$$H(s) = \frac{V_0(s)}{V_i(s)} = \frac{\dfrac{1}{Cs}}{R + \dfrac{1}{Cs}} = \frac{1}{RCs + 1} = \frac{1}{s + 1}$$

The Laplace transform of the input waveform is

$$V_i(s) = \mathscr{L}\{10 \cos 4t\, u(t)\} = \frac{10s}{s^2 + 16}$$

Thus, the transform of the output voltage $v_0(t)$ is

$$V_0(s) = H(s)\, V_i(s) = \frac{10s}{(s^2 + 16)(s + 1)}$$

If we expand the right-hand side in partial fractions, we obtain

$$V_0(s) = \frac{As + B}{s^2 + 16} + \frac{C}{s + 1} \tag{6.61}$$

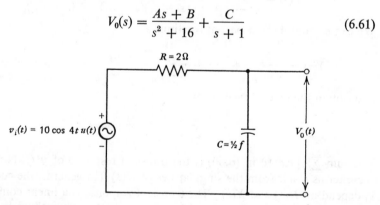

$$R = 2\,\Omega$$

$$v_i(t) = 10 \cos 4t\, u(t)$$

$$C = \tfrac{1}{2}\,f$$

$$V_0(t)$$

FIGURE 6.15

where

$$C = (s + 1)V_0(s)\big|_{s=-1} = \frac{10s}{s^2 + 16}\bigg|_{s=-1} = -\frac{10}{17}$$

Thus, clearing (6.61) of fractions yields

$$10s = (As + B)(s + 1) - \tfrac{10}{17}(s^2 + 16)$$
$$= As^2 + Bs + As + B - \tfrac{10}{17}s^2 - \tfrac{160}{17}$$

Equating coefficients of like powers of s yields

$$A - \tfrac{10}{17} = 0, \qquad A = \tfrac{10}{17}$$
$$\tfrac{10}{17} + B = 10, \qquad B = \tfrac{160}{17}$$

and so

$$V_0(s) = \frac{\tfrac{10}{17}s + \tfrac{160}{17}}{s^2 + 16} - \frac{\tfrac{10}{17}}{s + 1} \tag{6.62}$$

The first term on the right-hand side of (6.62) is the steady-state term, and the second term is the transient component. Taking inverse transforms, we obtain

$$v_0(t) = \mathscr{L}^{-1}\{V_0(s)\} = \mathscr{L}^{-1}\left\{\frac{-\tfrac{10}{17}}{s + 1} + \frac{\tfrac{10}{17}s}{s^2 + 16} + \frac{\tfrac{160}{17}}{s^2 + 16}\right\}$$

$$= \underbrace{-\tfrac{10}{17}e^{-t}}_{\text{transient part}} + \underbrace{\tfrac{10}{17}\cos 4t + \tfrac{40}{17}\sin 4t}_{\text{steady-state part}}, \qquad t > 0$$

One can also obtain the steady-state component directly via the Fourier transform. Recall that a time-invariant linear system forced with a sinusoid yields as an output a sinusoid of the same frequency, with only the phase and amplitude of the input sinusoid being modified. The manner in which the input is modified is completely specified by the system transfer function $H(j\omega)$. The steady-state response to a sinusoid of angular frequency ω has magntude $|H(j\omega)|$ and is shifted in phase by $\underline{/H(j\omega)}$. In our case,

$$H(j\omega) = \frac{1}{1 + j\omega}$$

For an angular frequency of $\omega = 4$,

$$H(j4) = \frac{1}{1 + j4} = \frac{1 - j4}{17} = \sqrt{\left(\frac{1}{17}\right)^2 + \left(\frac{4}{17}\right)^2}\;\underline{/\tan^{-1}\left(\frac{-4}{1}\right)}$$

$$= \frac{1}{\sqrt{17}}\;\underline{/-76°}$$

Hence, the steady-state response to an input of 10 cos 4t is

$$\frac{10}{\sqrt{17}} \cos (4t - 76°) = \frac{10}{\sqrt{17}} \cos 4t \cos 76° + \frac{10}{\sqrt{17}} \sin 4t \sin 76°$$

$$= \frac{10}{17} \cos 4t + \frac{40}{17} \sin 4t$$

which is the same result obtained by the Laplace transform process.

6.10   FREQUENCY RESPONSE OF LINEAR SYSTEMS

The last example suggests that there is an intimate connection between the system's transfer function $H(s)$ and the steady-state response of the system to a sinusoidal input: i.e., the system's frequency response. If we examine the previous example, we see that, in fact, the frequency response for a system with transfer function $H(s)$ is simply $H(s)$ evaluated along the $j\omega$ axis.

This result can also be seen by the following argument. Assume that $H(s)$ is the transfer function of a stable causal system. The transformed output for an input $x(t) = \sin \omega t$ is

$$Y(s) = X(s) H(s) = H(s)\left(\frac{\omega}{s^2 + \omega^2}\right) = \frac{N(s)}{D(s)}\left(\frac{\omega}{s^2 + \omega^2}\right)$$

If we expand $Y(s)$ in partial fractions, we have

$$Y(s) = H(s)\, \omega\left(\frac{1}{(s + j\omega)(s - j\omega)}\right)$$

$$= \frac{N(s)}{D(s)}\, \omega\, \frac{1}{(s + j\omega)(s - j\omega)}$$

$$= \frac{A}{s + j\omega} + \frac{B}{s - j\omega} + \frac{\hat{N}(s)}{D(s)}$$

where

$$A = H(s)\, \omega\, \frac{1}{s - j\omega}\bigg|_{s=-j\omega} = \frac{H(-j\omega)\, \omega}{-2j\omega} = -\frac{H(-j\omega)}{2j}$$

$$B = H(s)\, \omega\, \frac{1}{s + j\omega}\bigg|_{s=j\omega} = \frac{H(j\omega)\, \omega}{2j\omega} = \frac{H(j\omega)}{2j}$$

and $D(s)$ is the denominator polynomial of $H(s)$. Therefore

$$Y(s) = -\frac{H(-j\omega)}{2j(s + j\omega)} + \frac{H(j\omega)}{2j(s - j\omega)} + \frac{\hat{N}(s)}{D(s)}$$

Because $H(s)$ has singularities only in the left-hand half of the plane, the term $\dfrac{\hat{N}(s)}{D(s)}$ represents transient terms in the time domain which die away as $t \to \infty$. Thus, the steady-state output is

$$y_{ss}(t) = \mathscr{L}^{-1}\left\{ -\frac{H(-j\omega)}{2j(s + j\omega)} + \frac{H(j\omega)}{2j(s - j\omega)} \right\}$$

$$= -\frac{H(-j\omega)e^{-j\omega t}}{2j} + \frac{H(j\omega)e^{j\omega t}}{2j} \tag{6.63}$$

Let $H(j\omega)$ be written in polar form as

$$H(j\omega) = |H(j\omega)|\, e^{j\theta(j\omega)}$$

Then

$$H(-j\omega) = |H(-j\omega)|\, e^{j\theta(-j\omega)} = |H(j\omega)|\, e^{-j\theta(j\omega)}$$

Substituting for $H(j\omega)$ and $H(-j\omega)$ in (6.60) then yields

$$y_{ss}(t) = \frac{|H(j\omega)|\, e^{j\theta(j\omega)}e^{j\omega t} - |H(j\omega)|\, e^{-j\theta(j\omega)}e^{-j\omega t}}{2j}$$

$$= \frac{|H(j\omega)|\, [e^{j(\theta(j\omega)+\omega t)} - e^{-j(\theta(j\omega)+\omega t)}]}{2j}$$

$$= |H(j\omega)| \sin(\omega t + \theta(j\omega)) \tag{6.64}$$

Equation (6.64) states that if the input $x(t)$ is $\sin \omega t$, then the steady-state output $y(t)$ is $|H(j\omega)| \sin(\omega t + \theta(\omega))$. The functions $|H(j\omega)|$ and $\theta(\omega)$ are the amplitude and angle functions obtained from the transfer function $H(s)$ by substituting $j\omega$ for s. This is an alternate way of establishing that the frequency response of a linear-time invariant system is the magnitude and phase of $H(j\omega)$. This relationship between the frequency response of a system and the transfer function of a system makes it easy to obtain the steady-state response for a sinusoidal input.

6.11 LAPLACE TRANSFORM ANALYSIS OF CAUSAL PERIODIC INPUTS TO LINEAR SYSTEMS

In applications, it is often useful to consider the Laplace transform of a causal periodic function $f(t)$. Assume that $f(t)$ is periodic with period T and that $f(t)$ is causal: i.e., $f(t) = 0$, $t < 0$ and $f(t) = f(t + T)$, $t > 0$. We can write the Laplace transform of $f(t)$ as

$$\mathscr{L}\{f(t)\} = \int_0^\infty f(t)e^{-st}\,dt$$

$$= \sum_{n=0}^\infty \int_{nT}^{(n+1)T} f(t)e^{-st}\,dt$$

$$= \sum_{n=0}^\infty e^{-nTs} \int_0^T f(\tau)e^{-s\tau}\,d\tau, \qquad \text{for } \tau = t - nT$$

The properties of a geometric series allow us to evaluate the sum $\sum_{n=0}^\infty e^{-nTs}$ as

$$\sum_{n=0}^\infty e^{-nTs} = \frac{1}{1 - e^{-sT}}$$

Thus

$$\mathscr{L}\{f(t)\} = \frac{\displaystyle\int_0^T f(t)e^{-st}\,dt}{1 - e^{-sT}} \tag{6.65}$$

EXAMPLE 6.15.   Find the Laplace transform of the periodic wave-form shown in Figure 6.16. Substituting into (6.65) yields

$$F(s) = \frac{1}{1 - e^{-sT}}\left[\int_0^{T/2} 1 \cdot e^{-st}\,dt + \int_{T/2}^T -1 \cdot e^{-st}\,dt\right]$$

$$= \frac{1}{1 - e^{-sT}}\left[\frac{1 - e^{-sT/2}}{s} + \frac{e^{-sT} - e^{-sT/2}}{s}\right]$$

$$= \frac{1 - e^{-sT/2}}{s(1 + e^{-sT/2})}$$

Another way of arriving at (6.65) is to recognize that we can write a periodic function $f(t)$ as

$$f(t) = f_1(t) + f_1(t - T)\,u(t - T) + f_1(t - 2T)\,u(t - 2T) + \cdots \tag{6.66}$$

where

$$f_1(t) = f(t), \qquad t \in [0, T]$$
$$= 0, \qquad\qquad \text{otherwise}$$

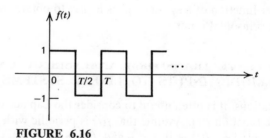

FIGURE 6.16

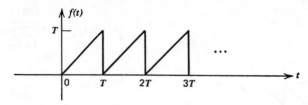

FIGURE 6.17

Taking the Laplace transform of (6.66) gives

$$\mathscr{L}\{f(t)\} = F_1(s) + F_1(s)e^{-sT} + F_1(s)e^{-2sT} + \cdots$$
$$= F_1(s)[1 + e^{-sT} + e^{-2sT} + \cdots]$$
$$= \frac{F_1(s)}{1 - e^{-sT}} \tag{6.67}$$

where
$$F_1(s) = \mathscr{L}\{f_1(t)\} = \int_0^T f(t)e^{-st}\,dt$$

EXAMPLE 6.16. Find the transform of the saw-tooth waveform shown in Figure 6.17. The period of $f(t)$ is T. Hence, using (6.67), we have

$$\mathscr{L}\{f(t)\} = \frac{1}{1 - e^{-sT}} \int_0^T t e^{-st}\,dt = \frac{1}{1 - e^{-sT}}\left[\frac{e^{-st}}{s^2}(-st - 1)\right]\Big|_0^T$$

$$= \frac{1 - (1 + sT)e^{-sT}}{s^2(1 - e^{-sT})}$$

$$= \frac{(1 + sT)(1 - e^{-sT}) - sT}{s^2(1 - e^{-sT})} = \frac{1 + sT}{s^2} - \frac{T}{s(1 - e^{-sT})}$$

EXAMPLE 6.17. Find the steady-state output current waveform of the *RLC* circuit shown for a periodic saw-tooth input voltage waveform of Figure 6.17 with $T = 1$. Assume an initial current of 1 amp flowing in the inductor as shown in Figure 6.18. The differential equation describing this system is

$$e(t) = R\,i(t) + L\frac{d\,i(t)}{dt} + \frac{1}{C}\int_0^t i(t')\,dt'$$

Taking Laplace transforms of both sides yields

$$3I(s) + (s\,I(s) - 1) + \frac{2}{s}I(s) = \frac{1 + s}{s^2} - \frac{1}{s(1 - e^{-s})}$$

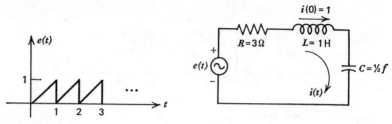

FIGURE 6.18

Solving for $I(s)$,

$$I(s) = \frac{s^2 + s + 1}{s(s + 1)(s + 2)} - \frac{1}{(s + 1)(s + 2)(1 - e^{-s})}$$

The first fraction of the preceding equation can be expanded in partial fractions to yield

$$\tfrac{1}{2} - e^{-t} + \tfrac{3}{2}e^{-2t}$$

To find the inverse of the second fraction, we have

$$\frac{1}{(s + 1)(s + 2)(1 - e^{-s})} = \left(\frac{1}{s + 1} - \frac{1}{s + 2}\right)\left(\frac{1}{1 - e^{-s}}\right)$$

$$= \frac{1}{(s + 1)(1 - e^{-s})} - \frac{1}{(s + 2)(1 - e^{-s})} \qquad (6.68)$$

Consider the expression

$$\frac{1}{(s + a)(1 - e^{-sT})} = \frac{1}{s + a}(1 + e^{-sT} + e^{-2sT} + e^{-3sT} + \cdots)$$

$$= \frac{1}{s + a} + \frac{e^{-sT}}{s + a} + \frac{e^{-2sT}}{s + a} + \cdots$$

Consider the inverse of the $(n + 1)$st term. The contribution of $1/(s + a)$ is e^{-at}. The exponential e^{-nsT} translates the time function e^{-at} to the right by nT and then truncates it to the left of $t = nT$. When this is done for each term, we obtain

$$f(t) = e^{-at} + e^{-a(t-T)}u(t - T) + e^{-a(t-2T)}u(t - 2T) + \cdots$$

Over the interval $[nT, (n + 1)T]$, we thus have for $f(t)$

$$f(t) = e^{-at}[1 + e^{aT} + e^{2aT} + \cdots + e^{naT}] \qquad (6.69)$$

We can sum the bracketed term in (6.69), using the properties of the geometric series, to obtain

$$f(t) = e^{-at} \left[\frac{(e^{aT})^{(n+1)} - 1}{e^{aT} - 1} \right] = \frac{e^{-a[t-(n+1)T]}}{e^{aT} - 1} - \frac{e^{-at}}{e^{aT} - 1},$$

$$t \in [nT, (n+1)T]$$

Returning to (6.68), we have that $T = 1$, so that over the interval $n < t < n + 1$,

$$\mathscr{L}^{-1} \left\{ \frac{1}{(s+1)(1-e^{-s})} - \frac{1}{(s+2)(1-e^{-s})} \right\}$$

$$= \left(\frac{e^{-(t-(n+1))}}{e-1} - \frac{e^{-t}}{e-1} \right) - \left(\frac{e^{-2(t-(n+1))}}{e^2-1} - \frac{e^{-2t}}{e^2-1} \right), \qquad n < t < n+1$$

The total solution is therefore

$$y(t) = \tfrac{1}{2} - e^{-t} + \tfrac{3}{2}e^{-2t} + \left(\frac{e^{-2t}}{e^2-1} - \frac{e^{-t}}{e-1} \right) + \frac{e^{-[t-(n+1)]}}{e-1} - \frac{e^{-2[t-(n+1)]}}{e^2-1}$$

$$= \left(\tfrac{3}{2}e^{-2t} - e^{-t} - \frac{e^{-t}}{e-1} + \frac{e^{-2t}}{e^2-1} \right) + \left(\frac{1}{2} + \frac{e^{-[t-(n+1)]}}{e-1} - \frac{e^{-2[t-(n+1)]}}{e^2-1} \right),$$

$$\underbrace{\qquad\qquad\qquad\qquad\qquad}_{\text{transient}} \qquad\qquad \underbrace{\qquad\qquad\qquad\qquad\qquad\qquad}_{\text{steady-state}}$$

$$n < t < n+1$$

Thus the steady-state current in the $(n + 1)$st interval is

$$y_{ss}(t) = \frac{1}{2} + \frac{e^{-[t-(n+1)]}}{e-1} - \frac{e^{-2[t-(n+1)]}}{e^2-1}, \qquad n < t < n+1$$

6.12   RELATIONSHIP OF THE Z-TRANSFORM TO THE FOURIER AND LAPLACE TRANSFORMS

We consider next the connection between the Fourier and Laplace transforms of a continuous function and the Z-transform of a sequence. Let $f(t)$ be a continuous function which is sampled at the time instants $\{\ldots, -T, 0, T, 2T, \ldots\}$. The sampled version can be represented as

$$f_s(t) = f(t) \sum_{n=-\infty}^{\infty} \delta(t - nT)$$

$$= \sum_{n=-\infty}^{\infty} f(nT)\,\delta(t - nT) \qquad (6.70)$$

where we have used the property that $f(t)\,\delta(t) = f(0)\,\delta(t)$ to equate

$f(t) \, \delta(t - nT)$ to $f(nT) \, \delta(t - nT)$. The sampled version of $f(t)$ has the Laplace transform

$$F_s(s) = \mathscr{L}_b\{f_s(t)\} = \mathscr{L}_b\left\{ \sum_{n=-\infty}^{\infty} f(nT) \, \delta(t - nT)\right\}$$

$$= \sum_{n=-\infty}^{\infty} f(nT) \, \mathscr{L}_b\{\delta(t - nT)\}$$

Now $\mathscr{L}_b\{\delta(t)\} = 1$, and by the time-shift theorem, $\mathscr{L}_b\{\delta(t - nT)\} = e^{-nsT}$. Thus we obtain

$$F_s(s) = \sum_{n=-\infty}^{\infty} f(nT)e^{-snT} \qquad (6.71)$$

If we now make the substitution $z = e^{-sT}$, then $F_s(s)$ becomes

$$F_s(s)\big|_{z=e^{-sT}} = \sum_{n=-\infty}^{\infty} f(nT)z^n = F(z) \qquad (6.72)$$

where $F(z)$ is the Z-transform of the sequence of samples of $f(t)$, $\{f(nT)\}$, $n = 0, \pm 1, \pm 2, \ldots$.

From this discussion, we see that the Z-transform may be viewed as the Laplace transform of the sampled time function $f(t)$ (with an appropriate change of variable), or quite independently as the generating function for a sequence $\{f_n\}$ which assumes the values $f(nT)$ for $n = \cdots, -2, -1, 0, 1, 2, \ldots$. We note that with $z = e^{-sT}$, the complex s plane maps into the complex z plane. Under this mapping, the imaginary axis, Re $(s) = 0$, maps onto the unit circle $|z| = 1$ in the z plane. Also, the left-hand half plane Re $(s) < 0$ corresponds to the exterior of the unit circle $|z| = 1$ in the z plane. This correspondence is shown schematically in Figure 6.19.

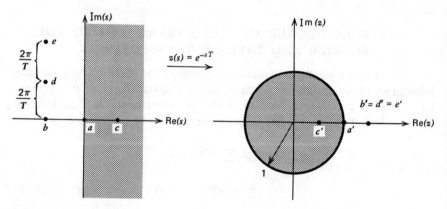

FIGURE 6.19

If we restrict s to the $j\omega$ axis in the s plane, then $F_s(s)$ becomes $F_s(j\omega)$, the Fourier transform of the sampled function $f_s(t)$. From (6.71), we obtain

$$F_s(j\omega) = \sum_{n=-\infty}^{\infty} f(nT)e^{-j\omega nT} = F(z)\Big|_{z=e^{-j\omega T}} \qquad (6.73)$$

Notice that $F_s(j\omega) = F_s(j\omega + 2\pi/T)$: i.e., $F_s(j\omega)$ is periodic with period $2\pi/T$. Now, if $s = j\omega$, in the z plane we have that $z = e^{-j\omega T}$. Therefore $F_s(j\omega)$ can be obtained by evaluating $F(z)$ where z is restricted to the unit circle. One period of $F_s(j\omega)$ can be obtained by evaluating $F(z)$ once around the unit circle $|z| = 1$.

6.13 SUMMARY

The Laplace transform is a generalization of the Fourier transform in which the frequency variable s is a complex variable, $\sigma + j\omega$. Through this generalization, we increase the class of functions which can be transformed at the cost of dealing with a more abstract, less physical representation. As with other transform methods, we can often solve a problem modeled by a linear differential equation by a table look-up.

The Laplace transform analysis of a system is useful in examining the system's stability directly by finding the roots of the system's transfer function. This transform method is also a convenient way to formulate transient problems in which initial energy storage is nonzero at time $t = 0$, because the initial conditions at $t = 0$ are easily incorporated into the transform equations.

PROBLEMS

1. Find the inverse Laplace transforms of the following transforms. Assume that $f(t)$ is causal.

a. $\dfrac{1}{(s + b)^4}$ 　　 $ans.: \dfrac{t^3 e^{-bt}}{6}$

b. $\dfrac{s}{(s^2 + a^2)^2}$ 　　 $ans.: \dfrac{t}{2a}\sin at$

c. $\dfrac{e^{-sT}}{(s + 1)^3}$ 　　 $ans.: \dfrac{(t - T)^2 e^{-(t-T)}}{2}u(t - T)$

d. $\dfrac{2}{(s^2 + 1)^2}$ 　　 $ans.: \sin t - t\cos t$

e. $\dfrac{1}{s^4 - a^4}$ *ans.:* $\dfrac{1}{2a^3}(\sinh at - \sin at)$

f. $\dfrac{1}{s(s^2 + a^2)^2}$ *ans.:* $\dfrac{1}{a^4}(1 - \cos at) - \dfrac{1}{2a^3}t\sin at$

g. $\dfrac{s}{s^4 + 4a^4}$ *ans.:* $\dfrac{1}{2a^2}\sin at \sinh at$

2. Find the Laplace transform of the following time functions.

a. $u(t - T)$ *ans.:* $\dfrac{e^{-sT}}{s}$

b. $f(t) = \begin{cases} \sin t, & 0 < t < \pi \\ 0, & \text{otherwise} \end{cases}$ *ans.:* $\dfrac{1 + e^{-\pi s}}{s^2 + 1}$

c. $\dfrac{1}{a^2}(1 - \cos at)u(t)$ *ans.:* $\dfrac{1}{s(s^2 + a^2)}$

d. $\dfrac{1}{2a^3}(\sin at - at \cos at)u(t)$ *ans.:* $\dfrac{1}{(s^2 + a^2)^2}$

e. $\left(\dfrac{1 - \cos bt}{t}\right)u(t)$ *ans.:* $\dfrac{1}{2}\left\{6\cot^{-1}\left(\dfrac{s}{b}\right) + s\ln\dfrac{s^2}{s^2 + b^2}\right\}$

3. A linear system is described by the following differential equation. This system is forced with an input as shown in the graph. Find the output of the system.

$$\frac{d^2y(t)}{dt^2} + \frac{3dy(t)}{dt} + 2y(t) = x(t), \qquad y(0) = 0, \qquad y'(0) = 1$$

ans.: $(e^{-t} - e^{-2t})u(t) + \frac{1}{2}\{1 - 2e^{-(t-1)} + e^{-2(t-1)}\}u(t - 1)$

4. Consider the Laplace transform $F(s)$ given by

$$F(s) = \frac{1}{(s - 3)(s - 2)(s + 1)}$$

Find four possible time functions that have this Laplace transform. State the region of convergence for each.

5. Find the output voltage in the following circuit for all t in response to an input voltage $e^{-|t|}$.

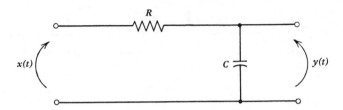

6. John Brightfellow has been asked to determine the contents of a black box (containing an electrical circuit). John decides to find the system's frequency response by forcing the system with a sinusoid. He applies a sinusoidal voltage and measures a direct current output. What is the circuit?

7. A voltage $e^{-t} u(t)$ is applied to a circuit. The response voltage is a ramp $t\, u(t)$. What is the circuit?

8. We wish to compare two systems, A and B, as shown below. The two systems are two methods of performing integration. In system A, the response $r(t)$ is integrated and then sampled to obtain a sequence $y_A(nT)$. In system B, the response $r(t)$ is first sampled and then these samples are summed (the discrete analog of integration) to obtain a sequence $y_B(nT)$. Assume that the input $x(t)$ is an impulse $\delta(t)$ and that $T = 1$ and $a = .2$. Obtain closed form expressions for $y_A(nT)$ and $y_B(nT)$. Compare these two sequences for $n = 0, 1, 2, 3, 4$. Are the systems equivalent? Suppose you could use a single delay element in the discrete integrator. Design a better discrete integrator (not limited to the specific input of this problem).

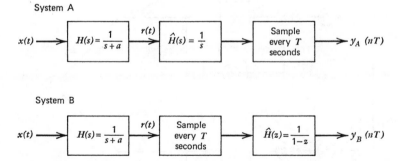

9. Find the impulse response of the following continuous-time system. Is the system stable?

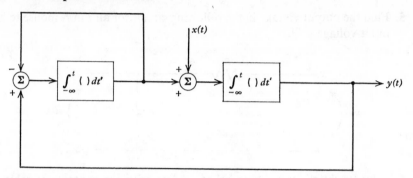

10. Is the feedback system shown below stable if the gain g is zero: i.e., with no feedback? Plot the locus of poles in the s plane for the overall system for both positive and negative values of g. For what range of g is the feedback system stable?

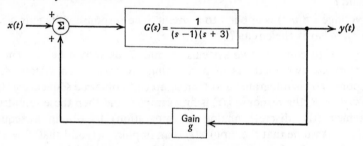

11. Assume that we have a transfer function of the form

$$H(s) = \frac{a_0 + a_1 s + \cdots + a_m s^m}{b_0 + b_1 s + \cdots + b_n s^n}$$

Parallel form:

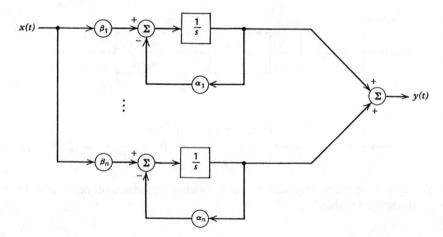

Cascade form:

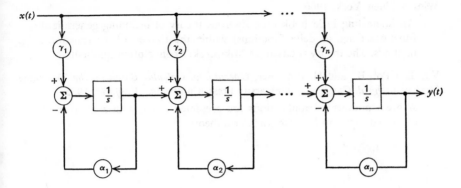

Assume that the poles of $H(s)$ are real and distinct. Show that $H(s)$ can be realized in the two forms shown above. Develop a method for computing the constants α_i, β_i, and γ_i.

12. What is the Laplace transform of the waveform shown below?

$$Ans.: \quad X(s) = \frac{1}{s(1 - e^{-sT})}$$

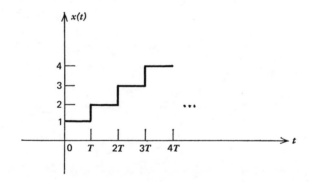

SUGGESTED READINGS

There are many good references on the Laplace transform and its use in systems analysis. The books in the first three references listed below present Laplace transform methods at a junior-senior level.

Cheng, D. K., *Analysis of Linear Systems*, Addison-Wesley, Reading, Mass., 1959.

Cooper, G. R. and C. D. McGillem, *Methods of Signal and System Analysis*, Holt, Rinehart, and Winston, New York, 1967.

Lathi, B. P., *Signals, Systems, and Communication*, Wiley, New York, 1967.

Erdélyi, A., *Operational Calculus and Generalized Functions*, Holt, Rinehart, and Winston, New York, 1966.

An interesting little book on a rigorous theory of including generalized functions (such as delta functions) within the framework of transform methods. The theory is based on Mikusinski's convolution quotients.

Van Der Pol, B., and H. Bremmer, *Operational Calculus Based on the Two-sided Laplace Integral*, Cambridge University Press, Cambridge, England, 1955.

A great number of applications are presented along with a number of advanced topics in Laplace transform theory.

7

Implementation of Continuous-Time Filtering Functions with Digital Filters

In the design of communication, control, and telemetry systems, we must often include filters to alter the various internal signals in some fashion. In communications, for example, we make extensive use of low-pass, high-pass, band-pass, band reject, and other spectrum-shaping filters. In control systems we often require compensation filters to achieve a desired system response.

In this chapter we treat the problem of realizing these continuous-time filters by using discrete-time filters programed on a digital computer. Figure 7.1a shows the desired filter H in block diagram form, with input $x(t)$ and output $y_1(t)$. Figure 7.1b shows its implementation with a digital filter D. Here, the continuous input $x(t)$ is first sampled and converted to a sequence of digital words by an analog-to-digital converter. The digitized samples are then passed through a digital filter, and finally the digital filter output sequence is reconverted to the continuous output $y_2(t)$ by the digital-to-analog converter. The problem is to design a digital filter D such that the two systems have identical input-output relations. Ideally, then, the same input, $x(t)$, to both filters would result in identical outputs, $y_1(t) = y_2(t)$. We shall determine a class of inputs and filters for which this is possible and develop two approaches to use in designing the appropriate digital filter.

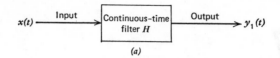

(a)

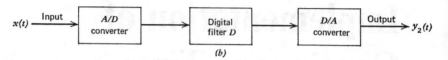

(b)

FIGURE 7.1

First, however, it is appropriate to question the motivation for this approach. One reason for studying the problem is purely pedagogic—we shall find that this study ties together much of the linear system material we have developed. We shall unite concepts from difference equations, convolution, Z-transforms, Fourier transforms, and Laplace transforms. Studying the interrelationships among these areas will strengthen our understanding of their utility.

A more important reason for this study is that the digital filter is an increasingly important type of linear system. A small special-purpose computer can be time-shared among several components in a chemical plant, rocket guidance system, or multichannel communication system. The cost of digital implementation is often considerably lower than that of its analog counterpart. The accuracy of a digital filter is dependent only on the computer word length, rather than on the tolerance of physical (e.g., RLC) elements, which drift with temperature changes and age. Complex filters of high order can be realized very easily with digital filters, without the constraints imposed by nonideal physical elements. Finally, digital filters can be modified simply by changing the coefficients in a digital computer program, in contrast to analog systems, which may have to be physically rebuilt.

7.1 DESIGN OF THE DIGITAL FILTER

We shall use the following model for our digital filter implementation (see Figure 7.2). The input $x(t)$ is sampled regularly at points T seconds apart, and each sample is converted into a digital word. This process forms the digital filter input sequence $\{x_k\}$, where $x_k \triangleq x(kT)$. The digital filter operates on the sequence $\{x_k\}$ to form the output sequence $\{w_k\}$. This output sequence is then converted to a train of impulses spaced T seconds apart with

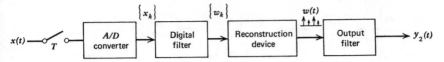

FIGURE 7.2

area equal to T times the respective sequence values. Call this impulse sequence $w(t)$.

$$w(t) = T \sum_{k=-\infty}^{\infty} w_k \, \delta(t - kT)$$

Now $w(t)$ is passed through an output filter with transfer function $F(\omega)$ to recover the continuous output $y_2(t)$. In this model, $F(\omega)$ includes the effects of both the D/A converter and any external filters.

Let $h(t)$ be the impulse response of the desired filter in Figure 7.1 and $\{d_k\}$ be the impulse response of the digital filter D. Then with $X(\omega)$, $H(\omega)$, $Y_1(\omega)$, $W(\omega)$, and $Y_2(\omega)$, the Fourier transforms of $x(t)$, $h(t)$, $y_1(t)$, $w(t)$, and $y_2(t)$, respectively, we have

$$Y_1(\omega) = H(\omega) \, X(\omega) \tag{7.1}$$

$$Y_2(\omega) = W(\omega) \, F(\omega) \tag{7.2}$$

where

$$\{w_k\} = \{x_k\} * \{d_k\}$$

and

$$w(t) = T \sum_{k=-\infty}^{\infty} w_k \, \delta(t - kT)$$

$$= T \sum_{k=-\infty}^{\infty} \left[\sum_{m=-\infty}^{\infty} x_{k-m} \, d_m \right] \delta(t - kT) \tag{7.3}$$

Our design problem is to choose the digital filter impulse response $\{d_k\}$ such that $y_1(t) = y_2(t)$, or equivalently $Y_1(\omega) = Y_2(\omega)$, for any input $x(t)$ (perhaps of a limited class of inputs). Having found the sequence $\{d_k\}$, we then compute the coefficients in the standard Z-transform specification of the rational digital filter $D(z)$.

$$D(z) = \sum_{k=-\infty}^{\infty} d_k z^k$$

$$= \frac{a_0 + a_1 z + \cdots + a_m z^m}{1 + b_1 z + \cdots + b_n z^n}$$

(Here we do not constrain the orders of numerator and denominator, m and n.) D can then be implemented with the difference equation

$$w_k = a_0 x_k + a_1 x_{k-1} + \cdots + a_m x_{k-m} - b_1 w_{k-1} - \cdots - b_n w_{k-n}$$

as shown schematically in Figure 7.3.

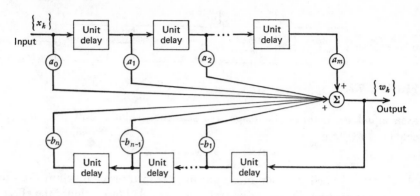

FIGURE 7.3

7.2 EVALUATION OF THE RESULTING TRANSFER FUNCTION

Before proceeding with a discussion of design methods, we wish first to derive an expression for the transfer function

$$H_{eq}(\omega) = \frac{Y_2(\omega)}{X(\omega)} \tag{7.4}$$

which we shall realize with the digital filter system of Figure 7.1b. From (7.3), we have

$$w(t) = T \sum_{k=-\infty}^{\infty} \left[\sum_{m=-\infty}^{\infty} x_{k-m}\, d_m \right] \delta(t - kT)$$

Taking the Fourier transform on both sides we have

$$W(\omega) = T \sum_{k=-\infty}^{\infty} \sum_{m=-\infty}^{\infty} x_{k-m}\, d_m e^{-jk\omega T}$$

$$= \sum_{m=-\infty}^{\infty} d_m e^{-jm\omega T}\, T \sum_{k=-\infty}^{\infty} x_{k-m} e^{-j(k-m)\omega T}$$

We recognize the first sum as the Z-transform $D(z)$ evaluated at $z = e^{-j\omega T}$: i.e., around the unit circle. With $n = k - m$ in the second sum, we apply (5.145) to obtain

$$T \sum_{n=-\infty}^{\infty} x_n e^{-jn\omega T} = \mathscr{F}\left\{ T \sum_{n=-\infty}^{\infty} x(nT)\, \delta(t - nT) \right\}$$

$$= \mathscr{F}\left\{ Tx(t) \sum_{n=-\infty}^{\infty} \delta(t - nT) \right\}$$

$$= \sum_{m=-\infty}^{\infty} X\left(\omega - \frac{m2\pi}{T} \right)$$

Thus, the second sum is just the Fourier transform of the impulse-sampled input,

$$T\,x_s(t) = T\,x(t) \sum_{n=-\infty}^{\infty} \delta(t - nT)$$

Notice that the process of sampling $x(t)$ has produced a time function $T\,x_s(t)$ whose spectrum is an aliased form of $X(\omega)$: i.e.,

$$\sum_{m=-\infty}^{\infty} X(\omega - m2\pi/T).$$

Finally, with $Y_2(\omega) = W(\omega)F(\omega)$, we obtain

$$Y_2(\omega) = D(z)\Big|_{z=e^{-j\omega T}} F(\omega) \sum_{m=-\infty}^{\infty} X\left(\omega - \frac{m2\pi}{T}\right) \qquad (7.5)$$

The three functions on the right-hand side of (7.5), whose product equals the Fourier transform of the output waveform $y_2(t)$, are shown in Figure 7.4. Here $D(z)\big|_{z=e^{-j\omega T}}$ is the periodic function obtained by evaluating the digital filter transfer function around the unit circle, $F(\omega)$ is the transfer function of the output filter, and the final term is the "wound-up" or aliased form of the Fourier transform of the input waveform $x(t)$.

We obtain a special case of considerable interest if the input waveform is bandlimited such that $X(\omega) = 0$ for $|\omega| \geq \pi/T$, and the output filter $F(\omega)$ is an ideal low-pass filter.

$$F(\omega) = \begin{cases} 1, & |\omega| \leq \dfrac{\pi}{T} \\[2mm] 0, & |\omega| > \dfrac{\pi}{T} \end{cases}$$

In this case, the product $F(\omega) \sum_{m=-\infty}^{\infty} X(\omega - m2\pi/T)$ is just $X(\omega)$, and

$$Y_2(\omega) = D(z)\big|_{z=e^{-j\omega T}} X(\omega)$$

Now the equivalent transfer function of the digital filter system is given by

$$H_{eq}(\omega) = \frac{Y_2(\omega)}{X(\omega)} = D(z)\Big|_{z=e^{-j\omega T}} \qquad (7.6)$$

Equations (7.5) and (7.6) are fundamental to the design and use of digital filters. Equation (7.5) holds in general and allows us to predict the output $y_2(t)$ of the system in Figure 7.2, while (7.6) holds for a bandlimited input and an output filter such that $F(\omega) \sum_{m=-\infty}^{\infty} X(\omega - m2\pi/T)$ equals $X(\omega)$. Other output filters are treated in the problems at the end of this chapter.

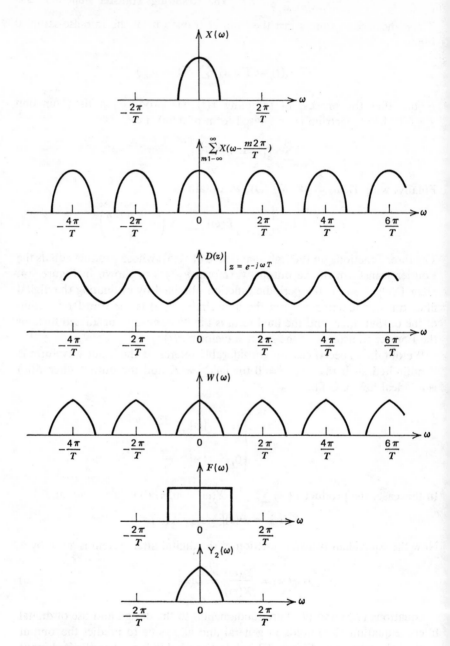

FIGURE 7.4

7.3   THE DIRECT Z-TRANSFORM METHOD

We begin our study with the direct Z-transform method, which is the most straightforward approach and yields satisfactory results in many cases. This design method is also referred to in the literature as the *impulse invariance* method. The central idea is to design the digital filter so that its impulse-response sequence is identical to the sampled impulse response of the desired continuous-time filter. With $h(t)$ the desired impulse response and $\{d_k\}$ the digital filter impulse response, we require

$$d_k = Th(kT), \qquad k = \cdots -1, 0, 1, 2, \ldots \tag{7.7}$$

where the factor T, equal to the sampling period, is incorporated to make the filter transmission independent of the sampling rate. The relation (7.7) is sketched in Figure 7.5.

For a bandlimited input waveform and an ideal low-pass output filter, we obtain the system transfer function

$$H_{eq}(\omega) = D(z)\big|_{z=e^{-j\omega T}}$$

$$= \sum_{k=-\infty}^{\infty} d_k e^{-j\omega kT}$$

$$= T \sum_{k=-\infty}^{\infty} h(kT)e^{-j\omega kT}$$

$$= T\mathscr{F}\left\{ \sum_{k=-\infty}^{\infty} h(kT)\,\delta(t - kT) \right\}$$

$$= T\mathscr{F}\left\{ h(t) \sum_{k=-\infty}^{\infty} \delta(t - kT) \right\}$$

$$= \sum_{m=-\infty}^{\infty} H\left(\omega - \frac{m2\pi}{T} \right), \qquad |\omega| \le \frac{\pi}{T} \tag{7.8}$$

where again we have applied the results of Chapter 5, Section 5.10, this time to the sampled time function $h_s(t) = h(t) \sum_{k=-\infty}^{\infty} \delta(t - kT)$. If the desired transfer function $H(\omega)$ is essentially zero for $|\omega| > \pi/T$, then the aliased version in (7.8) will yield a good approximation to $H(\omega)$.

For example, we might wish to design a digital filter which approximates an ideal low-pass filter with cut-off frequency equal to one fourth of the sampling frequency, $2\pi/T$ rad/sec. Here we have

$$H(\omega) = \begin{cases} 1, & |\omega| \le \dfrac{\pi}{2T} \\[2mm] 0, & |\omega| > \dfrac{\pi}{2T} \end{cases} \tag{7.9}$$

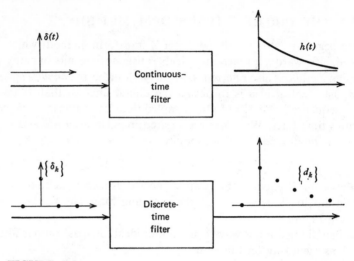

FIGURE 7.5

with the continuous-time impulse response

$$h(t) = \frac{1}{2T} \operatorname{sinc} \frac{\pi t}{2T}$$

Defining the digital filter impulse response from (7.7) as $d_k = Th(kT)$, we have

$$d_k = \frac{1}{2} \operatorname{sinc} \frac{k\pi}{2}, \qquad k = 0, \pm 1, \pm 2, \ldots$$

These functions are illustrated in Figure 7.6.

Two practical problems confront us when we attempt to implement this scheme: (1) In general, we require an infinite number of terms in the digital filter impulse response $\{d_k\}$, and (2) $\{d_k\}$ represents a noncausal filter, because in general, $h(t)$ is nonzero for negative t. To alleviate these problems, we must limit ourselves to taking only a finite number of terms in the sequence $\{d_k\}$ and we must add a delay such that the sequence members are zero for negative k. For example, we might include only the $2N + 1$ samples from $h(-NT)$ to $h(NT)$, and let

$$d_k = h[(k - N)T], \qquad k = 0, 1, 2, \ldots, 2N \qquad (7.10)$$

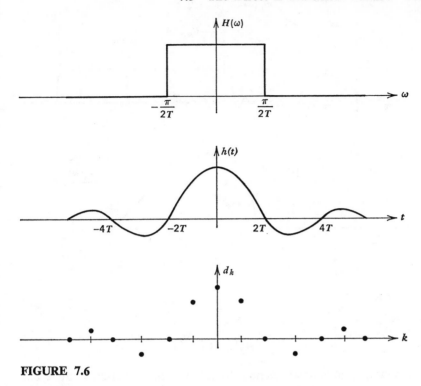

FIGURE 7.6

Mathematically, we multiply $h(t)$ by the gating or "window" function[1]

$$g_{2NT}(t) = \begin{cases} 1, & |t| \le NT \\ 0, & |t| > NT \end{cases}$$

and delay the resulting time function by NT seconds before sampling. In the frequency domain, these modifications are equivalent to convolving $H(\omega)$ with $(1/2\pi) G_{2NT}(\omega) = (NT/\pi)$ sinc ωNT, and multiplying by the delay phase term $\exp(-j\omega NT)$. If we call the resulting transform $H_2(\omega)$, we have

$$H_2(\omega) = \frac{1}{2\pi} [H(\omega) * G_{2NT}(\omega)] e^{-j\omega NT} \qquad (7.11)$$

The system transfer function $H_{eq}(\omega)$ from (7.8) is given by the aliased form

---

[1] The choice of an appropriate window function is an important step in the design of these filters. See, for example, Kuo, F. F. and J. F. Kaiser, *System Analysis by Digital Computer*, Wiley, New York, 1966, Chapter 7.

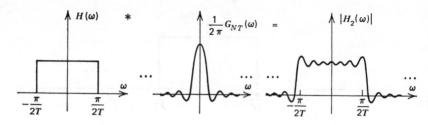

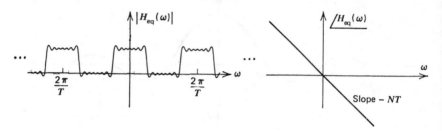

FIGURE 7.7

of $H_2(\omega)$.

$$H_{eq}(\omega) = \sum_{m=-\infty}^{\infty} H_2\left(\omega - \frac{m2\pi}{T}\right) \qquad |\omega| \leq \frac{\pi}{T} \qquad (7.12)$$

Given the desired transfer function $H(\omega)$, (7.11), and (7.12) allow us to evaluate the actual system transfer function $H_{eq}(\omega)$ for a particular value of N. These relations are sketched in Figure 7.7 for the low-pass filter $H(\omega)$ defined in (7.9).

EXAMPLE 7.1.   To treat a specific example for the low-pass filter approximation, let us take $N = 5$ with $H(\omega)$ as in (7.9). Now we have

$$d_k = Th[(k - 5)T]$$

$$= \frac{1}{2}\operatorname{sinc}\frac{(k - 5)\pi}{2}, \qquad 0 \leq k \leq 10$$

Thus

$$\{d_k\} = \left\{\frac{1}{5\pi}, 0, \frac{-1}{3\pi}, 0, \frac{1}{\pi}, \frac{1}{2}, \frac{1}{\pi}, 0, \frac{-1}{3\pi}, 0, \frac{1}{5\pi}\right\}$$
$$\qquad\quad \underset{k=0}{\uparrow} \qquad\qquad\qquad\qquad\qquad\qquad \underset{k=10}{\uparrow}$$

This digital filter impulse response is shown in Figure 7.8a. This filter can be implemented as in Figure 7.8b. The resulting transfer function is shown in Figure 7.8c.

The design method just discussed yields filters of the type known as *nonrecursive*—that is, with no feedback. Nonrecursive filters have a finite-length

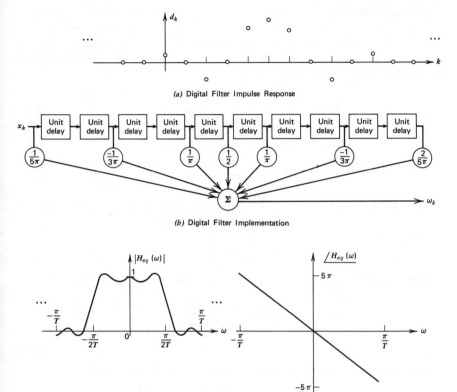

(a) Digital Filter Impulse Response

(b) Digital Filter Implementation

(c) Transfer Function

FIGURE 7.8

impulse-response sequence. Their structure, as in Figure 7.8b, is analogous to a tapped delay line, with the output formed as a weighted sum of delayed input samples. The method is quite general, and can be applied to the approximation of any desired transfer function. In general, however, many terms are required to obtain an $H_{eq}(\omega)$ which is a reasonable approximation to $H(\omega)$, especially in cases where $H(\omega)$ has discontinuities, as in the example above. In the next section, we consider the design of *recursive* filters. These filters are characterized by an infinite-length impulse-response sequence.

7.4 RECURSIVE FILTER DESIGN BY THE DIRECT METHOD

In the design of a recursive digital filter, we begin with a specification in terms of $H(s)$, the Laplace transform of the desired impulse response $h(t)$

This expression might represent, for example, a Chebyschev or maximally flat approximation to an ideal filter transfer function and would be obtained by standard network synthesis procedures. We shall require that $H(s)$ be rational—that is, a ratio of polynomials in s—with the numerator of lower order than the demoninator.

$$H(s) = \frac{\alpha_0 + \alpha_1 s + \cdots + \alpha_m s^m}{\beta_0 + \beta_1 s + \cdots + \beta_n s^n}, \qquad n > m \qquad (7.13)$$

We begin by expanding $H(s)$ in partial fractions.

$$H(s) = \frac{c_1}{s - p_1} + \frac{c_2}{s - p_2} + \cdots + \frac{c_n}{s - p_n} \qquad (7.14)$$

where $p_1 \cdots p_n$ are the poles of $H(s)$, and we assume for now that no pole is repeated. From this relation, we can write the impulse response $h(t)$ as a sum of functions

$$h(t) = h_1(t) + h_2(t) + \cdots + h_n(t)$$

where $h_i(t)$ is the inverse transform of the term $c_i/(s - p_i)$, namely

$$h_i(t) = \mathscr{L}^{-1}\left\{\frac{c_i}{s - p}\right\}$$

$$= c_i e^{p_i t} u(t) \qquad (7.15)$$

Now $\{d_k\}$ can be written as a corresponding sum of sequences

$$d_k = d_1(k) + d_2(k) + \cdots + d_n(k)$$

where

$$d_i(k) = T h_i(kT)$$

$$= \begin{cases} 0, & k < 0 \\ T c_i e^{p_i kT}, & k \geq 0 \end{cases}$$

Taking the Z-transform of $\{d_k\}$ term by term, we have

$$D(z) = D_1(z) + D_2(z) + \cdots + D_n(z)$$

where

$$D_i(z) = Z\{\{d_i(k)\}\}$$

$$= \sum_{k=-\infty}^{\infty} d_i(k) z^k$$

$$= \sum_{k=0}^{\infty} c_i T e^{p_i kT} z^k$$

$$= \frac{T c_i}{1 - z e^{p_i T}}, \qquad |z| < |e^{-p_i T}| \qquad (7.16)$$

We now have the required digital filter in partial fraction form, where each term $c_i/(s - p_i)$ in the partial fraction expansion of $H(s)$ has been replaced by the term $Tc_i/(1 - ze^{p_iT})$ in $D(z)$. Thus, each simple pole, p_i of $H(s)$ appears as a simple pole e^{-p_iT} of $D(z)$. Note that for Re $(p_i) < 0$, $|e^{-p_iT}| > 1$: hence, poles in the *left* half of the s plane map into poles *outside* the unit circle in the z plane, as we require for a stable causal filter. The design for nonrepeated roots is summarized as

$$H(s) = \frac{c_1}{s - p_1} + \frac{c_2}{s - p_2} + \cdots + \frac{c_n}{s - p_n} \qquad (7.17)$$

$$D(z) = \frac{Tc_1}{1 - ze^{p_1T}} + \frac{Tc_2}{1 - ze^{p_2T}} + \cdots + \frac{Tc_n}{1 - ze^{p_nT}}$$

$$= \frac{-Tc_1e^{-p_1T}}{z - e^{-p_1T}} - \frac{Tc_2e^{-p_2T}}{z - e^{-p_2T}} - \cdots - \frac{Tc_ne^{-p_nT}}{z - e^{-p_nT}}$$

$$\overset{\Delta}{=} \frac{\hat{c}_1}{z - \zeta_1} + \frac{\hat{c}_2}{z - \zeta_2} + \cdots + \frac{\hat{c}_n}{z - \zeta_n} \qquad (7.18)$$

where

$$\hat{c}_i = -Tc_ie^{-p_iT}, \qquad i = 1, 2, \ldots, n$$

$$\zeta_i = e^{-p_iT} \quad , \qquad i = 1, 2, \ldots, n \qquad (7.19)$$

Suppose now that $H(s)$ has a repeated root, which gives rise to terms of the form $c/(s - p)^r$ in the partial fraction expansion. Taking the inverse Laplace transform, we find the corresponding impulse-response component to be

$$h_i(t) = \mathscr{L}^{-1}\left\{\frac{c}{(s - p)^r}\right\}$$

$$= \frac{c}{(r - 1)!} t^{r-1}e^{pt}u(t)$$

The sampled version is

$$d_i(k) = \begin{cases} \dfrac{cT}{(r - 1)!} (kT)^{r-1}e^{pkT}, & k \geq 0 \\ 0, & k < 0 \end{cases}$$

with the Z-transform[2]

$$D_i(z) = \frac{c}{(r - 1)!} T^r \sum_{k=0}^{\infty} k^{r-1}(e^{pT}z)^k$$

$$= \frac{cT^r}{(r - 1)!}\left(z\frac{d}{dz}\right)^{r-1} \frac{1}{1 - ze^{pT}}$$

[2] See Section 4.4, page 179, equation (4.35).

TABLE 7.1

In this table, take $a \overset{\Delta}{=} e^{pT}$

$H_i(s)$	$D_i(z)$
$\dfrac{c}{s - p}$	$\dfrac{Tc}{1 - az}$
$\dfrac{c}{(s - p)^2}$	$T^2 \dfrac{caz}{(1 - az)^2}$
$\dfrac{c}{(s - p)^3}$	$\dfrac{T^3 c}{2}\left[\dfrac{az}{(1 - az)^2} + \dfrac{2a^2z^2}{(1 - az)^3}\right]$ $= T^3 \dfrac{caz(1 + az)}{2(1 - az)^3}$
$\dfrac{c}{(s - p)^4}$	$\dfrac{T^4 c}{6}\left[\dfrac{az}{(1 - az)^2} + \dfrac{6a^2z^2}{(1 - az)^3} + \dfrac{6a^3z^3}{(1 - az)^4}\right]$ $= T^4 \dfrac{caz(1 + 4az + a^2z^2)}{6(1 - az)^4}$
$\dfrac{c}{(s - p)^5}$	$\dfrac{T^5 c}{24}\left[\dfrac{az}{(1 - az)^2} + \dfrac{14a^2z^2}{(1 - az)^3} + \dfrac{36a^3z^3}{(1 - az)^4} + \dfrac{24a^4z^4}{(1 - az)^5}\right]$ $= T^5 \dfrac{caz(1 + 11az + 11a^2z^2 + a^3z^3)}{24(1 - az)^5}$ $= T^5 \dfrac{caz(1 + az)(1 + 10az + a^2z^2)}{24(1 - az)^5}$

The first few of these transforms are listed in Table 7.1 for $r = 1$ (the isolated pole we treated first), 2, 3, 4, and 5. Again, to design our digital filter by the direct Z-transform method, we first find the partial fraction expansion of $H(s)$ and then write $D(z)$ as a term-by-term substitution. Each pole of $H(s)$ at $s = p$ gives rise to a pole of $D(z)$ at $z = e^{-pT}$ of the same multiplicity. In Table 7.1, the substitution $a = e^{pT}$ was made for notational simplicity. Both the partial fraction and combined forms of $D_i(z)$ are given.

EXAMPLE 7.2. Let us apply this technique to the design of a low-pass filter. First, we shall find the digital equivalent of the RC network of Figure 7.9. We select the cut-off frequency to be one fourth the sampling frequency, as in Example 7.1.

$$\omega_0 \overset{\Delta}{=} \frac{1}{RC} = \frac{1}{4}\frac{2\pi}{T} = \frac{\pi}{2T}$$

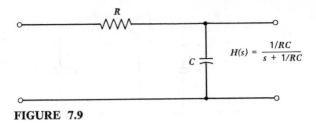

FIGURE 7.9

Thus,

$$H(s) = \frac{\dfrac{\pi}{2T}}{s + \dfrac{\pi}{2T}}$$

and hence the digitized version of $H(s)$ is

$$D(z) = \frac{T \cdot \dfrac{\pi}{2T}}{1 - e^{-\pi/2}z}$$

$$= \frac{1.57}{1 - .21z}$$

The equivalent difference equation is found from

$$(1 - .21z)W(z) = 1.57X(z)$$

Taking inverse transforms, we have

$$w_k = 1.57x_k + .21w_{k-1}$$

which might be realized as in Figure 7.10.

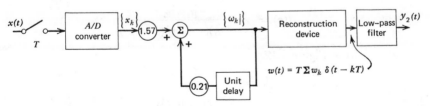

FIGURE 7.10

EXAMPLE 7.3. If we cascade two idential RC filters separated by an isolating amplifier as in Figure 7.11, we obtain the transfer

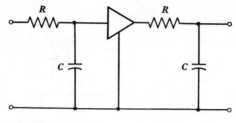

FIGURE 7.11

function

$$H(s) = \left[\frac{\dfrac{1}{RC}}{s + \dfrac{1}{RC}} \right]^2 = \frac{\omega_0^2}{(s + \omega_0)^2}$$

The equivalent digital filter transfer function $D(z)$ is found from Table 7.1 as

$$D(z) = T^2 \frac{\omega_0^2 e^{-\omega_0 T} z}{(1 - z e^{-\omega_0 T})^2}$$

Again, with $\omega_0 = \frac{1}{4} \times (2\pi/T)$, we have

$$D(z) = \frac{T^2 (\pi/2T)^2 e^{-\pi/2} z}{(1 - e^{-\pi/2} z)^2} = \frac{.51z}{(1 - .21z)^2}$$

Note that $D(z)$ cannot be found simply by cascading the two identical digital filters of the previous example.

EXAMPLE 7.4. As a final example, let us construct the digital equivalent of a three-pole maximally flat (Butterworth) filter. We know that the continuous filter with unit dc gain has the transfer function

$$H(s) = \frac{-p_1 p_2 p_3}{(s - p_1)(s - p_2)(s - p_3)}$$

where, from Figure 7.12,

$$p_1 = -\omega_0$$

$$p_2 = -\omega_0 \frac{(1 - j\sqrt{3})}{2}$$

$$p_3 = -\omega_0 \frac{(1 + j\sqrt{3})}{2} = p_2^*$$

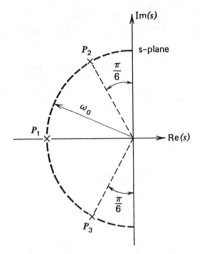

FIGURE 7.12

Expanding in partial fractions, we have

$$H(s) = \frac{c_1}{s - p_1} + \frac{c_2}{s - p_2} + \frac{c_3}{s - p_3} \qquad (7.20)$$

with

$$c_1 = \frac{-p_1 p_2 p_3}{(p_1 - p_2)(p_1 - p_3)} = \omega_0$$

$$c_2 = \frac{-p_1 p_2 p_3}{(p_2 - p_1)(p_2 - p_3)} = \frac{2\omega_0}{-3 + j\sqrt{3}}$$

$$c_3 = \frac{-p_1 p_2 p_3}{(p_3 - p_1)(p_3 - p_2)} = c_2^* = \frac{2\omega_0}{-3 - j\sqrt{3}}$$

Now the digital filter is found by applying the transformation

$$\frac{c}{s - p} \to \frac{cT}{1 - e^{pT}z}$$

term by term to (7.20)

$$D(z) = \frac{T\omega_0}{1 - ze^{-\omega_0 T}} - \frac{2T\omega_0}{3 - j\sqrt{3}} \cdot \frac{1}{1 - ze^{(-\omega_0 T/2)(1 - j\sqrt{3})}}$$

$$- \frac{2T\omega_0}{3 + j\sqrt{3}} \cdot \frac{1}{1 - ze^{(-\omega_0 T/2)(1 + j\sqrt{3})}} \qquad (7.21)$$

We wish, of course, to realize our digital filter using only real-valued coefficients. We can do so by combining the last two terms of $D(z)$ to give

$$D(z) = \frac{\omega_0 T}{1 - ze^{-\omega_0 T}} - \omega_0 T \frac{1 - ze^{-(\omega_0 T/2)}\left(\cos\alpha + \frac{1}{\sqrt{3}}\sin\alpha\right)}{1 - 2ze^{-\omega_0 T/2}\cos\alpha + z^2 e^{-\omega_0 T}}$$

where

$$\alpha \triangleq \frac{\omega_0 T\sqrt{3}}{2}$$

If we take $\omega_0 = \pi/2T$, as in the previous two examples, we have

$$\alpha = \frac{\sqrt{3}\,\pi}{4}\text{ rad} = 78°$$

and

$$D(z) = \frac{1.57}{1 - .21z} - \frac{1.57 - .55z}{1 - .19z + .21z^2} \tag{7.22}$$

As written above, $D(z)$ could be realized in the parallel form shown in Figure 7.13. If a cascade or direct form is preferred, the two terms could be combined, yielding

$$D(z) = \frac{.58z + .21z^2}{(1 - .21z)(1 - .19z + .21z^2)} \tag{7.23}$$

which might be realized in the cascade form of Figure 7.14.

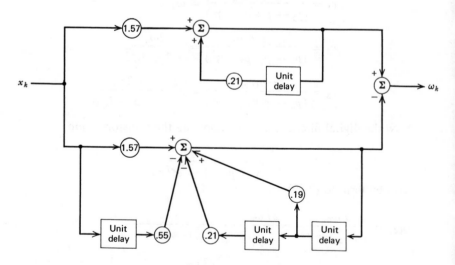

FIGURE 7.13

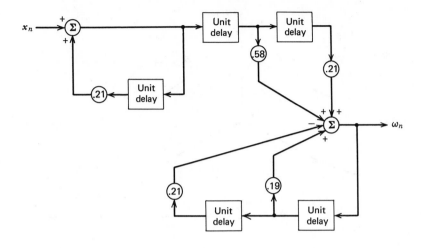

FIGURE 7.14

As can be seen in the last example, the direct Z-transform implementation involves a certain amount of computation. Because the digital filter corresponding to the cascade $H_1(s)H_2(s)$ is not the cascade of the digital filters $D_1(z)$ and $D_2(z)$ corresponding to $H_1(s)$ and $H_2(s)$ individually, we must first obtain a partial fraction expansion of $H(s)$. Second, we apply Table 7.1 to generate $D(z)$ term by term, and last, we must recombine the terms of $D(z)$ if we wish to realize our digital filter in cascade form.

Let us now compare the desired frequency response and the actual digital filter response. Figure 7.15 shows the measured transfer function for the digital filter of Figure 7.10 from Example 7.2 compared with the desired $H(j\omega)$. Note that the dc transmission[3] is $D(1) = 1.99$, rather than $H(0) = 1$. Deviations in amplitude and phase can be seen at other frequencies. Figures 7.16 and 7.17 show the predicted and measured transfer functions for the filters of Examples 7.3 and 7.4, where the same sampling and reconstruction systems as in Figure 7.10 have been used. Again we see a discrepancy in the measured transfer functions.

One source of this discrepancy is expressed by (7.8). We should expect to obtain not the original transfer function $H(\omega)$, but rather its aliased version $\sum_{n=-\infty}^{\infty} H(\omega - n2\pi/T)$. In fact, we find in Figures 7.16 and 7.17 that this is just what we do measure. Comparing $\sum_{n=-\infty}^{\infty} H(\omega - n2\pi/T)$ with the measured results of Example 7.2, however, points to an additional

---

[3] $\{w_k\}$ is found from the convolution of $\{x_k\}$ with $\{d_k\}$. With $x_k = 1$, all k, w_k is given by $w_k = \sum_m d_m x_{k-m} = \sum_m d_m = D(z)\big|_{z=1}$.

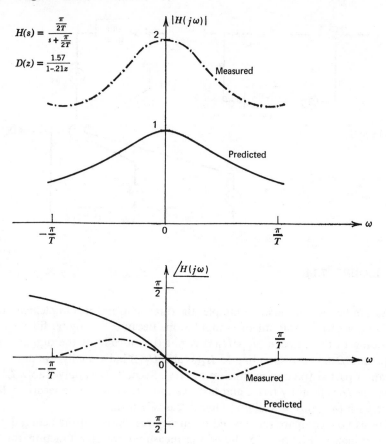

$$H(s) = \frac{\frac{\pi}{2T}}{s + \frac{\pi}{2T}}$$

$$D(z) = \frac{1.57}{1 - .21z}$$

FIGURE 7.15

error. These results are sketched in Figure 7.18. Here, the real and imaginary parts of the functions have been plotted, rather than the magnitude and phase. These plots show that the measured transfer function exceeds the predicted curve by the constant value .785 for all ω. To understand this discrepancy, we return to (7.8).

$$
\begin{aligned}
H_{\text{eq}}(\omega) &= \sum_{k=-\infty}^{\infty} d_k e^{-j\omega kT} \\
&= T\mathscr{F}_{-}\left\{ \sum_{k=-\infty}^{\infty} h(kT)\, \delta(t - kT) \right\} \\
&= T\mathscr{F}\left\{ h(t) \sum_{k=-\infty}^{\infty} \delta(t - kT) \right\}
\end{aligned}
$$

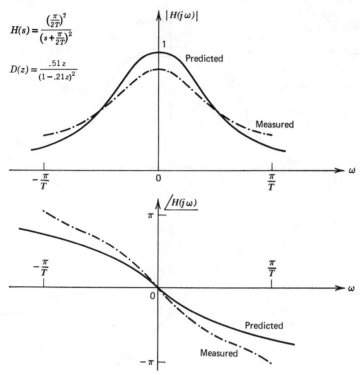

$$H(s) = \frac{\left(\frac{\pi}{2T}\right)^2}{\left(s + \frac{\pi}{2T}\right)^2}$$

$$D(z) = \frac{.51\,z}{(1 - .21z)^2}$$

FIGURE 7.16

For the filter of Example 7.2,

$$h(t) = \omega_0 e^{-\omega_0 t} u(t)$$

and $h(t)$ is discontinuous at $t = 0$: hence, we should *not* have used the relation

$$h(t)\,\delta(t) = h(0)\,\delta(t)$$

in the $k = 0$ term above, because this relation holds only if h is continuous at 0.

To correct this error, we can take, as before in Chapter 2,

$$\delta(t) = \lim_{a \to 0} \frac{1}{a}\, p\left(\frac{t}{a}\right)$$

where

$$\int_{-\infty}^{\infty} p(t)\, dt = 1$$

Here, however, we restrict $p(t)$ to be an *even* function

$$p(-t) = p(t)$$

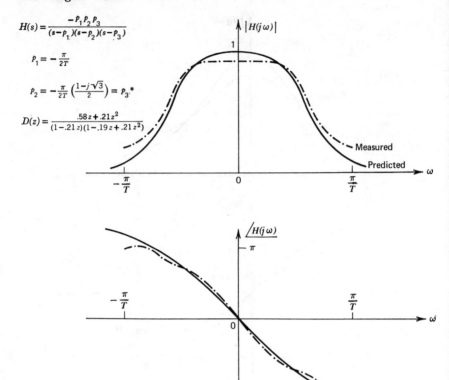

$$H(s) = \frac{-P_1 P_2 P_3}{(s-P_1)(s-P_2)(s-P_3)}$$

$$P_1 = -\frac{\pi}{2T}$$

$$P_2 = -\frac{\pi}{2T}\left(\frac{1-j\sqrt{3}}{2}\right) = P_3{}^*$$

$$D(z) = \frac{.58z + .21z^2}{(1-.21z)(1-.19z+.21z^2)}$$

FIGURE 7.17

For example, $p(t)$ might be a rectangular or Gaussian function centered at $t = 0$. Now if $f(t)$ has a jump discontinuity at $t = 0$ but is otherwise continuous in the neighborhood of 0, and the testing function $\theta(t)$ is continuous at 0, we can argue that

$$\lim_{a\to 0}\int_{-\infty}^{\infty} f(t)\frac{1}{a}p\left(\frac{t}{a}\right)\theta(t)\,dt = \lim_{a\to 0}\frac{f(0^+)+f(0^-)}{2}\int_{-\infty}^{\infty}\frac{1}{a}p\left(\frac{t}{a}\right)\theta(t)\,dt$$

and thus that we should take $f(t)\delta(t)$ to be

$$f(t)\,\delta(t) = \frac{f(0^+)+f(0^-)}{2}\delta(t) \qquad (7.24)$$

where

$$f(0^+) = \lim_{\Delta\to 0} f(t + |\Delta|)$$

$$f(0^-) = \lim_{\Delta\to 0} f(t - |\Delta|)$$

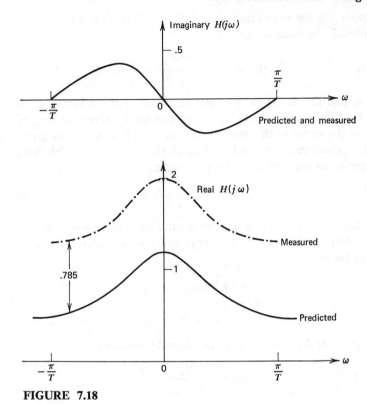

FIGURE 7.18

are the limits of $f(t)$ at 0 from the right and left, respectively. (Note that this result is consistent with our previous result if $f(t)$ is continuous at 0.)

Now, applying (7.24) to (7.8), we have the corrected relation

$$H_{eq}(\omega) = T\mathscr{F}\left\{\sum_{k=-\infty}^{\infty} h(kT)\,\delta(t - kT)\right\}$$

$$= T\mathscr{F}\left\{h(t)\sum_{k=-\infty}^{\infty}\delta(t - kT) + \frac{h(0)}{2}\,\delta(t)\right\}$$

$$= T\frac{h(0)}{2} + \sum_{n=-\infty}^{\infty} H\left(\omega - \frac{n2\pi}{T}\right) \tag{7.25}$$

where we have written

$$h(t)\,\delta(t) = \frac{h(0)}{2}\,\delta(t)$$

to account for the discontinuity of $h(t)$ at $t = 0$. For the filter treated in Example 7.2, the constant term is

$$\frac{T}{2} h(0) = \frac{T}{2} \omega_0 = \frac{T}{2} \frac{\pi}{2T} = \frac{\pi}{4} = .785$$

which is just the deviation we have measured. For the filters of Examples 7.3 and 7.4, $h(0) = 0$, and so this effect was not observed. In fact, $h(0) = 0$ whenever the order of the denominator of $H(s)$ is at least two greater than the order of its numerator (in Equation (7.13), $n \geq m + 2$). This fact follows from the initial value theorem on page 316:

$$h(0) = \lim_{s \to \infty} sH(s)$$

If the denominator is of order one greater than the numerator order, the constant term $(T/2)h(0)$ can be eliminated from the transfer function $Y_2(\omega)/X(\omega)$ by taking

$$d_k = \begin{cases} \dfrac{T}{2} h(0), & k = 0 \\ Th(kT), & k \neq 0 \end{cases} \qquad (7.26)$$

which gives for $H(s) = c/(s - p)$ the digital equivalent

$$D(z) = \frac{Tc}{1 - az} - \frac{Tc}{2}$$

$$= \frac{Tc}{2} \frac{1 + az}{1 - az}$$

with $a = e^{-pT}$ (compare with the first pair of Table 7.1). Let us take the filter of Example 7.2.

$$H(s) = \frac{\dfrac{\pi}{2T}}{s + \dfrac{\pi}{2T}}$$

from which

$$D(z) = \frac{T}{2} \frac{\pi}{2T} \frac{1 + .21z}{1 - .21z}$$

$$= .785 \frac{1 + .21z}{1 - .21z}$$

The measured transfer function $Y_2(\omega)/X(\omega)$ with this $D(z)$ is now equal to $\sum_{n=-\infty}^{\infty} H(\omega - n2\pi/T)$ for $|\omega| < \pi/T$.

Let us summarize the direct Z-transform method. Beginning with the Laplace transfer function, $H(s)$, of the desired continuous-time filter, we first find the partial fraction expansion of $H(s)$, and then apply the equivalent pairs of Table 7.1 term by term to generate the digital filter transfer function $D(z)$: e.g.,

$$\frac{c}{s - p} \to \frac{Tc}{1 - az}, \quad a = e^{pT} \tag{7.27}$$

If the desired impulse response, $h(t)$, is nonzero at $t = 0$, the pair

$$\frac{c}{s - p} \to \frac{Tc}{2} \frac{1 + az}{1 - az}, \quad a = e^{pT} \tag{7.28}$$

should be used; if $h(0) = 0$, the two pairs will give the same results. The input $x(t)$ must be bandlimited to $1/2T$ Hz, where T is the sampling interval. The Fourier transfer function which will be realized is

$$\frac{Y_2(\omega)}{X(\omega)} = H_{eq}(\omega) = \sum_{n=-\infty}^{\infty} H\left(\omega - \frac{n2\pi}{T}\right), \quad |\omega| \le \frac{2\pi}{T} \tag{7.29}$$

In general, this expression will be a satisfactory approximation to $H(\omega)$ for filters whose denominators are of considerably higher order than their numerators and whose highest break frequency is a small fraction of the sampling frequency, $1/T$ Hz. In any case, the actual transfer function can be evaluated from (7.29) before any design efforts are carried out.

7.5 THE BILINEAR TRANSFORM METHOD

To avoid the problems caused by frequency aliasing in the direct Z-transform method, we shall develop an alternate approach in this section. Our aim is to design a rational function $\tilde{H}$, representing a desired continuous-time transfer function $H(s)$, and then to transform $\tilde{H}$ into a rational digital filter transfer function, $D(z)$, which, in the system of Figure 7.2, will result in the desired overall continuous-time response. As we saw previously, $H(s)$ must be bandlimited to $1/2T$ Hz before a direct continuous-to-discrete system translation can be made. The bilinear transform method is motivated by the need to find an appropriate rational (and hence nonbandlimited) function $\tilde{H}$ corresponding to the desired bandlimited transfer function $H(s)$.

We shall deal with three complex-valued functions, $\tilde{H}(\cdot)$, $H(\cdot)$, and $D(\cdot)$, and three complex variables, $\tilde{s}$, s, and z. An appropriate rational function $\tilde{H}(\tilde{s})$ is first designed by standard Butterworth, Chebyshev, or other method. Next, the entire imaginary $\tilde{s}$ axis is mapped onto the segment of the imaginary s axis, $-\pi/T \le \text{Im}(s) \le \pi/T$, with a transformation $s = \zeta(\tilde{s})$. $H(s)$ is defined to be identically $\tilde{H}(\tilde{s})$ with the substitution $\tilde{s} = \zeta^{-1}(s)$. $H(s)$ will be the continuous-time transfer function of our system. Finally, z is related to s by $z \triangleq e^{-sT}$ as before, and $D(z)$ is defined to be identical to $H(s)$ with e^{-sT} replaced by z. This substitution maps the same segment of the imaginary s axis onto the unit circle in the z plane. These relations are sketched in Figure 7.19 below.

Our objectives are to find the transform ζ which insures that $D(z)$ is rational (as it must be if we are to implement it as in Figure 7.3), and to define an appropriate $\tilde{H}(\tilde{s})$ which yields the $H(s)$ we wish. The transform we propose to use is defined by the relations

$$\tilde{s} = \frac{2}{T} \tanh \frac{sT}{2} \qquad (7.30)$$

and

$$s = \frac{2}{T} \tanh^{-1} \frac{\tilde{s}T}{2} \qquad (7.31)$$

where, as before, T is the sampling interval. If we substitute $z = e^{-sT}$ in (7.30), we obtain

$$\tilde{s} = \frac{2}{T} \tanh \frac{sT}{2}$$

$$= \frac{2}{T} \frac{e^{sT/2} - e^{-sT/2}}{e^{sT/2} + e^{-sT/2}}$$

$$= \frac{2}{T} \frac{1 - e^{-sT}}{1 + e^{-sT}}$$

$$= \frac{2}{T} \frac{1 - z}{1 + z} \qquad (7.32)$$

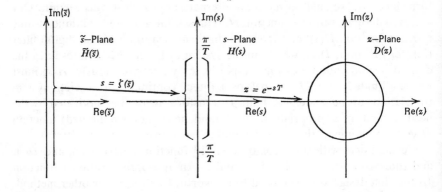

FIGURE 7.19

with the inverse[4]

$$z = \frac{1 - \dfrac{\tilde{s}T}{2}}{1 + \dfrac{\tilde{s}T}{2}} \qquad (7.33)$$

Now we define $H(s)$ and $D(z)$ by

$$H(s) = \tilde{H}\left(\frac{2}{T} \tanh \frac{sT}{2}\right) \qquad (7.34)$$

and

$$
\begin{aligned}
D(z) &= H(s)\big|_{e^{-sT}=z} \\
&= \tilde{H}\left(\frac{2}{T} \tanh \frac{sT}{2}\right)\bigg|_{e^{-sT}=z} \\
&= \tilde{H}\left(\frac{2}{T} \cdot \frac{1-z}{1+z}\right) \qquad (7.35)
\end{aligned}
$$

Thus, our digital filter transfer function $D(z)$ is derived from $\tilde{H}(\tilde{s})$ simply by replacing the argument $\tilde{s}$ with the expression $(2/T)[(1-z)/(1+z)]$. If $\tilde{H}(\tilde{s})$ is a rational function in $\tilde{s}$, it follows immediately that $D(z)$ is a rational function in z. The resulting continuous-time transfer function for the overall system of Figure 7.2 is

$$
\begin{aligned}
H(s) &= D(e^{-sT}) \\
&= \tilde{H}\left(\frac{2}{T} \tanh \frac{sT}{2}\right) \qquad (7.36)
\end{aligned}
$$

The mapping in (7.31) accomplishes what we wish it to do: it transforms the rational function $\tilde{H}(\tilde{s})$ into the rational function $D(z)$. Now how should $\tilde{H}(\tilde{s})$ be chosen to yield a desired $H(s)$ in (7.36)? Let us begin with an idealized transfer function, $H_1(s)$, which we wish to approximate. H_1 might, for example, be an ideal low-pass transfer function. Mapping from the s plane to the $\tilde{s}$ plane, we transform $H_1(s)$ into the equivalent ideal filter $\tilde{H}_1(\tilde{s})$ by applying (7.30).

$$\tilde{H}_1(\tilde{s}) = H_1\left(\frac{2}{T} \tanh^{-1} \frac{\tilde{s}T}{2}\right)$$

Now a rational function $\tilde{H}(\tilde{s})$ can be designed with standard techniques to approximate $\tilde{H}_1(\tilde{s})$ as closely as desired. Normally, $H_1(s)$ will be specified

---

[4] The bilinear relations (7.32) and (7.33) give the name "bilinear" to this method: they have the property of translating circles and straight lines in one complex plane into circles and straight lines in another complex plane.

on the $j\omega$ axis: i.e., for $s = j\omega$. The equivalent $\tilde{s}$ in this case is

$$\tilde{s} = \frac{2}{T} \tanh \frac{j\omega T}{2}$$

$$= \frac{2}{T} \frac{e^{j(\omega T/2)} - e^{-j(\omega T/2)}}{e^{j(\omega T/2)} + e^{-j(\omega T/2)}}$$

$$= j \frac{2}{T} \tan \frac{\omega T}{2}$$

Thus, with $\tilde{s} = j\tilde{\omega}$, we relate the imaginary s and $\tilde{s}$ axes by

$$\tilde{\omega} = \frac{2}{T} \tan \frac{\omega T}{2} \tag{7.37}$$

and

$$\omega = \frac{2}{T} \tan^{-1} \frac{\tilde{\omega} T}{2} \tag{7.38}$$

and on these axes, we obtain $\tilde{H}_1$ from H_1 by applying the transformation

$$\tilde{H}_1(j\omega) = H_1\left(j \frac{2}{T} \tan \frac{\omega T}{2} \right) \tag{7.39}$$

The relationship between ω and $\tilde{\omega}$ represents a distortion or *warping* of the frequency axis, as shown in Figure 7.20. In mapping $H_1(s)$ into $\tilde{H}_1(\tilde{s})$,

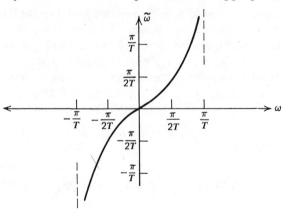

FIGURE 7.20

we prewarp the frequency axis using (7.37). This procedure then compensates for the distortion which is later introduced in mapping the rational approximation $\tilde{H}(\tilde{s})$ into $H(s)$.

We shall now summarize the bilinear transform design method and present some examples of its use (see Figure 7.21).

Step 1. Define an appropriate ideal continuous-time transfer function $H_1(j\omega)$.

Step 2. Define

$$\tilde{H}_1(j\tilde{\omega}) = H_1\left(j\frac{2}{T}\tan^{-1}\frac{\omega T}{2}\right)$$

mapping the segment $|\omega| \le \pi/T$ of the $j\omega$ axis onto the entire $j\tilde{\omega}$ axis.

Step 3. Using standard synthesis methods, design the rational function $\tilde{H}(\tilde{s})$ to approximate $\tilde{H}_1(j\tilde{\omega})$ as closely as desired on the $j\tilde{\omega}$ axis.

Step 4. The appropriate digital filter $D(z)$ is given by $D(z) = \tilde{H}\left(\frac{2}{T}\frac{1-z}{1+z}\right)$, and the resulting continuous-time transfer function is equal to

$$H(j\omega) = D(e^{-j\omega T})$$
$$= \tilde{H}\left(j\frac{2}{T}\tan\frac{\omega T}{2}\right) \qquad (7.40)$$

These steps are diagramed in Figure 7.21 for the case where $H_1(j\omega)$ is the transfer function of an ideal low-pass filter with upper half power frequency $1/4 \times 2\pi/T$, and where $\tilde{H}_1$ is approximated with a three-pole maximally flat (Butterworth) filter. For numerical details of this design, see Example 7.7.

Examples 7.5 through 7.8 which follow treat various designs for a low-pass filter with cut-off frequency equal to $1/4$ of the sampling frequency. The reader should compare these results with those of Examples 7.1 through 7.4, which treated the same design problem using the direct method. As before, the system input $x(t)$ must be bandlimited to half the sampling frequency so that the digital filter input sequence $\{x_k\}$ is not distorted by aliasing effects.

EXAMPLE 7.5. We define the ideal transfer function, as before, as

$$H_1(j\omega) = \begin{cases} 1, & |\omega| \le \dfrac{\pi}{2T} \\ 0, & |\omega| > \dfrac{\pi}{2T} \end{cases}$$

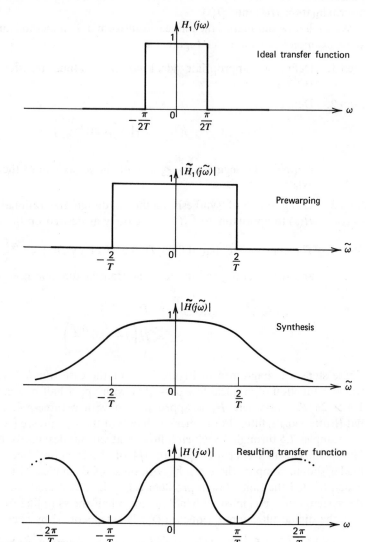

FIGURE 7.21

Then we prewarp the frequency axis by setting the $\tilde{s}$ plane cut-off frequency at

$$j\tilde{\omega}_0 = j\frac{2}{T}\tan\left(\frac{T}{2}\frac{\pi}{2T}\right)$$

$$= j\frac{2}{T}\tan\left(\frac{\pi}{4}\right)$$

$$= j\frac{2}{T}$$

This method yields the $\tilde{s}$ plane ideal transfer function

$$\tilde{H}_1(j\tilde{\omega}) = \begin{cases} 1, & |\tilde{\omega}| \leq \dfrac{2}{T} \\[2ex] 0, & |\tilde{\omega}| > \dfrac{2}{T} \end{cases}$$

First we take a single-pole approximation to $\tilde{H}_1$ with

$$\tilde{H}(j\tilde{\omega}) = \frac{1}{1 + j\dfrac{\tilde{\omega}}{\tilde{\omega}_0}} = \frac{1}{1 + j\dfrac{\tilde{\omega}T}{2}}$$

This approximation is transformed into the digital filter

$$D(z) = \tilde{H}\left(\frac{2}{T}\frac{1-z}{1+z}\right)$$

$$= \frac{1}{1 + \dfrac{2}{T}\dfrac{1-z}{1+z}\dfrac{T}{2}}$$

$$= \frac{1+z}{2}$$

Now the overall continuous-time transfer function is

$$H_{eq}(\omega) = D(z)\big|_{z=e^{-j\omega T}} = \tfrac{1}{2}(1 + e^{-j\omega T})$$

The magnitude of this function is plotted in Figure 7.22. Note that $H(s) = (1/2)(1 + e^{-sT})$ is periodic, with period $2\pi/T$ in the imaginary s direction. Because $x(t)$ is bandlimited, only the central period of $H(s)$ contributes to the filtering function.

EXAMPLE 7.6.   If we take the same $H_1(j\omega)$ and $\tilde{H}_1(j\tilde{\omega})$ as in the above example, but in this case approximate $\tilde{H}_1(j\tilde{\omega})$ by two identical cascaded

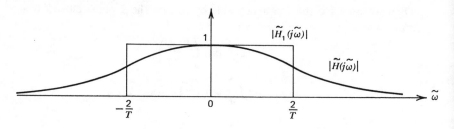

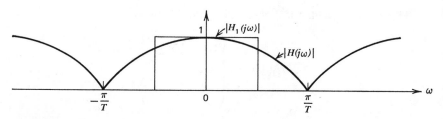

FIGURE 7.22

sections, we now have

$$\tilde{H}(j\tilde{\omega}) = \left[\frac{1}{1 + j\dfrac{\tilde{\omega}T}{2}} \right]^2$$

$$D(z) = \frac{(1 + z)^2}{4}$$

and

$$H_{eq}(\omega) = D(z)\big|_{z=e^{-j\omega T}} = \tfrac{1}{4}(1 + e^{-j\omega T})^2$$

The magnitude of this transfer function is sketched in Figure 7.23. Note that cascaded sections of $\tilde{H}(\tilde{s})$ are transformed into cascaded sections of $D(z)$. This property will often facilitate the design procedure.

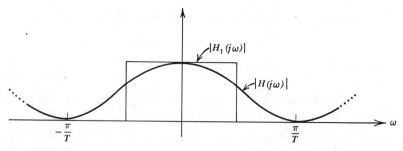

FIGURE 7.23

EXAMPLE 7.7. Again we take $H_1(j\omega)$ and $\tilde{H}_1(j\tilde{\omega})$ as in Example 7.5. Here, however, we approximate $\tilde{H}_1(j\tilde{\omega})$ with a three-pole Butterworth function. As in Example 7.4, we have (with $\tilde{\omega}_0 = 2/T$)

$$\tilde{H}(\tilde{s}) = \frac{-p_1 p_2 p_3}{(\tilde{s} - p_1)(\tilde{s} - p_2)(\tilde{s} - p_3)}$$

with

$$p_1 = -\frac{2}{T}$$

$$p_2 = -\frac{2}{T}\frac{1 - j\sqrt{3}}{2} = -\frac{1}{T}(1 - j\sqrt{3})$$

$$p_3 = p_2^* = -\frac{1}{T}(1 + j\sqrt{3})$$

Substituting for p_1, p_2, and p_3, we obtain the rational approximation

$$\tilde{H}(\tilde{s}) = \frac{\dfrac{8}{T^3}}{\left(\tilde{s} + \dfrac{2}{T}\right)\left(\tilde{s} + \dfrac{1 - j\sqrt{3}}{T}\right)\left(\tilde{s} + \dfrac{1 + j\sqrt{3}}{T}\right)}$$

$$= \frac{8}{T^3\left(\tilde{s} + \dfrac{2}{T}\right)\left(\tilde{s}^2 + \dfrac{2\tilde{s}}{T} + \dfrac{4}{T^2}\right)}$$

The equivalent digital filter is given by

$$D(z) = \tilde{H}\left(\frac{2}{T}\frac{1 - z}{1 + z}\right) = \frac{(1 + z)^3}{2(3 + z^2)}$$

The actual continuous-time transfer function which we realize is given by

$$H(j\omega) = \frac{(1 + e^{-j\omega T})^3}{2(3 + e^{-j2\omega T})}$$

This transfer function is sketched in Figure 7.24 (see also Figure 7.21).

As can be seen from these examples, the bilinear transform method is easily implemented, especially in cases where the desired transfer function $H_1(s)$ takes the values 0 or 1 along the $j\omega$ axis (e.g., for low-pass, band-pass, etc. filters). In this case, we merely map the cut-off frequencies from one plane to another. The designed transfer function $\tilde{H}$ can be made as close to the ideal

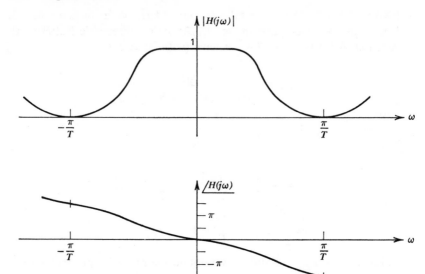

FIGURE 7.24

characteristic H_1 as desired. Workers in the field of physical network synthesis have developed an extensive body of theory to attack this very problem. This theory is particularly applicable because the realized filter $H(j\omega)$ is, except for a distortion of the $j\omega$ axis, identical to the designed filter $\tilde{H}(j\tilde{\omega})$.

> **EXAMPLE 7.8.** It should be apparent to the reader that the bilinear transform filter design procedure is well suited to computer implementation. A program written for digital filter design was used to generate a six-pole Chebychev low-pass filter with half-power frequency $\pi/2T$ (as in the previous examples) and 0.1 ripple in the pass band. The appropriate s plane poles are shown in Figure 7.25. The corresponding digital filter has the transfer function
>
> $$D(z) = \frac{.18 + .37z + .18z^2}{1.0 - .06z + .88z^2} \times \frac{.20 + .40z + .20z^2}{1.0 - .50z + .64z^2} \times \frac{.24 + .48z + .24z^2}{1.0 - 1.11z + .41z^2}$$
>
> The transfer functions $\tilde{H}(j\tilde{\omega})$ and $H(j\omega)$ for this filter are shown in Figure 7.26. Note the horizontal distortion that occurs in mapping $\tilde{H}(j\tilde{\omega})$ into $H(j\omega)$.

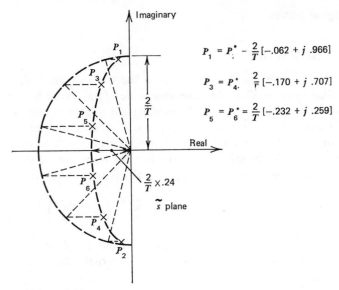

$$P_1 = P_2^* - \frac{2}{T}[-.062 + j\ .966]$$

$$P_3 = P_4^* \cdot \frac{2}{T}[-.170 + j\ .707]$$

$$P_5 = P_6^* = \frac{2}{T}[-.232 + j\ .259]$$

FIGURE 7.25

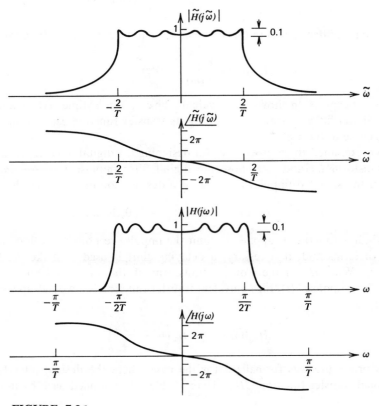

FIGURE 7.26

7.6 SUMMARY

In this chapter we have investigated the use of digital filters in continuous-time systems. Our objective was to design a digital filter which, in terms of the system of Figure 7.2, would yield a desired continuous-time transfer function. We obtained two expressions for the output $y_2(t)$ of this system. In general, with $D(z)$ the digital filter transfer function, $F(\omega)$ the output filter transfer function, $X(\omega)$ the Fourier transform of the input $x(t)$ and T the sampling interval, the output $y_2(t)$ has the Fourier transform

$$Y_2(\omega) = D(z)\Big|_{z=e^{-j\omega T}} F(\omega) \sum_{n=-\infty}^{\infty} {}^{'} X\left(\omega - \frac{n2\pi}{T}\right)$$

If the input is bandlimited to $1/2T$ Hz, or half the sampling frequency, and the output filter has an ideal low-pass characteristic, then we obtain the simpler form

$$Y_2(\omega) = D(z)\Big|_{z=e^{-j\omega T}} X(\omega)$$

The digital filter system in this case has the equivalent transfer function

$$H_{eq}(\omega) = \frac{Y_2(\omega)}{X(\omega)} = D(z)\Big|_{z=e^{-j\omega T}}$$

This last expression allows us to evaluate the continuous-time performance of any digital filter by simply evaluating its transfer function $D(z)$ around the unit circle in the z plane.

We examined three procedures for designing the digital filter. In the direct Z-transform method, we chose the digital filter impulse response $\{d_k\}$ to equal the sampled impulse response of a desired continuous-time filter $h(t)$

$$d_k = Th(kT) \qquad k = \ldots -1, 0, 1, 2, \ldots$$

A window function is applied to limit the impulse response to a finite number of terms and, if necessary, a delaying shift is used to make the filter causal. With $H_2(\omega)$ the Fourier transform of the windowed and shifted impulse response $h(t)$, the equivalent transfer function is a wound-up version of $H_2(\omega)$.

$$H_{eq}(\omega) = \sum_{n=-\infty}^{\infty} H_2\left(\omega - \frac{n2\pi}{T}\right)$$

This procedure was formalized for the case where the desired filter has a rational transfer function, $H(s)$. Here, Table 7.1 is applied term by term to the partial fraction expansion form of $H(s)$. Because no truncation or delay is required, the equivalent transfer function is a wound-up version of $H(\omega)$ itself.

$$H_{eq}(\omega) = \sum_{n=-\infty}^{\infty} H\left(\omega - \frac{n2\pi}{T}\right)$$

The bilinear method is motivated by a desire to avoid this aliasing of the transfer function $H(\omega)$. An ideal transfer function $H_1(s)$ is first mapped onto the $\tilde{s}$ plane, where the synthesis of a rational approximation $\tilde{H}(\tilde{s})$ is performed. The digital filter transfer function is obtained from this rational approximation by

$$D(z) = \tilde{H}\left(\frac{2}{T}\frac{1-z}{1+z}\right)$$

The actual transfer function which is realized by this method is

$$H(s) = \tilde{H}\left(\frac{2}{T}\tanh\frac{sT}{2}\right)$$

Along the $j\omega$ axis, we obtain

$$H_{eq}(\omega) = H(j\omega) = \tilde{H}\left(j\frac{2}{T}\tan\frac{\omega T}{2}\right)$$

The nonlinear mapping from the s plane to the $\tilde{s}$ plane, introducing a distortion of the frequency axis, is most easily performed for piecewise-constant transfer functions. Therefore, this method is particularly well suited to the approximation of ideal low-pass, high-pass, band-pass, and similar transfer functions.

PROBLEMS

1. A zero-order hold (ZOH) is perhaps the most common form of output filter, and is usually programed into the digital-to-analog converter. This device puts out a constant voltage, equal to the last digital input, for one sampling period as shown at the top of page 396:
   a. Verify that the model above is appropriate for the input-output time functions $w(t)$ and $y(t)$.
   b. What is the impulse response of the ZOH?
   c. What is the transfer function $F(\omega)$ for the ZOH?
   d. Do you think the ZOH would make a good output filter?

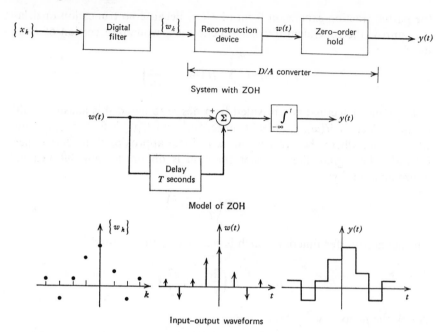

System with ZOH

Model of ZOH

Input–output waveforms

2. The linear point connector (LPC) is another output filter. The LPC output $y(t)$ is formed by connecting samples of the digital filter output sequence with straight line segments.

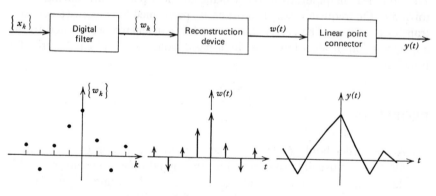

a. Derive a model of the LPC.
b. What is the impulse response of the LPC? Is this a realizable system? If not, how could you modify it to make it realizable?
c. What is the transfer function $F(\omega)$ for the LPC?
d. Compare the attenuation of the ZOH and LPC output filters at the crossover signal frequency, π/T rad/sec.

3. Compare the operation of the integrator circuits below. The first is a continuous-time integrator; the second is a digital filter implementation which approximates an integral by the sum $w_k = \sum\limits_{m=-\infty}^{k} x_m$.

Continuous–time system

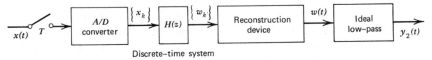

Discrete–time system

  a. What is the transfer function $H(z)$ of the digital filter?

  b. What is the continuous-time transfer function for each system? You may consider the input $x(t)$ to be bandlimited to π/T rad/sec.

  c. Under what conditions is the digital integrator a good approximation to the continuous-time integrator?

4. Evaluate the transfer function $H_{eq}(\omega)$ for the nonrecursive filter of Example 7.1 if N is taken to be 7 rather than 5, and the digital filter impulse response contains 15 nonzero terms. Repeat for $N = 9$, with 19 nonzero terms in $\{d_k\}$. (You may wish to write a computer program to perform this evaluation.)

5. What transfer function $H_{eq}(\omega)$ is obtained if a triangular window function,

$$f(t) = \begin{cases} 1 - \dfrac{|t|}{NT}, & |t| \le NT \\ 0, & |t| > NT \end{cases}$$

is used in the design of nonrecursive filters by the direct method? In this case, we set

$$d_k = \begin{cases} T\left(1 - \dfrac{|k - N|}{N}\right) h(kT), & |k| \le N \\ 0, & |k| > N \end{cases}$$

Compare this filter with equations (7.11) and (7.12) of Section 7.3. What qualitative remarks can you make regarding use of this window function vis-a-vis the rectangular window—e.g., sharpness of the transition region or overshoot in the transfer function $H_{eq}(\omega)$?

6. Carry out the design of a 4-pole approximation to an ideal low-pass filter having a cut-off frequency of $(\pi/2T)$ rad/sec: i.e., one-fourth the sampling frequency. Use both the direct and bilinear methods, and compare the equivalent transfer functions you obtain. Which filter has the sharper cut-off?

7. A seismic transponder signal is to be transmitted from a landed space-craft back to Earth. Although the signal has frequency components up to 10 Hz, the scientists responsible for analyzing the data are interested in spectral components only up to 5 Hz. Therefore, to reduce the re-quired data rate, it is planned to low-pass filter the signal, eliminating energy above 5 Hz; sample the filtered signal at 10 Hz; and send this sampled data back to Earth at a rate of 10 samples per second.

One way to implement this plan is to sample the original signal at 20 samples per second (twice the highest frequency), operate on the samples with a digital filter having a cut-off frequency of $f_{\text{samp}}/4$, and then subsample the digital filter output, transmitting only every other output sample. The continuous time signal is then reconstructed at the receiver by use of an ideal low-pass filter.

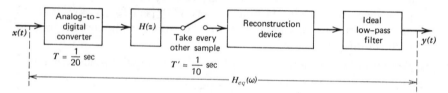

Assuming that an ideal low-pass filter is available, will this method work? If $H(z)$ is the digital filter transfer function, what would be the equivalent continuous-time transfer function, $H_{\text{eq}}(\omega)$, from input to output? What would be the output from an input of $x(t) = \cos \omega t$, where $\omega/2\pi$ is not necessarily less than 20 Hz?

8. Although a bandlimited signal can in theory be reconstructed exactly from samples taken at slightly greater than twice the highest frequency, in practice it is difficult to realize a good approximation to the required ideal low-pass filter. At the same time, we would often like to take as few samples as possible—for example, in the preceding próblem, where we wish to minimize a required transmission rate.

To help resolve this problem, it has been proposed that a digital interpolation filter be used to effectively double the sampling frequency by inserting samples between those taken from the original signal. If $s(t)$ is the bandlimited signal and $\{s(kT)\}$ is the sequence of samples taken at a rate of $1/T = 2 \times f_{\text{max}}$, then a new sequence $\{x_k\} = \cdots s(-T), 0,$ $s(0), 0, s(T), 0, s(2T), \ldots$ would be formed and used as the input to a digital filter with transfer function $H(z)$. $H(z)$ is to be chosen such that the output sequence $\{w_k\}$ has the form

$$\{w_k\} = \cdots s(-T), \quad s(-T/2), \quad s(0), \quad s(T/2), \quad s(T), \quad s(3T/2), \cdots$$

If this scheme is successful, we shall have inserted a guard band of $2\pi/T$ rad/sec with no additional signal energy into the spectrum of

$$w(t) = T \sum w_k \, \delta\left(t - \frac{kT}{2}\right)$$

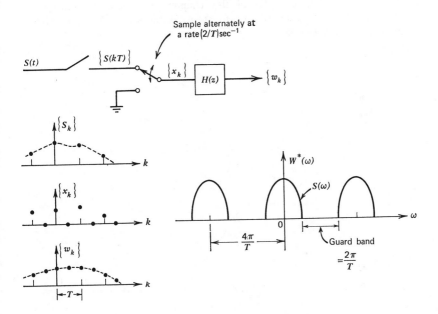

This will make the reconstruction filter much easier to build. Could such a digital filter be built? If so, how would you do it? If not, why not?

SUGGESTED READINGS

Churchill, R. V., *Complex Variables and Applications*, McGraw-Hill, New York, 1960.
    The bilinear transform is treated on pages 74 ff.
Guillemin, E. A., *Synthesis of Passive Networks*, Wiley, New York, 1957.
Skwirzynski, J. K., *Design Theory and Data for Electrical Filters*, Van Nostrand, New York, 1965.
Van Valkenburg, M. E., *Modern Network Synthesis*, Wiley, New York, 1960.

These three references treat the problem of designing a rational filter $H(s)$ for approximating a given transfer function. The most relevant topics are those dealing with pole-zero locations for maximally flat, Chebyshev, and other approximations to ideal characteristics. The transformation of low-pass filters into high-pass, band-pass and other transfer functions is also of interest. Guillemin and Van Valkenburg's texts are classics in the theory of network synthesis; Skwirzynski's reference is directed more toward detailed design.

Kuo, F. F., and J. F. Kaiser, *System Design by Digital Computer*, Wiley, New York, 1966.

Chapter 7 treats the topics of this chapter and presents additional examples. The selection of window functions for the design of nonrecursive filters is discussed in detail.

Appendix A
Tabulation of the sinc
Function

The following table lists values of sinc(x) for x between 0. and 20. in steps of .05. The definition of sinc (x) is

$$\text{sinc} (x) \triangleq \frac{\sin (x)}{x}$$

x	sinc (x)	x	sinc (x)	x	sinc (x)
.0	1.000000	1.25	.759187	2.50	.239388
.05	.999583	1.30	.741198	2.55	.218699
.10	.998334	1.35	.722758	2.60	.198269
.15	.996254	1.40	.703892	2.65	.178124
.20	.993346	1.45	.684629	2.70	.158288
.25	.989615	1.50	.664996	2.75	.138785
.30	.985067	1.55	.645021	2.80	.119638
.35	.979708	1.60	.624733	2.85	.100869
.40	.973545	1.65	.604160	2.90	.082499
.45	.966590	1.70	.583332	2.95	.064550
.50	.958851	1.75	.562277	3.00	.047040
.55	.950340	1.80	.541026	3.05	.029988
.60	.941070	1.85	.519608	3.10	.013413
.65	.931056	1.90	.498052	3.15	−.002668
.70	.920310	1.95	.476389	3.20	−.018241
.75	.908851	2.00	.454648	3.25	−.033290
.80	.896695	2.05	.432859	3.30	−.047801
.85	.883859	2.10	.411052	3.35	−.061761
.90	.870363	2.15	.389255	3.40	−.075159
.95	.856226	2.20	.367498	3.45	−.087983
1.00	.841470	2.25	.345810	3.50	−.100223
1.05	.826117	2.30	.324219	3.55	−.111872
1.10	.810188	2.35	.302754	3.60	−.122922
1.15	.793707	2.40	.281442	3.65	−.133366
1.20	.776699	2.45	.260312	3.70	−.143198

x	sinc (x)	x	sinc (x)	x	sinc (x)
3.75	−.152416	5.95	−.054967	8.15	.117362
3.80	−.161015	6.00	−.046569	8.20	.114723
3.85	−.168993	6.05	−.038194	8.25	.111830
3.90	−.176350	6.10	−.029862	8.30	.108695
3.95	−.183085	6.15	−.021592	8.35	.105327
4.00	−.189200	6.20	−.013401	8.40	.101737
4.05	−.194697	6.25	−.005308	8.45	.097938
4.10	−.199579	6.30	.002668	8.50	.093939
4.15	−.203851	6.35	.010514	8.55	.089754
4.20	−.207518	6.40	.018210	8.60	.085395
4.25	−.210585	6.45	.025742	8.65	.080873
4.30	−.213061	6.50	.033095	8.70	.076203
4.35	−.214954	6.55	.040253	8.75	.071397
4.40	−.216273	6.60	.047203	8.80	.066467
4.45	−.217027	6.65	.053931	8.85	.061429
4.50	−.217228	6.70	.060425	8.90	.056294
4.55	−.216888	6.75	.066673	8.95	.051077
4.60	−.216019	6.80	.072663	9.00	.045790
4.65	−.214635	6.85	.078386	9.05	.040449
4.70	−.212749	6.90	.083831	9.10	.035065
4.75	−.210377	6.95	.088990	9.15	.029653
4.80	−.207534	7.00	.093855	9.20	.024227
4.85	−.204236	7.05	.098417	9.25	.018798
4.90	−.200500	7.10	.102671	9.30	.013382
4.95	−.196344	7.15	.106611	9.35	.007990
5.00	−.191784	7.20	.110231	9.40	.002635
5.05	−.186841	7.25	.113528	9.45	−.002668
5.10	−.181532	7.30	.116498	9.50	−.007910
5.15	−.175877	7.35	.119138	9.55	−.013078
5.20	−.169895	7.40	.121447	9.60	−.018159
5.25	−.163606	7.45	.123423	9.65	−.023142
5.30	−.157031	7.50	.125066	9.70	−.028016
5.35	−.150190	7.55	.126377	9.75	−.032771
5.40	−.143104	7.60	.127357	9.80	−.037395
5.45	−.135794	7.65	.128008	9.85	−.041880
5.50	−.128280	7.70	.128333	9.90	−.046215
5.55	−.120583	7.75	.128335	9.95	−.050392
5.60	−.112726	7.80	.128018	10.00	−.054402
5.65	−.104728	7.85	.127387	10.05	−.058236
5.70	−.096611	7.90	.126448	10.10	−.061888
5.75	−.088396	7.95	.125206	10.15	−.065349
5.80	−.080103	8.00	.123669	10.20	−.068615
5.85	−.071754	8.05	.121844	10.25	−.071677
5.90	−.063368	8.10	.119739	10.30	−.074532

x	sinc (x)	x	sinc (x)	x	sinc (x)
10.35	$-.077174$	12.55	$-.001304$	14.75	.055459
10.40	$-.079598$	12.60	.002668	14.80	.053260
10.45	$-.081801$	12.65	.006603	14.85	.050943
10.50	$-.083780$	12.70	.010490	14.90	.048515
10.55	$-.085532$	12.75	.014321	14.95	.045982
10.60	$-.087054$	12.80	.018086	15.00	.043352
10.65	$-.088345$	12.85	.021777	15.05	.040631
10.70	$-.089405$	12.90	.025385	15.10	.037827
10.75	$-.090232$	12.95	.028902	15.15	.034947
10.80	$-.090827$	13.00	.032320	15.20	.031999
10.85	$-.091191$	13.05	.035631	15.25	.028991
10.90	$-.091324$	13.10	.038829	15.30	.025930
10.95	$-.091229$	13.15	.041905	15.35	.022825
11.00	$-.090908$	13.20	.044854	15.40	.019683
11.05	$-.090363$	13.25	.047668	15.45	.016512
11.10	$-.089599$	13.30	.050343	15.50	.013320
11.15	$-.088618$	13.35	.052873	15.55	.010116
11.20	$-.087426$	13.40	.055251	15.60	.006907
11.25	$-.086027$	13.45	.057475	15.65	.003701
11.30	$-.084426$	13.50	.059539	15.70	.000507
11.35	$-.082629$	13.55	.061440	15.75	$-.002668$
11.40	$-.080642$	13.60	.063173	15.80	$-.005816$
11.45	$-.078472$	13.65	.064737	15.85	$-.008931$
11.50	$-.076126$	13.70	.066128	15.90	$-.012003$
11.55	$-.073610$	13.75	.067344	15.95	$-.015026$
11.60	$-.070933$	13.80	.068383	16.00	$-.017993$
11.65	$-.068102$	13.85	.069245	16.05	$-.020897$
11.70	$-.065126$	13.90	.069928	16.10	$-.023731$
11.75	$-.062014$	13.95	.070432	16.15	$-.026488$
11.80	$-.058773$	14.00	.070757	16.20	$-.029161$
11.85	$-.055413$	14.05	.070904	16.25	$-.031746$
11.90	$-.051944$	14.10	.070873	16.30	$-.034236$
11.95	$-.048374$	14.15	.070665	16.35	$-.036625$
12.00	$-.044714$	14.20	.070283	16.40	$-.038908$
12.05	$-.040973$	14.25	.069729	16.45	$-.041081$
12.10	$-.037160$	14.30	.069005	16.50	$-.043138$
12.15	$-.033287$	14.35	.068114	16.55	$-.045075$
12.20	$-.029363$	14.40	.067059	16.60	$-.046888$
12.25	$-.025397$	14.45	.065845	16.65	$-.048574$
12.30	$-.021400$	14.50	.064475	16.70	$-.050128$
12.35	$-.017383$	14.55	.062954	16.75	$-.051548$
12.40	$-.013355$	14.60	.061287	16.80	$-.052831$
12.45	$-.009325$	14.65	.059478	16.85	$-.053975$
12.50	$-.005305$	14.70	.057533	16.90	$-.054977$

403

x	sinc (x)	x	sinc (x)	x	sinc (x)
16.95	−.055837	18.00	−.041721	19.05	.010451
17.00	−.056552	18.05	−.039725	19.10	.012975
17.05	−.057123	18.10	−.037641	19.15	.015454
17.10	−.057547	18.15	−.035475	19.20	.017880
17.15	−.057826	18.20	−.033232	19.25	.020250
17.20	−.057959	18.25	−.030919	19.30	.022557
17.25	−.057947	18.30	−.028541	19.35	.024796
17.30	−.057790	18.35	−.026105	19.40	.026962
17.35	−.057490	18.40	−.023617	19.45	.029049
17.40	−.057049	18.45	−.021084	19.50	.031053
17.45	−.056468	18.50	−.018512	19.55	.032969
17.50	−.055750	18.55	−.015908	19.60	.034794
17.55	−.054896	18.60	−.013278	19.65	.036522
17.60	−.053911	18.65	−.010629	19.70	.038150
17.65	−.052797	18.70	−.007967	19.75	.039676
17.70	−.051558	18.75	−.005300	19.80	.041094
17.75	−.050197	18.80	−.002634	19.85	.042403
17.80	−.048719	18.85	.000023	19.90	.043600
17.85	−.047127	18.90	.002667	19.95	.044682
17.90	−.045427	18.95	.005291	20.00	.045647
17.95	−.043624	19.00	.007888		

Appendix B
Evaluation of Geometric
Series

In this appendix, we derive closed-form expressions for various geometric series such as

$$\sum_{n=n_1}^{n_2} a^n = a^{n_1} + a^{n_1+1} + \cdots + a^{n_2-1} + a^{n_2} \qquad n_2 \geq n_1$$

These series appear recurrently in our study of discrete-time systems.
We begin with $n_1 = 0$ and obtain,

$$\sum_{n=0}^{n_2} a^n = 1 + a + a^2 + \cdots + a^{n_2}$$

$$= \begin{cases} \dfrac{1 - a^{n_2+1}}{1 - a}, & a \neq 1 \\ n_2 + 1, & a = 1 \end{cases}$$

The closed form expression for $a \neq 1$ can be verified by multiplying both sides by $1 - a$ to obtain an identity. For the case $a = 1$, we note that there are $n_2 + 1$ terms in the series, with each term equal to 1.
Now, using these expressions, we obtain the relations

$$\sum_{n=n_1}^{n_2} a^n = \sum_{0}^{n_2} a^n - \sum_{0}^{n_1-1} a^n$$

$$= \begin{cases} \dfrac{1 - a^{n_2+1}}{1 - a} - \dfrac{1 - a^{n_1}}{1 - a}, & a \neq 1 \\ n_2 + 1 - n_1, & a = 1 \end{cases}$$

$$= \begin{cases} \dfrac{a^n - a^{n_1+1}}{1 - a}, & a \neq 1 \\ n_2 - n_1 + 1, & a = 1 \end{cases}$$

where we take $0 \leq n_1 \leq n_2$.

To deal with an infinite sum, we let $n_2 \to \infty$ above. Now, if $|a| < 1$, we have

$$\lim_{n \to \infty} a^n = 0$$

and

$$\sum_{n=n_1}^{\infty} a^n = \lim_{n_2 \to \infty} \sum_{n_1}^{n_2} a^n$$

$$= \lim_{n_2 \to \infty} \left[\frac{a^{n_1}}{1 - a} - \frac{a^{n_2+1}}{1 - a} \right]$$

$$= \frac{a^{n_1}}{1 - a}, \quad \begin{array}{l} |a| < 1 \\ n_1 \geq 0 \end{array}$$

The corresponding sum for $a \geq 1$ does not converge. Letting $n_1 = 0$ and $n_1 = 1$, we obtain the special cases

$$\sum_{n=0}^{\infty} a^n = \frac{1}{1 - a}, \quad |a| < 1$$

$$\sum_{n=1}^{\infty} a^n = \frac{a}{1 - a}, \quad |a| < 1$$

Note that n_1 and n_2 are positive numbers in these expressions. To generalize the results for any n_1 and n_2, we first let $n_1 < 0 \leq n_2$. Now we have

$$\sum_{n=n_1}^{n_2} a^n = \sum_{n=n_1}^{-1} a^n + \sum_{n=0}^{n_2} a^n$$

With $m = -n$ in the first sum on the right, we obtain the following sums over positive indices, which can be evaluated using the previous results:

$$\sum_{n=n_1}^{n_2} = \sum_{m=1}^{-n_1} \left(\frac{1}{a} \right)^m + \sum_{n=0}^{n_2} a^n$$

$$= \frac{\left(\frac{1}{a} \right) - \left(\frac{1}{a} \right)^{-n_1+1}}{1 - \frac{1}{a}} + \frac{1 - a^{n_2+1}}{1 - a}$$

$$= \frac{1 - a^{n_1}}{a - 1} + \frac{1 - a^{n_2+1}}{1 - a}$$

$$= \frac{a^{n_1} - a^{n_2+1}}{1 - a}, \quad a \neq 1$$

as before.

With $n_1 \leq n_2 \leq 0$, we again substitute $m = -n$ to obtain

$$\sum_{n=n_1}^{n_2} a^n = \sum_{m=-n_2}^{-n_1} \left(\frac{1}{a}\right)^m, \qquad 0 \geq -n_2 \geq -n_1$$

$$= \frac{\left(\dfrac{1}{a}\right)^{-n_2} - \left(\dfrac{1}{a}\right)^{-n_1+1}}{1 - \dfrac{1}{a}},$$

$$= \frac{a^{n_2} - \left(\dfrac{1}{a}\right)a^{n_1}}{1 - \dfrac{1}{a}}$$

$$= \frac{a^{n_1} - a^{n_2+1}}{1 - a}, \qquad a \neq 1$$

Thus, for $a \neq 1$ and $n_1 \leq n_2$, with no other restrictions on n_1 and n_2, we have

$$\sum_{n=n_1}^{n_2} a^n = \frac{a^{n_1} - a^{n_2+1}}{1 - a} \qquad a \neq 1$$

For $a = 1$, we observe that there are $n_2 - n_1 + 1$ terms in the sum for any $n_1 \leq n_2$, and obtain the value $n_2 - n_1 + 1$ for the sum. Alternatively, we can apply L'Hospital's rule to obtain

$$\lim_{a \to 1} \sum_{n=n_1}^{n_2} a^n = \lim_{a \to 1} \frac{a^{n_1} - a^{n_2+1}}{1 - a}$$

$$= \frac{\dfrac{d}{da}(a^{n_1} - a^{n_2+1})\Big|_{a=1}}{\dfrac{d}{da}(1 - a)\Big|_{a=1}}$$

$$= \frac{n_1 - (n_2 + 1)}{-1}$$

$$= n_2 - n_1 + 1$$

The results of this appendix are summarized in Table B.1.

TABLE B.1

---

1. $\displaystyle\sum_{n=n_1}^{n_2} a^n = \begin{cases} \dfrac{a^{n_1} - a^{n_2+1}}{1 - a}, & a \neq 1 \\[2ex] n_2 - n_1 + 1, & a = 1 \end{cases}$

2. $\displaystyle\sum_{n=0}^{\infty} a^n = \dfrac{1}{1 - a}, \quad |a| < 1$

3. $\displaystyle\sum_{n=1}^{\infty} a^n = \dfrac{a}{1 - a}, \quad |a| < 1$

4. $\displaystyle\sum_{n=n_1}^{\infty} a^n = \dfrac{a^{n_1}}{1 - a}, \quad |a| < 1$

Note: n_1 and n_2 may be either positive or negative in the expressions above, with $n_1 \leq n_2$.

Author Index

Subject Index